Cdre (retired) **Michele Cosentino** joined the Italian Navy in 1974 and sp[...] submarines and surface combatants as Chief Marine Engineer Officer. [...] included the Italian Navy Staff, the NATO HQ (Brussels), the Italian Min[...] OCCAR-EA Central Office (Bonn). Since 1987 he has contributed to Ita[...] and defence magazines and is the author of several books.

Aidan Dodson is Professor of Egyptology at the University of Bristol, as well as having parallel research interests in naval history; he also worked in defence procurement for 25 years, including as project leader for the offshore patrol vessel HMS Clyde. He is the author of more than 300 articles and reviews plus some twenty books, including *The Kaiser's Battlefleet* (2016) and *Before the Battlecruiser* (2018), with *Spoils of War: the Fates of the ex-Enemy Fleets after the two World Wars* currently going to press.

David Hobbs, a former naval pilot, is a naval historian and the author of 21 books including *The British Pacific Fleet*. He flew both fixed- and rotary-wing aircraft and served in a number of aircraft carriers including HMS *Victorious*. After leaving the Royal Navy he was curator of the Fleet Air Arm Museum for eight years.

Ian Johnston has written shipbuilding histories of William Beardmore's Naval Construction Works at Dalmuir and the John Brown yard at Clydebank; his book *Ships For All Nations* was republished by Seaforth in 2015. Other recent major books include *Clydebank Battlecruisers* (Seaforth, 2011), and *The Battleship Builders* (Seaforth 2013) with Ian Buxton. Ian was the associate producer of the Channel 4 series 'The Battleships'.

Hans Lengerer is an acknowledged authority on the Imperial Japanese Navy, and has written a number of books on the subject, including the *Technical and Operational History* series. His book *Imperial Japanese Warships Illustrated* was published in the autumn of 2016, followed by *Japanese Hybrid Warships* in October 2018, and Vol I of *The Aircraft Carriers of the Imperial Japanese Navy and Army* in January 2019. He co-authored, with Lars Ahlberg, *The Yamato Class and Subsequent Planning*, published in 2016, and Vol I of a planned trilogy on IJN battleships is due to appear in February 2019.

Stephen McLaughlin retired in 2017 after working for 35 years as s a librarian at the San Francisco Public Library. In addition to contributing regularly to *Warship*, he is the author of *Russian and Soviet Battleships* (US Naval Institute Press, 2003) and is co-editor of an annotated version of the controversial *Naval Staff Appreciation of Jutland* (Seaforth Publishing, 2016).

Kathrin Milanovich has been researching the history of the Imperial Japanese Navy and has contributed a number of recent articles to *Warship*.

After a career in marine engineering, **Brian Newman** took degrees from the University of Newcastle-upon-Tyne. His research interests include marine engine building historically considered, facilities provision in shipbuilding, and the fitting-out process.

Conrad Waters is a barrister by training but a banker by profession. He is the author of numerous articles on modern naval matters and editor of the annual *World Naval Review* (Seaforth Publishing). He also edited Seaforth's *Navies in the 21st Century*, shortlisted for the 2017 Mountbatten Award. He is currently writing a history of the Royal Navy's 'Town' class cruisers of the Second World War.

Jon Wise has written numerous articles for various publications, including *Warship*. His third book on naval matters, *The Royal Navy in South America, 1920–1970*, was published by Bloomsbury in 2014.

WARSHIP 2019

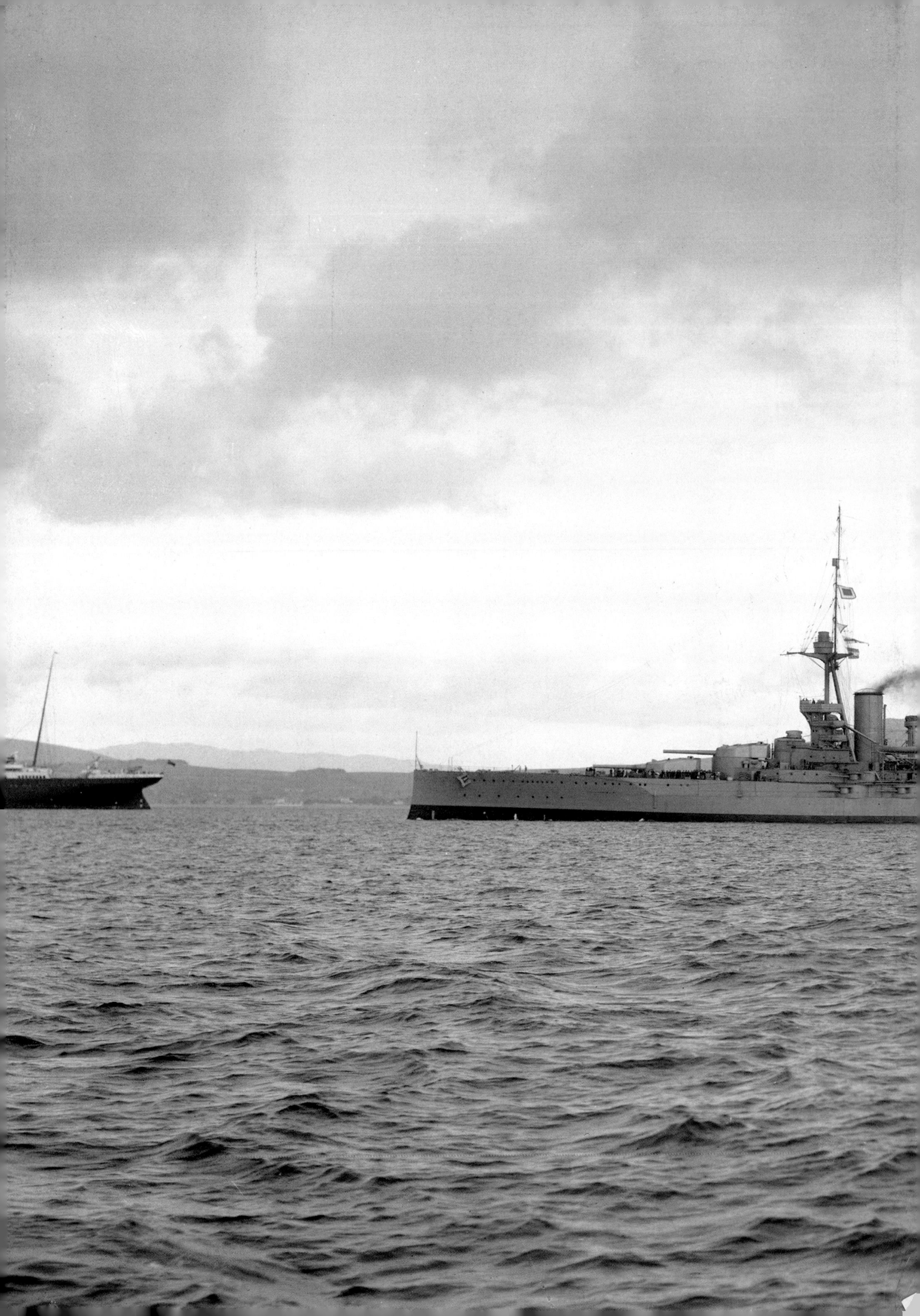

WARSHIP 2019

Editor: **John Jordan**

Assistant Editor: **Stephen Dent**

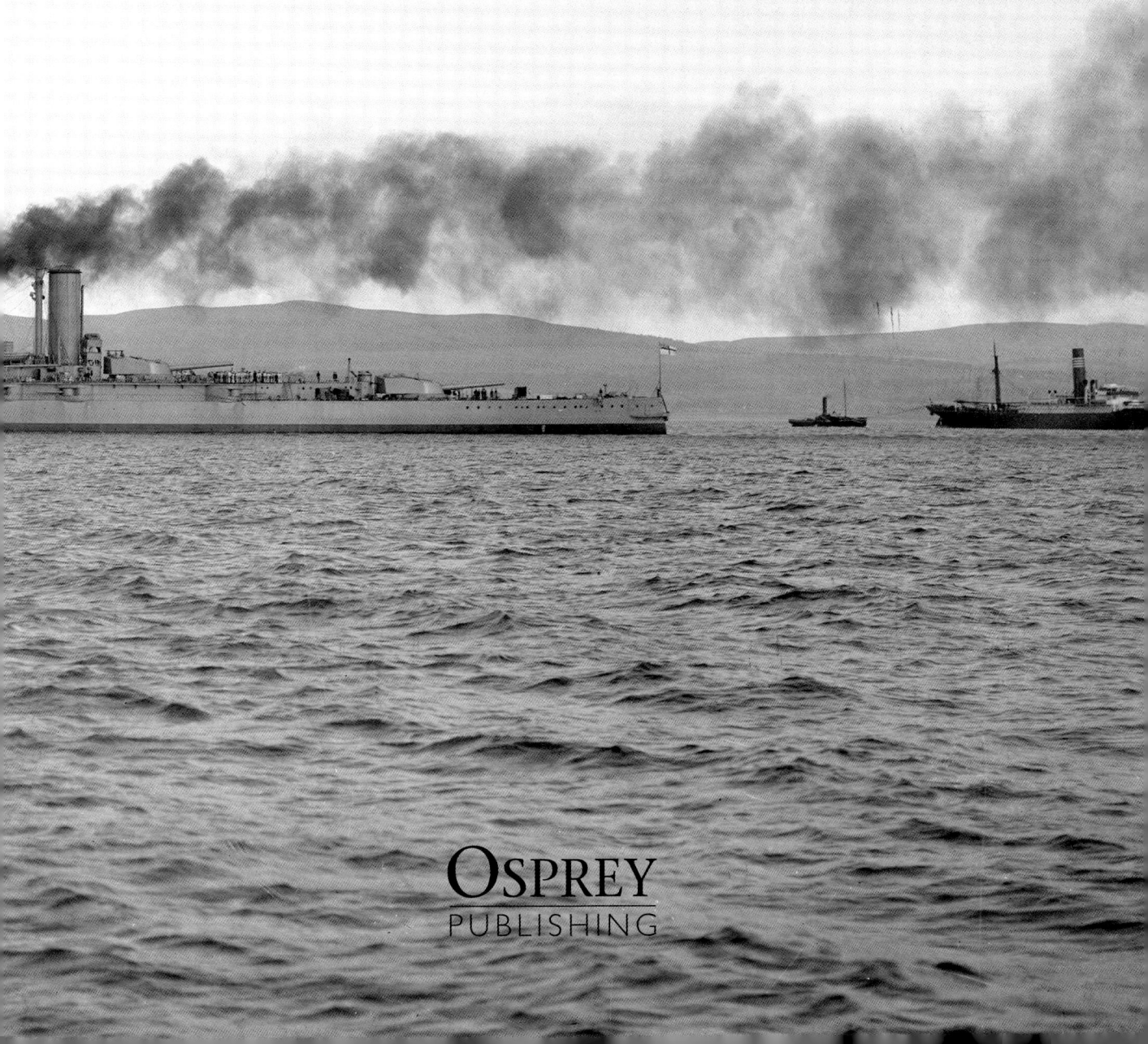

OSPREY
PUBLISHING

Title pages: The battlecruiser HMS *Tiger* on 4 October 1914, anchored at the Tail of the Bank in the Firth of Clyde, prior to her sea trials. A century's worth of erroneous description of the ship's machinery is critically examined by Dr Brian Newman in this edition of *Warship*. (NRS, UCS 1-118-418-163)

OSPREY PUBLISHING
Bloomsbury Publishing Plc
PO Box 883, Oxford, OX1 9PL, UK
1385 Broadway, 5th Floor, New York, NY 10018, USA
E-mail: info@ospreypublishing.com
www.ospreypublishing.com

OSPREY is a trademark of Osprey Publishing Ltd

First published in Great Britain in 2019

© Osprey Publishing Ltd 2019

For legal purposes the Acknowledgements within the individual articles constitute an extension of this copyright page.

All rights reserved. No part of this publication may be reproduced or transmitted in any form or by any means, electronic or mechanical, including photocopying, recording, or any information storage or retrieval system, without prior permission in writing from the publishers.

A catalogue record for this book is available from the British Library.

ISBN: HB 978-1-4728-3595-6; eBook 978-1-4728-3594-9; ePDF 978-1-4728-3593-2; XML 978-1-4728-3596-3

19 20 21 22 23   10 9 8 7 6 5 4 3 2 1

Typeset by Stephen Dent
Originated by PDQ Digital Media Solutions, Bungay, UK
Printed and bound in India by Replika Press Private Ltd.

Osprey Publishing supports the Woodland Trust, the UK's leading woodland conservation charity.

To find out more about our authors and books visit www.ospreypublishing.com. Here you will find extracts, author interviews, details of forthcoming events and the option to sign up for our newsletter.

CONTENTS

Editorial	6

Feature Articles

Armed Merchant Cruiser: The Conversion of HMS *Kanimbla*, 1939 — 8
Peter Cannon describes the conversion of the Australian coastal passenger liner *Kanimbla* into an AMC.

The French Battleship *Brennus* — 29
Philippe Caresse looks at the tortured history of this powerful vessel, which never fired a gun in anger.

The Genesis of the Six-Six Fleet — 47
Hans Lengerer examines the rationale behind the fleet that won the Russo-Japanese War.

The Rise of the Brown Curtis Turbine — 58
Ian Johnston assesses the development of the Curtis turbine, built under licence by John Brown & Co.

Battlecruiser *Tiger*: The Arrangement of the Main Engines — 69
Dr Brian Newman investigates the Brown Curtis turbine machinery installed in HMS *Tiger*, and establishes the actual configuration.

In *Avrora*'s Shadow: The Russian Cruisers of the *Diana* Class — 81
Stephen McLaughlin tells the story of the Russian Imperial Navy's first attempt at a true light cruiser.

Project 1030: A Nuclear Attack Submarine for the Italian Navy — 98
Michele Cosentino looks at the Italian Navy's ambition to build a force SSNs and SSBNs.

The 340mm Coast Defence Battery at Cape Cépet — 110
John Jordan describes the development and history of this key element in the defences of Toulon.

Powder Magazine Explosions on Japanese Warships — 118
Kathrin Milanovich looks in detail at magazine explosions involving major vessels, including the efforts of the various Investigation Committees set up to determine the causes.

Beyond the Kaiser: The IGN's Destroyers and Torpedo Boats after 1918 — 129
Aidan Dodson reviews the careers and ultimate fates of the vessels in service or building in 1918.

Early British Iron Armour — 145
David Boursnell looks at the manufacture and testing of early British iron armour plate.

Australia's First Destroyers — 153
Mark Briggs tells the story of the six destroyers of the 'River' class built for Australia.

North Sea Partners: The British and Dutch Navies in the Cold War Era — 167
Jon Wise looks at Anglo-Dutch collaboration on a series of postwar naval programmes and initiatives.

USS *Lebanon* (AG-2): A Jack of Several Trades — 183
A D Baker III focuses on USS *Lebanon*, taken up from trade as a collier and subsequently used for a variety of tasks.

Warship Notes — 188

Reviews — 201

Warship Gallery — 217
Przemyslaw Budzbon presents a set of previously unpublished photographs of the inter-war Polish Navy.

EDITORIAL

This year's annual is unusual in having no fewer than fourteen feature articles covering the full span of navies, periods and types of ship. The normal figure for the number of features is ten or eleven, but submissions this year have generally been shorter, and with the additional 16 pages we were granted by Osprey Publishing last year we have been able to accommodate all the material submitted.

The lead article this year has as its subject the conversion of the motor vessel *Kanimbla*, ordered in July 1934 by the Melbourne-based shipping line of McIlwraith McEacharn from the renowned passenger liner builders of Harland & Wolff in Belfast, into an Armed Merchant Cruiser. Peter Cannon, whose grandfather served in the ship, looks at the 'pre-fitting' programme undertaken by the British Admiralty for selected merchantmen before the war to provide the necessary stiffening for gun mountings, and at the extensive modifications that needed to be made to fit them for service as AMCs when war was declared. This is a fascinating study of a much-neglected but important aspect of Britain's trade defence strategy.

The article by Philippe Caresse that follows is the first of a series that looks in detail at the French battleships of the 'Fleet of Samples', the *Flotte d'échantillons*. Philippe leads off with the battleship *Brennus*, which was conceived during the early 1880s, suspended by Admiral Aube, and completed to a radically modified design in 1893. She incorporated a number of striking new technical features, but poor stability meant that she had to be completely rebuilt before she was fit for service. During the late 1890s and early 1900s she generally performed the role of fleet flagship, but her powerful 34cm guns were destined never to be fired in anger, and she decommissioned shortly before the First World War.

Two articles in this year's annual feature the Imperial Japanese Navy. A keynote article by Hans Lengerer addresses the thinking behind the so-called 'Six-Six Fleet', largely comprising ships built in British shipyards, that would inflict defeat on the Russian Admiral Rozhestvensky at Tsushima in 1905, while Kathrin Milanovich investigates the powder magazine explosions that plagued the Japanese Navy before and during the First World War.

A second pair of articles looks at different aspects of the Brown Curtis turbine that powered many of Britain's fastest battlecruisers. Ian Johnston's article, which makes extensive use of contemporary letters and memoirs, studies the adoption of the American Curtis turbine by John Brown's and the features that distinguished it from the better-known Parsons turbine. One interesting fact that emerges is that, for every Curtis turbine installed by John Brown in a Royal Navy warship, the company had to pay a royalty to Charles Parsons because of the licensing contract it had signed with the latter.

The accompanying article by Brian Newman is less a history than a detective story. The precise configuration of the Brown Curtis turbine installation of the battlecruiser *Tiger* has long been a matter for conjecture and dispute. Using photographs of the machinery taken in the workshop at John Brown's, Brian has established beyond doubt the composition and layout of the machinery. His conclusions contradict the arrangement seen in early plans of the ship and descriptions of the machinery by numerous 'authorities', including a former Director of Naval Construction, Sir Philip Watts. This is a cautionary tale for all those who embark on research using secondary (and even primary) sources: the Russian proverb 'Trust, but verify' seems particularly appropriate here.

The Russian cruiser *Avrora* is famous largely for her distinctive contribution to the Revolution of 1917 and for her subsequent preservation as a museum ship at Saint Petersburg (formerly Leningrad). In his latest article for *Warship*, Stephen McLaughlin looks at the tortuous design process of this ship and her two sisters, *Diana* and *Pallada*, and their very different career paths following their completion in 1901–03.

The history of nuclear propulsion for submarines in the US, Soviet, British and French navies during the postwar years has been well documented. The attempt by the Italian Navy to join this elite nuclear club is less well-known, and is now recounted in detail by Michele Cosentino, who recently came across documents and drawings related to the design of a nuclear attack submarine for the *Marina Militare* when researching another project at the Italian Navy Historical Office (IHNO). The proposed attack submarine, designated Project 1030, would have employed a reactor derived from that mounted in the US submarines of the *Skipjack* class but, in keeping with Italian Navy traditions, would have been able to operate an underwater vehicle for Special Forces operations.

In a departure from *Warship*'s normal coverage, the Editor has focused on the powerful French coastal battery of Cape Cépet, built during the interwar period using 34cm battleship guns in naval-style twin turrets to defend the approaches to Toulon, while articles by Aidan Dodson and David Boursnell continue themes introduced in earlier editions of *Warship*. After detailing the fates of the Kaiser's battleships and cruisers following the German defeat in the Great War of 1914–18, Aidan turns his attention to the destroyers and torpedo boats, while David follows up his article on the production of armour plate in the UK during the early 1900s with a study of the development and manufacture of iron armour during the second half of the 19th century.

Following the federation of the Australian colonies in 1901, naval defence would become a controversial political issue. The following year the Commonwealth

EDITORIAL

A view of the flight deck of the French aircraft carrier *Béarn* during the 1930s. Two Levasseur PL.101 reconnaissance aircraft are being brought up to the flight deck using the centre and after lifts. Note the heavy clamshell doors which ensured the flight deck remained in operation when the lifts were lowered to hangar level. The Editor's article, due to published in next year's annual, will look at some of the more unusual aspects of the ship's design, not all of which were successful. (Private collection, courtesy of Philippe Caresse)

government entered into a ten-year naval agreement with Britain whereby Australia paid £200,000 annually to the Royal Navy to base a fixed number of warships in Australian waters. However, as Mark Briggs' article states, this arrangement did not sit well with the heightened sense of national identity that had emerged in Australia after federation, and in 1909 it was decided that Australia would purchase eight destroyers of the 'River' class, modified for increased endurance. Two ships were to be delivered in the UK and a third vessel was to be built in Britain, taken apart and reassembled at Cockatoo Dockyard, Sydney, to provide the experience for further local construction from the keel up. In the event only three further hulls would be built in Australia, and construction of these relatively sophisticated vessels proved to be challenging and costly. Mark's article again highlights the difficult choice that has to be made by the lesser naval powers between purchasing 'off-the-shelf' designs abroad and the more costly alternative of developing a local industrial base.

The features section concludes with an article by Jon Wise about the close, but not always smooth-running relationship between the British Royal Navy and the Royal Netherlands Navy since 1945 and our regular drawing feature by A D Baker III. Jon's article focuses on the politics of joint projects for ships, weaponry and sensors and the inevitable tensions between affordability and capability. Dave Baker features an unusual vessel, the USS *Lebanon*, which began her career as a naval collier and was subsequently employed for a variety of tasks, including the towing of gunnery targets for the fleet.

The annual concludes with the customary comprehensive complement of Warship Notes, reviews of the major naval books published this year, and a Gallery that features some rare photos of the Polish Navy during the 1930s with a particular focus on the German aggression of early September 1939 and its consequences.

Warship 2020 already has a full complement of features promised or delivered. The author of this year's Gallery, Przemysław Budzbon, will publish a major article the influence of the salvaged British *L 55* on the design of the early Soviet submarines. Hans Lengerer will follow up his article on the Six-Six Fleet with one on the Eight-Eight Fleet in which he looks at the ambitious plans for the projection of Japanese naval power during the second decade of the 20th century. Philippe Caresse's series on the French *Flotte d'échantillons* will continue with the battleship *Charles Martel*, the lead ship of the 1890 programme, while the Editor will publish a major article on France's first aircraft carrier, *Béarn*, built using an incomplete battleship hull during the 1920s. David Hobbs will follow up his Warship Note on berthing the reconstructed *Victorious* with a major feature detailing the carrier's modernisation during the 1950s, while Stephen McLaughlin will turn his attention to the unusual design of the Italian ironclad battleships *Italia* and *Lepanto*.

John Jordan
April 2018

ARMED MERCHANT CRUISER:
The Conversion of HMS *Kanimbla*, 1939

British interwar contingency plans for the protection of Empire trade at sea called for the rapid conversion of selected passenger ships into armed merchant cruisers to augment the Royal Navy's inadequate force of regular warships. **Peter Cannon** looks at the conversion of the Australian coastal passenger liner *Kanimbla* into one of the most successful auxiliary warships of her kind upon the outbreak of war in 1939.

During the late nineteenth century, Britain developed a policy of employing passenger vessels as auxiliary cruising warships to supplement the regular vessels protecting her global empire's trade. Armed Merchant Cruisers (AMC) served successfully during the First World War, and experience of that conflict shaped future Admiralty contingency planning. The Royal Navy's (RN) immediate post-war planning, with Japan the only credible opponent, called for 70 cruisers to fight a war in the Far East: 25 with the battle fleet and the remaining 45 allocated to trade defence. The latter would be supplemented by 74 AMCs expeditiously converted at ports in the United Kingdom, Gibraltar, Malta, India, South Africa, Canada and Australia when hostilities were considered imminent.[1]

Beginning in late 1919, the Admiralty facilitated arrangements with patriotically-minded ship owners to begin incorporating structural stiffening to support both Low Angle (LA) anti-surface and High Angle (HA) anti-aircraft guns during the construction of 50 suitably-sized passenger ships. Furthermore, the register of potential AMCs included a pool of unprepared vessels from which to select a further 24 in the event of war. Conversions would be equipped predominantly from stockpiled equipment removed from decommissioned warships during the drastic downsizing of the peacetime fleet.

Australia, having recently assumed the status of full partner in the business of Empire naval defence through the creation of a blue-water navy, participated in the scheme from the outset. By 1921 seven ships building in

HMS *Kanimbla* sailing for her workup period on 27 November 1939. The photo was taken off Sydney by No 6 Squadron RAAF. Only the forward four (No 1 and No 2) guns have been provided with shields, while the after No 3 and poop guns remain unshielded. Also of note is the toned-down civilian paint scheme and short-lived air recognition roundel on the poop deck. (Royal Australian Navy: HMAS *Cerberus* Museum)

UK and Australian yards were allocated for stiffening and a commitment made to convert three ships for the RN. As the whereabouts of ships engaged in international trade at the outbreak of war was unpredictable, either Australian- or British-registered vessels trading in Australian waters would be requisitioned when required. These ships would be liable for service anywhere in the world under Admiralty orders. The Australian Government also undertook to furnish two additional locally-owned ships for service in the Royal Australian Navy (RAN) utilising its own surplus equipment.[2]

Policy

The Admiralty first promulgated comprehensive AMC instructions in 1926. Despite administrative refinement up to and beyond the eve of mobilisation, the broad outline remained constant. AMCs were to be employed on escort and convoy duties defending Allied shipping as well as blockade patrols intercepting and examining vessels for contraband. The recent wartime experience of German commerce raiders, as well as the paramount role of economic warfare in British maritime strategy, constituted the AMC's chief *raisons d'être*. They would serve as commissioned auxiliaries flying the White Ensign and initially complete to three different standards of equipment: emergency, semi-complete and complete, dependent upon the time required to reach their war stations. Emergency equipment was envisaged to require two and a half to three weeks of dockyard work, with five weeks needed to convert a ship to semi-complete standard. Complete equipment involved eight to ten weeks and was to include full director firing equipment. Earlier ships would be more thoroughly outfitted when they could be relieved. Most overseas conversions, including those planned for Australia, would be to semi-complete standard.

The programme to pre-position ordnance, ammunition and naval stores for 74 conversions at UK and overseas arming ports was originally planned for completion in 1935. While the RAN accumulated equipment for its two AMCs, the only equipment for the three Imperial ships initially shipped to Australia was their guns. Six 3in guns were received in Sydney in June 1926 with their mountings arriving in January 1929. Twenty-one 6in guns and mountings arrived between October 1926 and January 1927, all having been shipped from the RN Armament Depot at Gosport to the RAN's Spectacle Island armament repository. The outstanding equipment and technical documentation was only despatched during the late 1930s, and was still arriving as the ships were being converted.

Evolving war plans for the Far East included up to 23 British AMCs in Australian waters, primarily to defend the shipping focal point off south-western Australia from Fremantle and institute a patrol line of Darwin-based AMCs and regular cruisers between Java and Australian waters. The three Imperial conversions were allocated to Fremantle in 1926 and Darwin from 1932. The 1939 war orders, revised to reflect the European war then considered likely, allotted them to the East Indies, but hostilities would see them deploy to the China Station.

TSMV *Kanimbla*

The Twin Screw Motor Vessel (TSMV) *Kanimbla* was ordered in July 1934 by the Melbourne-based shipping line of McIlwraith McEacharn from the renowned passenger liner builders of Harland & Wolff in Belfast. At 488 feet long, 10,984 gross tons and with a design speed of 19 knots, her luxurious appointments were unprecedented on the Australian coastal trade. She also fell into the category of intermediate-sized vessels Admiralty specifications desired. Experience had shown larger ships to be uneconomical, insufficiently manoeuvrable and overly large targets, while smaller vessels often suffered from limited endurance and unsatisfactory sea-keeping qualities in heavy weather.[3]

The Australian Naval Board suggested the addition of *Kanimbla* to the Imperial register of stiffened ships before her keel was even laid. The Admiralty thereafter negotiated the requisite £2,500 of stiffening work with the owners during construction. She completed on 26 April 1936. Arriving in Australian waters the following month, she was employed servicing major ports between Fremantle and the Queensland port of Cairns.

Requisitioning

On 24 August 1939, as hostilities appeared imminent, the Admiralty and Board of Trade assumed powers to requisition British vessels for auxiliary service. The following day, two British-registered steamers operating

TSMV *Kanimbla*

Type:	Twin Screw Motor Vessel
Owner:	McIlwraith McEacharn Ltd, 94–96 Bridge Street, Melbourne
Port of Registry:	Melbourne, Australia
Builder:	Harland & Wolff Ltd, Belfast, Yard No 955
Laid Down:	July 1934
Launched:	15 December 1935
Delivery Date:	26 April 1936
Characteristics	
Length pp:	460ft
Length oa:	488ft 8in
Beam:	66ft 4in
Frames:	190
Frame Spacing:	2ft 6in
Watertight Bulkheads:	8
Gross Tonnage:	10,984.56 tons
Holds:	5
Tween Deck Holds:	9
Total Insulated Cargo:	9,537cu/ft (grain); 8,635 cu/ft (bale)
Passengers:	250 First Class, 198 Cabin Class
Complement:	17 officers, 150 men

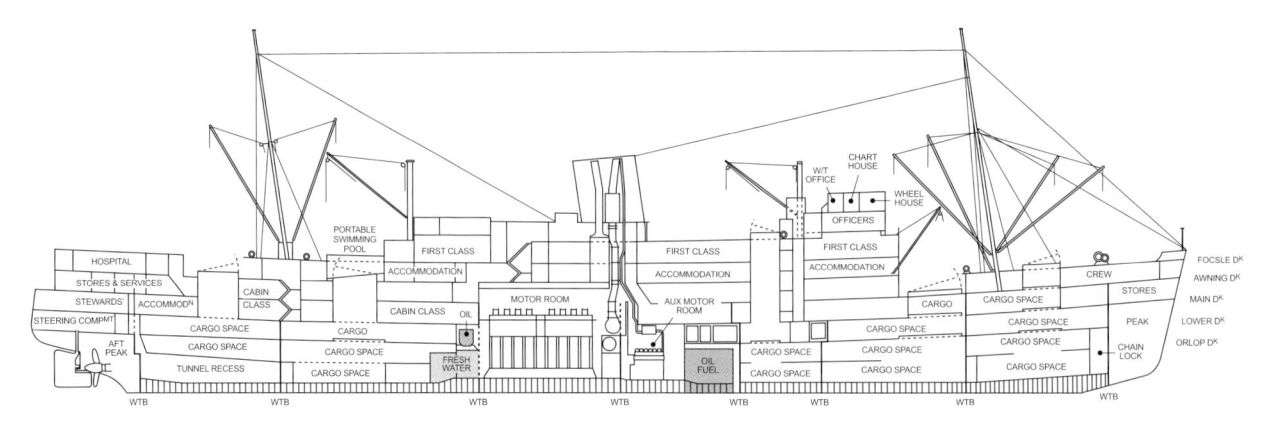

Fig 1: TSMV *Kanimbla* as completed, June 1936, for Australian coastal passenger service. (© John Jordan 2018)

in Australian waters, *Moreton Bay* and *Arawa*, were ordered to proceed to Sydney while the Australian Government was requested to requisition the locally-registered *Kanimbla*.[4] The Naval Board subsequently arranged to have all three vessels taken in hand for conversion to AMCs on Imperial account immediately upon arrival. International maritime law prevented the secret arming of merchantmen, and the ships were formally commissioned as men-of-war, under naval personnel subject to military discipline, prior to the commencement of work.

Kanimbla was visiting Cairns on 26 August when her owners received a letter from the Naval Board requesting the ship's delivery to Sydney for naval service on 4 September under the Australian Naval Defence Act. The ship arrived at the allotted time, only hours after the declaration of war, before discharging passengers, cargo and the majority of her crew. She was handed over to the Navy the following day before securing alongside Garden Island Naval Dockyard at 0700 on 6 September 1939, where she immediately commissioned into the RN as HMS *Kanimbla*. The Admiralty was thereafter responsible for the payment of charter rates to her owners for the duration of her service as an HM ship.

Conversion

The limited Australian naval engineering infrastructure, centred upon Sydney, was heavily engaged in preparing the fleet for war as well as equipping troop transports and defensively armed merchant ships. The two AMCs for the RAN were deferred as non-essential, and existing capacity was directed towards the Imperial ships. *Kanimbla*'s coastal passenger competitors *Manoora* and *Westralia* would be taken up a month later.[5] *Moreton Bay* was allocated to the Cockatoo Island shipyard and

TSMV *Kanimbla* on builder's trials in the Irish Sea, 21 April 1936. (National Archives of Australia)

Kanimbla as AMC

Commissioned:	6 September 1939
Form of Charter:	T98: Australian register but requisitioned for Imperial Govt
Converted to AMC:	Garden Island Naval Dockyard, Sydney
Pennant Number:	F23
Decommissioned:	31 May 1943

Characteristics

Loaded Displacement:	13,200 tons
Maximum Loaded Draft:	24ft 3in
GM at full load:	4.6ft
Paravane Gear:	4 x Mark VII
Echo Sounding Gear:	Marconi Type 429
Compass:	Standard magnetic
Boats (1940):	1 x 30ft, 1 x 32ft service cutters
	1 x 35ft motor boat,
	1 x 30ft motor cutter
	3 x 30ft Fleming 'B' Type lifeboats
Fresh Water:	1,358 tons capacity
Daily Consumption:	35 tons
Complement:	
13 Dec 1939	32 officers, 259 men
9 Apr 1942	32 officers, 296 men

Conversion

Timeline

Requisitioned for Naval Service:	26 August 1939
Placed at Naval Disposal:	5 September 1939
Taken in Hand for Conversion:	6 September 1939
Docking:	16–21 October 1939
Gunnery Trials:	13 November 1939
Refitting Completed:	23 November 1939
Work Up Period:	27 November – 8 December 1939
Deployed for War Station:	13 December 1939

Permanent Ballast and Buoyant Material 1939

Rock:	1,280 tons
Timber and bagged cork:	22,500 cu/ft
Timber stowage battens:	4,000 cu/ft
Total timber and bagged cork:	26,500 cu/ft
14-gauge, 46 gallon drums:	2,000
18-gauge, 47 gallon drums:	6,008
20-gauge, 44 gallon drums:	5,364
Total drums:	13,642

Arawa to the less capable Mort's Dockyard in Balmain. *Kanimbla*, under Lt-Cdr Geoffrey Branson, RN (Retd), until the arrival of Commander Frank Getting, RAN, on 4 October was allocated to Garden Island. The fleet base was capable of structural and machinery refitting, gun mounting, electrical and optical work, boat building and rigging as well as the manufacture of a considerable range of naval stores. With the construction of the yard's graving dock still in the planning phase, all underwater work would take place in Cockatoo Island's Sutherland graving dock.

At first glance, it might appear that fitting an early 20th-century passenger vessel as an auxiliary cruiser would involve little aside from bolting on a few guns, drafting in some sailors and sending her to sea, but the reality was far more involved. Guns and their crews had to be controlled, ammunition safely stowed and accessible, battle damage prepared for and the crew accommodated; unnecessary equipment had to be landed and a myriad of naval stores, required to operate as a warship, had to be embarked. Generic guideline drawings for dockyards had been provided since 1926, but converting the ships required co-ordination, under the direction of the Commodore-in-Charge, Sydney, between the three dockyards as well as periodic Admiralty advice to tailor alterations for each ship.

The work was a major undertaking by dockyard workers, initially assisted by a skeleton crew signed on by *Kanimbla*'s owners, as well as successive drafts of naval personnel. The ship was fitted out alongside Garden Island, the neighbouring Woolloomooloo Bay finger wharf, Cockatoo Island and finally Man-o-War Anchorage adjacent to the Sydney Harbour Bridge. The majority of equipment and stores came and went by lighter. *Kanimbla*'s general layout, seen in Figures 1–3, was typical of a ship of her kind and the narrative will concentrate on the specific areas of capability required to prepare her for war service.

Stripping and Structural Alterations

As soon as the ship came under dockyard control, the task of disembarking vast quantities of equipment, bedding and stores not required for naval service was begun. This also involved stocktaking and inventorying practically every item in the ship and providing storage ashore. Existing victualling provisions were taken on charge and quantities of naval stores, medical supplies, etc were embarked throughout the conversion process. A full Lloyds Register condition survey was also conducted.[6] More than 200 tons were removed including cabins, furniture, hold insulation, insulated hatches, boats and davits as well as most of the cargo-working derricks and some existing bridge structure. While many cabins would remain, workmen wasted no time in knocking down non-structural bulkheads and clearing away ablution facilities and other fixtures to provide space for messes and buoyancy ballast.

The ship contained considerable amounts of high-quality woodwork which added to the risk of fire and splinters. While much was removed, it was not a comprehensive effort and a far greater amount of flammable panelling, furniture and corticine decking than found in a regular warship was retained. A flying bridge, encompassing a naval-style open compass platform was built above the navigating bridge from which the ship would be conned at sea. Guard rails were replaced with hinged stanchions with two wires to facilitate weapon firing arcs. A crow's nest was constructed on the foremast, hatches were plated over and limited protective plating was provided. Furthermore, magazines, shell rooms and ammunition arrangements were erected by teams of boilermakers, welders, carpenters and other tradesmen working in shifts. Elements of this work will be described below.

Main Armament

Kanimbla's LA armament was installed in line with the original 1926 equipment scheme. The 40-year-old, 45-calibre Mk VII breech loading gun was the oldest model of British 6in gun to see service in the Second World War. By 1939, 42 AMCs were earmarked to receive these weapons on pedestal-type PIII mountings removed from broken-up turn-of-the-century cruisers and pre-dreadnought battleships. Later supplemented by limited numbers of more modern weapons as they became obtainable, it was the only model available when stockpiling began and the only one supplied to overseas arming ports.

The Mk VII was of wire-wound construction and had a Welin interrupted-screw breech. The mounting was hand worked, platforms being provided on the left and right sides for the gunlayer and trainer respectively to elevate and train via manual handwheels. The breechworker, sightsetter, rammer, two projectile and two cartridge loading numbers followed the weapon on the deck as it trained. There was no loading tray and projectiles were manhandled into the open chamber for ramming. Telescopes were fitted with night illumination circuits, the gunlayer's telescope being manipulated by the manual gun sights set for the range in use by the sightsetter. Maximum range was a relatively modest 14,200 yards firing standard 4crh 100lb shells with silk-bagged 23lb 2oz cordite SC103 propellant charges. Seven rounds per minute was achievable with a well-trained crew.[7]

Kanimbla's layout required seven mountings to achieve the mandated four-gun broadside. Port and Starboard No 1 guns (P1 and S1), mounted on the forecastle, could fire from dead ahead to 75° abaft the beam; P2 and S2 in the forward well deck were restricted to 65° before and 63° abaft the beam; P3 and S3 on B deck fired from 60° before the beam to dead astern and the single No 4 centreline gun on the poop deck was a given a 280° firing arc. Gun positions and permanent stiffening were marked on builder's drawings held in the Master's safe, while portable material had been despatched to Australia with the ship and stored at Garden Island. Installation of these pillars, channel stiffeners, angle, web and other plates during construction would have obstructed cabin spaces and interfered with the ship's commercial operation but they were now fitted prior to embarking ordnance. Merchant seamen remaining in the ship were surprised to see the teak decks in these positions removed to reveal packing rings (for 20° elevation) ready to receive gun mountings.

Guns and mountings from Spectacle Island were transported to Garden Island by Cockatoo's floating crane *Titan* on 29 September and hoisted aboard over the next three days. Only the four forward guns were provided with turn-of-the-century cruiser-pattern gunshields.

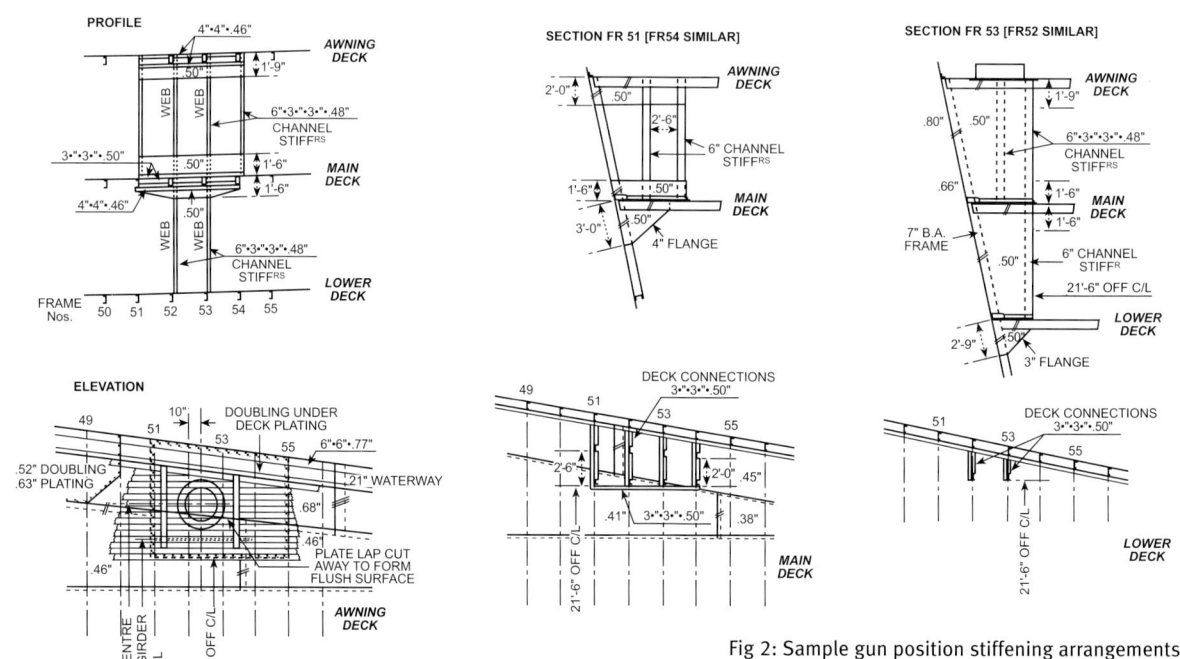

Fig 2: Sample gun position stiffening arrangements: port and starboard No 2 6in guns, forward well deck. (© John Jordan 2018)

Many mountings had been removed from battleship casemates, and some AMCs completed without any shields. Guardrails, winches, ventilators and other equipment obstructing training arcs were removed, and depression rails installed to inhibit inadvertent aiming at the ship's structure and blast damage.[8]

Low Angle Fire Control

Director firing systems had become standard equipment in capital ships, cruisers and finally modern destroyers by the end of the First World War, and by 1922 the Admiralty had determined that AMCs would receive an austere, destroyer-style arrangement. It was intended to utilise more capable surplus systems as they became available but the majority of conversions received a configuration of equipment similar, but less capable than the system introduced in the now-obsolescent V&W classes of destroyer during 1918.

The system was centred upon a pedestal-mounted 'light type' director and a 9-foot FQ2 coincidence rangefinder atop two separately constructed bandstands in the fore control position abaft the compass platform. Small main and secondary transmitting stations (TS) were arranged in former cabins in the forward and after superstructures respectively, with the after control position directly above the after TS. The director, essentially a gun sighting telescope on a trainable mounting, was designed to electrically transmit a bearing, corrected for deflection and drift by the director layer and sightsetter, to receivers at the guns. A pistol operated by the layer, assisted by Henderson gyro gear to account for roll, allowed coordinated electric broadside firing. However, the AMC outfit would be a 'training only' system, and the director firing components were removed prior to shipping to Australia in June 1939. Furthermore, elevation transmission, facilitated in heavier, more complex systems supported by a full fire control table in the TS, was not provided.

Range was mechanically transmitted from the rangefinder to a receiver in the TS backed by voicepipe communication. Estimated enemy course and speed was applied to either the port or starboard Dumaresq rate and deflection instrument, mounted on the bulwarks of fore control, to estimate the rate at which the range was changing as well as deflection. This, as well as spotting corrections, was passed to the TS by voicepipe. Range and range rate was then manually applied to a Vickers range clock to determine gun range. Deflection estimates were passed to the director layer by voicepipe from the TS for setting on the director. Control personnel were also provided with two deflection ready reckoners.

To fight the main armament from the TS, *Kanimbla's* electrical fire control installation comprised five main circuits backed by voicepiping between all key positions. Cabling was led from TS junction boxes to the mountings by separate wiring runs to the port and starboard No 1 guns, the remaining guns on their respective sides being supplied thereafter in series. No 4 gun was wired at the end of the port run. Ship's mains 220V high power was stepped down to supply the 22V fire control equipment from a low power room constructed adjacent to the TS. Battery banks, accessible via changeover switches and capable of at least three hours of contin-

HMS *Kanimbla* fitting out alongside No 8 Berth, Woolloomooloo Bay in Sydney in late October/early November 1939. The main armament is on board and lighters of timber, used both for buoyant material as well as for stowage of drums, are secured alongside. (Author's collection)

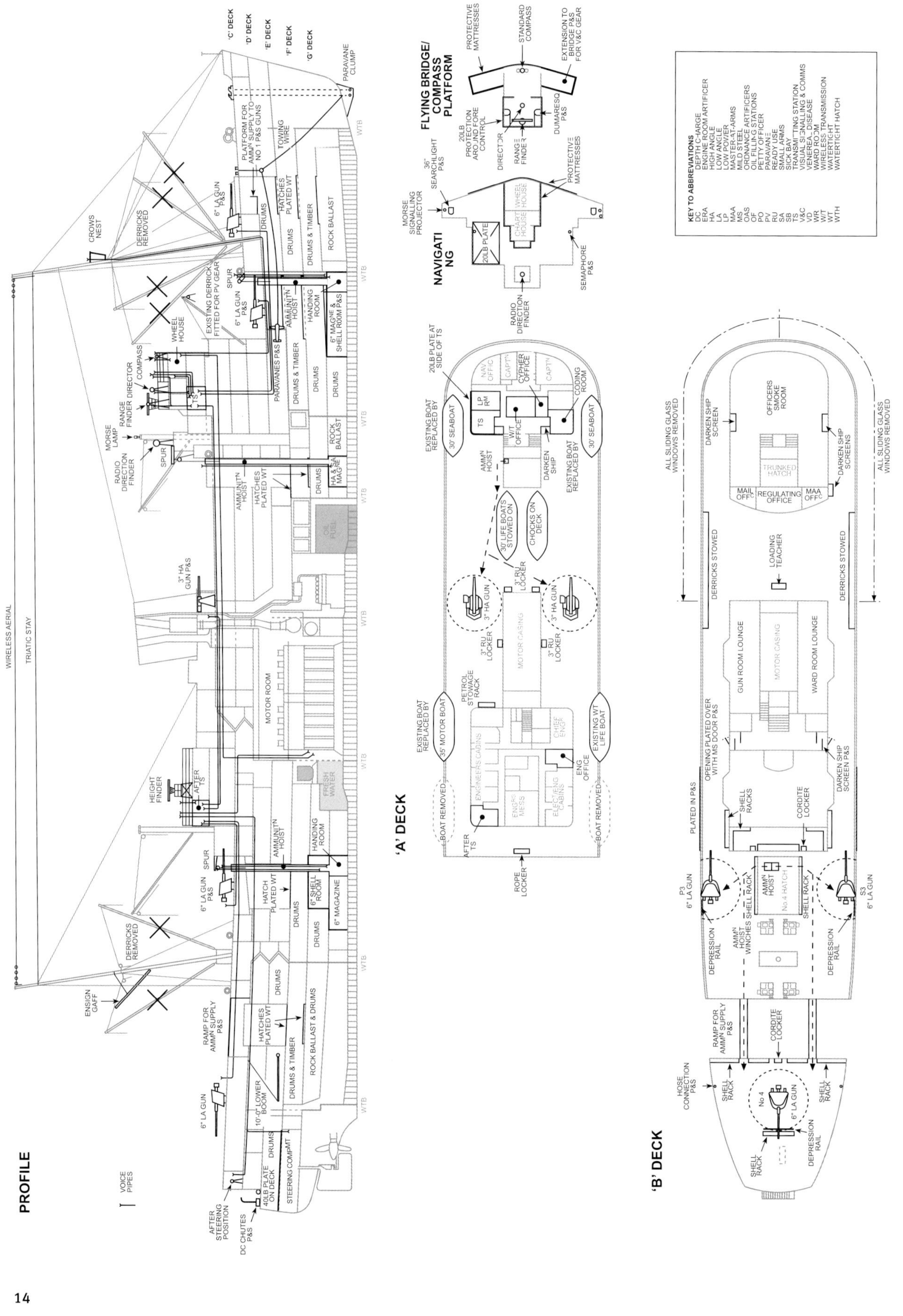
PROFILE
'A' DECK
'B' DECK
NAVIGATING
FLYING BRIDGE/ COMPASS PLATFORM
KEY TO ABBREVIATIONS
DC DEPTH CHARGE
ERA ENGINE ROOM ARTIFICER
HA HIGH ANGLE
LA LOW ANGLE
LP LOW POWER
MAA MASTER-AT-ARMS
MS MILD STEEL
OAS ORDNANCE ARTIFICERS
OF OIL FILLING STATIONS
PO PETTY OFFICER
PV PARAVANE
RU READY USE
SA SMALL ARMS
SB SICK BAY
TS TRANSMITTING STATION
V&C VISUAL SIGNALLING & COMMS
VD VENEREAL DISEASE
WR WARD ROOM
WT WIRELESS TRANSMISSION
W/T WATERTIGHT
WTH WATERTIGHT HATCH

Fig 3: Plans of HMS *Kanimbla* as completed, December 1939, and submitted to the Admiralty 8 March 1940. (© John Jordan 2018)

Technical Data

Machinery

Main Engines:	2 x Harland & Wolff, Burmeister & Wain, 4-stroke diesels with airless injection & supercharging; 4,250bhp, 108rpm
Screws:	2 x 3-bladed, 16ft 9in propellers
Speed:	
TSMV (22 Apr 1936)	19.05kts, trial
AMC (26 Sep 1940)	16.7kts, 105rpm, 10,023bhp
Power Generation:	4 x Harland & Wolff, Burmeister & Wain 6-cylinder engines and dynamos with airless injection; 430bhp, 270rpm, 300kW, 1340A, 225VDC; 1,600amp (2,200amp max) action load at sea
	1 x Laurence Scott, 50kW, 228A petrol-paraffin emergency dynamo
Auxiliary Boiler:	1 x Clarkson Thimble Tube Boiler, petrol- or oil-fired, 100lb/in^2
Fire Suppression:	Grinnell automatic sprinkler system throughout accommodation and public rooms; fire detection apparatus and CO2 fire extinguishing equipment in holds nos 1, 2, 3, 4 & 5; 40 x fire hydrants

Speed & Endurance

	Speed	RPM	Fuel Consumption	Endurance on Full Bunkers (760 tons)
'Authorised Full Power'	16kts	100	38 tons per day	7,680 miles, 20 days
'With all Convenient Despatch'	12kts	75	21 tons per day	10,422 miles, 36 days
'At Most Economical Speed'	8kts	50	11.5 tons per day	12,672 miles, 66 days

Turning Data (14 Jul 1941)

80 revs, 20° starboard wheel, 9 min 30 sec (360° turn), turning circle mean diameter 1,100 yards

Armament 1939

7 x 6in BL Mk VII guns on PIII mountings
2 x 3in QF Mk I guns on Mk II mountings
2 x .303in Lewis Mk I machine guns on single mountings
8 x Depth Charge Mk VII, 2 x DC chutes
41 x .303in rifles
60 x .303in Lee Enfield No 1, Mk III rifles
60 x No 1, Pattern 1907 bayonets
30 x No 1, Mk VI revolvers (6in barrel)

Armament Weights

6in BL Mk VII guns:	50.78 tons
3in Mk I guns:	2.00 tons
.303in Lewis Mk I MG:	0.03 tons
Depth Charge Mk VII:	1.61 tons
Cartridges, 6in:	13.60 tons
Projectiles, 6in combat:	31.84 tons
Projectiles, 6in practice:	10.1 tons
Cartridges, 3in QF:	8.63 tons
Cartridges, Small Arms, ball:	1.20 tons
Demolition Stores:	0.12 tons
Total:	119.92 tons

Fire Control

Director:	Light Type Director (training only)
Rangefinders:	1 x 9-foot Type FQ–2 rangefinder (fore control)
	1 x 9-foot Type FQ–2 rangefinder converted for height finding (after control)
Searchlights:	2 x 36in projectors

Ammunition 1939

6in Cartridge:	924 x cartridges, filled 23lb 2oz cordite SC103
	112 x cartridges, filled 11lb 8oz cordite SC103
6in Projectile:	77 x CPC shell, 4crh, TNT burster
	420 x HE shell, 4crh, TNT burster
	203 x CPC shell, 4crh, powder burster
	224 x practice shell
3in QF, 20 cwt:	300 x HE shell, 16lb, TF
	100 x starshell, 12.5lb
	8 x target smoke shell
	60 x HA practice (HETF in lieu)
Fuzes for HA guns:	420 x percussion, DAI No 45P Mk VII***
	100 x time No 198 Mk I
	300 x time No 198 Mk II
Vent Tubes:	924 x .4in percussion tubes
Small Arms:	5,000 x .303in cartridges Mk VII, rifle, chargers
	4,000 x .303in cartridges Mk VII, rifle, cartons
	1,000 x .303in tracer, GI
	4,000 x .455in cartridges, revolver, ball
Pyrotechnics:	12 x smoke floats Type RIII
	14 x igniters, smoke float Mk IV
	50 x 1in signal cartridges, red
	50 x 1in signal cartridges, white
	50 x 1in signal cartridges, green
Demolition Stores:	40 x TNT demolition blocks Mk I, 1.25lb
	20 x demolition primers Mk I, CE
	1 x safety fuze pistols
	50 x cartridges for safety fuze pistol, Mk I
	20 x electric detonators No 21, Mk VI
	20 x detonators No 25, Mk I
	30 x electric fuzes No 19, Mk III
	84 x safety fuzes No 9, Mk III*, feet

uous operation, were co-located to ensure supply in the event of a mains failure.

The ship's gun mountings were hurriedly modified at Spectacle Island just prior to embarkation with director operating gear and destroyer-pattern, follow-the-pointer style 'small type' training receivers despatched from Chatham during late 1939. The receivers, mounted in front of the trainers' position, received the director bearing, and a manual convergence adjustment enabled the alignment of each separate gun to the director's line of sight at the range in use.

Despite earlier intentions, only the main TS received a range and deflection transmitter, the after TS resorting to voice communications if required to fight the guns with the assistance of a second range clock. A range and deflection receiver was mounted on the left side of each gun for the sightsetter as a secondary method to voice-pipe communication. The sightsetter adjusted the layer's sight accordingly before the weapon was laid for elevation. Another receiver resided in fore control for confirmation that the control officer's spotting corrections were being applied correctly. If director bearing failed, the trainer would aim by telescope, while the additional loss of range and deflection would see the gunlayers estimating all fire control calculations for their individual guns in local control.

Grahams Navyphones, an obsolete exchange-type system wired in series with a call-up lamp and bell for noisy positions, had been fitted in older ships with separate exchanges for different services. Eleven phones, for fire control only, were grouped in *Kanimbla* with every position being able to hear and be heard by all others. Each TS had a direct-wired phone to speak to the other and a second to talk to the guns, with the after TS accessing the circuit via a junction box on a P3 mounting. Phones were a secondary communication method, voice-pipes being primary, and were situated in protected positions adjacent to the guns. Each gun had a fire gong, activated by a push in the TS, for coordinating salvo and broadside fire, while another circuit controlled by a separate push on the compass platform rang cease-fire bells at the guns and the TS. The five aforementioned circuits were supplemented by twelve alarm rattlers, situated at various positions around the upper decks, acting as an alert system also activated by a compass platform push.

Illumination for range and deflection dials, telescope night sights and loading lamps above the breech, were provided by on-mount batteries. *Kanimbla's* guns were not fitted for electrical firing and the ship only carried percussion vent tubes. Filled with gunpowder pellets, vent tubes were inserted by the breechworker into the breech mechanism to ignite the cordite charges.[9]

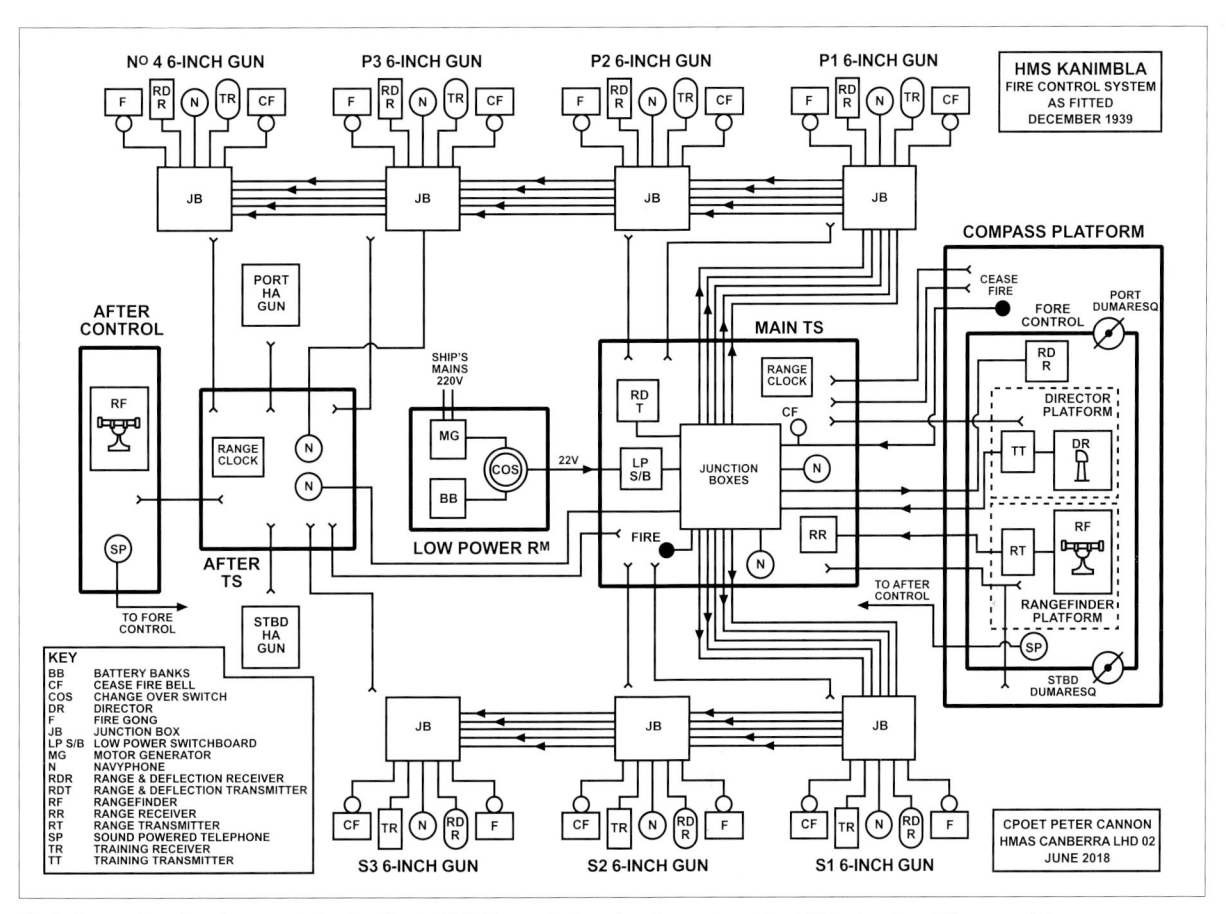

Fig 4: Comprehensive documentation for *Kanimbla's* fire control system has not survived. This drawing utilises surviving correspondence and Admiralty equipment manuals to facilitate a system overview. (© John Jordan 2018)

A system of levers, actuated by a hand lever operated by the gunlayer, was used to trigger a percussion striker and fire the tube. The final aspect of the fire control system was the addition of a 36in searchlight projector on each wing of the navigating bridge to illuminate targets at night. They were sited within earshot of the compass platform with no control arrangements provided.

Anti-Aircraft Armament and Control

The HA equipment reserved for AMCs during the 1920s comprised single 3in guns recently superseded as the fleet's primary anti-aircraft weapons by the 4in Mk V. While a few RN AMCs received the 4in, the Australian allocation remained two 3in Mk I guns on Mk II mountings per ship. The Mk I was the first British gun designed for the anti-aircraft role, entering service in 1913 and firing high explosive 16lb shells at between 12 and 14 rounds per minute to a maximum ceiling of 25,200 feet.[10] They were also provided with star shells to illuminate surface targets during a night encounter. The guns were percussion-initiated, sliding breechblock weapons manually controlled by a gunlayer and trainer with open barrage sights and a range drum operated by the gunlayer. They were supported by a breechworker, loader and two supply numbers. Positions had been prepared port and starboard amidships up on the boat deck, their packing rings having been fitted beneath the planking of the former tennis courts. As with the 6in guns, the HA guns were fitted with depression railings.

The Mk I was outclassed by the aerial threats of the early 1920s, let alone those of the late 1930s, on account of its weak ballistic properties and the small effective burst zone of its projectile. Furthermore, *Kanimbla*'s fire control arrangements were equally unsatisfactory.

View of *Kanimbla*'s open flying bridge/compass platform from the crow's nest. Behind the standard magnetic compass – the ship was not fitted with gyros – is the chart table and the two fire control platforms mounting the Light Type director and FQ-2 rangefinder. (Con Cannon)

Kanimbla's forecastle showing S1 6in gun and crew, plated-over No 1 hatch with inset ammunition supply hatch for both No 1 guns as well as numerous single ready-use 6in shell racks secured to the deck. A 6in cordite locker is immediately forward of the ammunition winch at bottom left. The photo was taken in Hong Kong early in 1940. (Author's collection)

ARMED MERCHANT CRUISER: THE CONVERSION OF HMS *KANIMBLA*, 1939

Kanimbla's after guns were provided with gunshields in Hong Kong early in 1940. This late 1941/early 1942 view shows the turn-of-the-century British cruiser-pattern shields as well as the safety depression rails and ready-use shell racks fitted in Sydney. (Author's collection)

The Standard Temporary System of fire control, removed and replaced in capital ships and cruisers of the *Ceres* and later classes by the High Angle Control System (HACS), was too heavy and complex for AMCs. However, the Admiralty prevaricated with the allocation of the similarly obsolescent Single Gun System, being removed in the mid-1930s from ships smaller than *Ceres*, as it struggled to address fleet anti-aircraft defence in its entirety. In the end, only a handful of RN AMCs received the Single Gun Unit (SGU).

Kanimbla shipped a 9-foot FQ2 rangefinder modified for height finding mounted on a bandstand in after control. A fixed voicepipe was fitted to the after TS and thereafter cumbersome flexible voicepiping ran between the TS and each gun for passing orders and height. Two sound powered telephones, wired together, provided communications between fore and after control. The ship therefore had no way of calculating or transmitting fuze settings for controlled anti-aircraft fire; local barrage control, ostensibly aided by hand-held deflection calculators, was the only option. Nor was the close-range armament any improvement, the entire outfit consisting of the original 1926 provision of two single .303in Lewis machine guns, one per bridge wing.[11]

The after 6in Mk VII gun on the poop deck. Electrical fire control wiring and voicepiping can be seen at the front of the PIII mounting pedestal. Above the breech at the rear of the gun is the night loading lamp with the breech mechanism lever behind the gun. Below the breech is the gunlayer's platform; note the hand elevating wheel at chest height, the telescope at head height as well as the adjacent range dial and smaller deflection dial for manual setting of the gun sights. (Con Cannon)

Ammunition Arrangements

Ammunition supply plans, prepared during 1938, were modified by dockyard and ship's officers during initial conversion inspections. Forward and after watertight cordite magazines and shell rooms were required for the main armament and a single midships magazine would supply fixed-cartridge anti-aircraft ammunition. The forward magazine and shell room were co-located in No 2 hold with the magazine to port and shell room to starboard. The after magazine resided between the shaft tunnels in No 4 hold with the shell room directly above while the HA magazine was built in No 3 hold. All were kept below the waterline for optimum protection and to enhance ship stability. Each 6in magazine provided cordite cartridges in cardboard Clarkson cases through flash-tight scuttles to adjacent handling rooms and were fitted with seawater spraying and flooding arrangements. All compartments were of steel construction as opposed to the wooden structures planned for emergency conversions.

Hatches Nos 2, 3 and 4 were plated over at each deck level up to and including E deck above the waterline as well as the weather deck to aid watertight subdivision. Steel trunking from cordite handing and shell rooms, with additional trunking servicing the HA magazine, was built to protect the ammunition up to small hatches inset at weather deck level adjacent to the guns they served. Existing 220V electric cargo winches acted as hoists in a similar arrangement to older cruisers with open gun mountings. Two forecastle winches delivered cordite and shells to the well deck adjacent to the No 2 guns, supplying them as well as the No 1 guns via a loading platform beneath No 1 hatch. Winches aft of No 4 hatch lifted ammunition to the No 3 guns while ammunition parties traversed two gangways constructed between the promenade and poop decks to supply No 4 gun from the same hoists. HA ammunition was hoisted up to the boat deck through No 3 hatch by an existing winch fitted with lead blocks and walked to the guns.

Ready-use ammunition was provided on deck in the immediate vicinity of each gun. Cordite for the 6in guns was stowed in a single Pattern IV locker per weapon while the projectiles, less susceptible to fire and weather, rested on numerous single shot racks affixed to the deck and hatch covers. Two Pattern 80 lockers per HA gun were secured to the motor casing on the boat deck. While reasonably satisfactory, in the case of prolonged firing of either the main or anti-aircraft armament, it was unlikely that ammunition resupply arrangements were capable of sustaining a high rate of fire from the exceedingly deep magazines once ready-use ammunition was expended.

Depth Charges

Nine of the 15 AMCs lost during the First World War fell victim to German submarines, and by the end of the conflict several vessels carried depth charges as a limited deterrent. Depth charges were issued to AMCs and cruisers primarily to keep a U-boat submerged while exiting the area. However, unlike their regular counterparts, AMCs were not fitted with a sonar set to detect approaching torpedoes.[12] Garden Island fitted *Kanimbla* with two simple First World War-pattern depth charge chutes, or 'roll off racks', with hand release gear along with a forged steel davit for loading. Only two charges

The port 3in Mk I gun on the site of the former boat deck tennis court. Note the safety depression rails around the mounting. Also visible is the port sea boat at its davit, the newly-constructed port bridge wing with 36in searchlight, splinter matting to the rear of the flying bridge, and crow's nest on the foremast, as well as the FQ-2 rangefinder atop its platform. (Lt-Cdr Donald Dykes, RANR[S])

Kanimbla's anti-submarine armament of two depth charge chutes, discharged by hand release gear through openings cut in the rear bulwark of D deck. The photo was taken in April 1940. (Con Cannon)

ARMED MERCHANT CRUISER: THE CONVERSION OF HMS *KANIMBLA*, 1939

View of the boat deck behind the direction finding antenna; the photo was taken in June 1940. A motor cutter and two merchant service lifeboats are stowed on deck and being used to air bedding. Looking aft are the two 3in HA guns, while another lifeboat to port and a service motor boat to starboard are on davits. (Lt-Cdr Donald Dykes, RANR[S])

were able to be dropped at a time, rolled over the stern through two openings cut in the bulwark on D deck. Six more charges were immediately accessible lashed to the deck. No dedicated communications between the chutes and the bridge were fitted until depth charge throwers were shipped in April 1941.

Aircraft and Torpedoes

Unlike the Germans, Admiralty intentions to equip British AMCs with reconnaissance aircraft had been scrapped by 1939 as an uneconomical use of scarce resources. It was also considered that torpedo tubes were not required in the trade defence role. The RAN fitted their two ships with Seagull V amphibians handled by ship's derricks, and some RN ships later received surplus catapults and cranes landed from cruisers. However, the RAN also assessed the odds of an AMC ever being in an advantageous position to use torpedoes as unlikely.

Accommodation

Semi-complete conversion guidelines included the fitting of the open mess deck system, otherwise known as broadside messes, for the ship's company. The main junior ratings' accommodation was aft on D deck, where a total of 39 cabin class and first class cabins, furniture and facilities were removed to install eighteen mess tables with forms (stools) as well as communal sculleries for 170 men. In common with the days of sail, each table constituted a 'mess' where the men ate, relaxed and slung their hammocks. Another two smaller messes for artisans and non-watchkeeping day workers were fashioned from cabins farther forward. The Petty

Kanimbla's Light Type director with windscreen atop its platform on the flying bridge; one end of the adjacent FQ-2 rangefinder can be seen above. The horizontal slewing handwheel can be seen on the right-hand side of the unit with the vertical training handwheel directly below. The standard firing pistol for electrical firing has been removed from below the deflection handwheel, which can be seen on the left hand side of the mounting. The director telescope is on top of the unit and cross-levelling gear is mounted to the outside of the windscreen. (Author's collection)

Officers and Chiefs also berthed in this area, either one or two to a cabin, and were provided with small recreational messes. Supply and secretariat personnel were accommodated in former merchant crew cabins below on E deck. The large cabin class lounge and smoking area on C deck was also split into separate recreation spaces for junior and senior ratings.

The officers received individual, unmodified first class cabins on C deck while the engineers utilised their merchant counterparts' accommodation in the after superstructure. Despite there being no shortage of well-appointed cabins, midshipmen were berthed in pairs as their superiors diligently guarded against such unheard-of excesses spoiling the young gentlemen. To this day, *Kanimbla's* recreation spaces remain unsurpassed, in the RAN at least, with the spacious first class dining saloon on E Deck converted into a wardroom and two annexes. Additionally, midshipmen shared the port-side first class lounge on B deck as a gunroom lounge while the starboard space served as the wardroom lounge. All officers also shared the large forward B deck first class area as a smoking room. All of these spaces retained their outfit of first class furniture. Further cabins were converted into various executive, engineering, communications and secretariat office spaces and other more naval requirements, such as venereal disease isolation and prison cells, were also provided for. Whether an oversight or otherwise, of particular note regarding the level of comfort awaiting *Kanimbla's* new ship's company was the fact that she would be the only Australian ship ever to sail for active service with a swimming pool![13]

Overall, the ship had been designed to carry paying passengers in far greater numbers than her naval complement and she deployed with more than adequate 'hotel' services as well as a large and well-equipped sickbay. These attributes greatly enhanced her flexibility as a cruising warship serving in isolated areas. Standards in the spacious broadside messes comfortably exceeded those of regular warships while the senior sailors, and to a greater extent the officers were berthed in relative luxury.

Ship's Boats

The Imperial AMCs required two service cutters with naval pattern releasing gear to act as seaboats as well as two additional motor boats. *Kanimbla* had carried eight large Fleming-type self-propelled lifeboats, but a September 1939 investigation into their conversion for naval purposes found the suggestion impractical. Difficulty was experienced in gathering sufficient boats,

Petty Officer Douglas Miller firing one of *Kanimbla*'s two .303in Lewis machine guns from the port bridge wing; the photo was taken early in 1940. (Con Cannon)

The port 3in HA gun crew drilling on *Kanimbla*'s boat deck during a northern Pacific patrol in early 1940. (Con Cannon)

Kanimbla's port-side 30ft service cutter being manned by No 1 Boarding Party and sea boat's crew in April 1940. The boats were hoisted 75 feet from the water to the boat deck by hand. (Con Cannon)

but the high risk of damage, given the need to recover them in all types of weather with very high ships' sides, made it imperative.

Kanimbla was initially issued with a 32-foot, sloop-rigged drop keel cutter, of the same pattern embarked in HMA cruisers, from reserves in Sydney, and a similar 30-foot cutter was appropriated during October. The cutters, naval davits and releasing gear replaced the forward life boats for boardings, man overboard work and general evolutions. A 35-foot motor boat was slung in existing davits on the port side of the boat deck and a converted 30-foot motor cutter to starboard. Two lifeboats (but not their davits) were removed from the boat deck aft, while boats and davits were removed from B deck to make way for the No 3 guns. Two Fleming boats were retained on chocks on the boat deck for lifesaving purposes and would later provide excellent service landing soldiers in the Persian Gulf.

Ballasting and Buoyancy

Special consideration was given to AMC ballasting and buoyancy arrangements. Ballasting was required to facilitate the optimum draught and trim for speed and stability as a gun platform. The ships were designed to carry cargo deep in the hull and its absence, coupled with added top weight, increased their metacentric height, causing them to ride higher in the water. This reduced the depth of water protection for the magazines and oil fuel tanks as well as offering an even bigger target than the ships already were. Permanent dry ballast would alleviate the problem, as would the more effective measure of filling and pressing fuel tanks with seawater. However, this would affect the ship's range, a most important consideration in a cruiser, and *Kanimbla* would combine both methods.

Adequate damaged stability was essential. Lacking cargo, ballast and a warship's watertight sub-division, experience had shown AMCs liable to capsize if seriously damaged below the waterline, with disastrous results for the crew. Some ships were liable to succumb with two adjacent watertight compartments flooded. The solution was to install empty sealed drums in non-essential compartments along the waterline, where damage was likely to occur, both to preserve stability through their inherent buoyancy and to reduce free surface water effects in flooded compartments. In concert with ballasting, the drums would improve the ship's motion and produce a steadier gun platform.

Kanimbla's fitting out suffered significant delays as a result of confusion surrounding her precise ballasting and buoyancy requirements; every ship was unique and Admiralty advice for British-built vessels was essential. While guidance plans based on P&O's *Carthage* arrived in early September, the Australian authorities required more specific advice. Materials were not ordered until after receipt of detailed instructions on 7 October 1939, rendering the projected completion date of 19 October unachievable.[14] Second-hand drums were not considered suitable, due to the doubtful nature of their previous contents and the necessity to clean and test, so new steel drums were hastily manufactured by the Rheem Manufacturing Company.

The task of lightering, embarking and then stowing blue metal stone ballast and buoyancy material deep in the ship was a protracted one, with local stevedores loading and trimming ballast from 9 October. Holds 1 and 3 were completely filled with 420 and 410 tons respectively, while No 5 hold was not quite filled, 450 tons being shipped to achieve the required trim. The embarkation of drums, and the timber dunnage to secure them, began on 18 October while the ship was dry-docked and continued alongside Woolloomooloo until 3 November. The work also involved the stripping out of a number of cabins and fittings aft on E deck. A total of 13,642 drums – a combination of 44-, 46- and 47-gallon patterns – were secured by 4,000 cubic feet of wooden battens. Another 22,500 cubic feet of timber and bagged cork completed the buoyancy outfit. *Kanimbla's* full load displacement as fitted was approximately 13,000 tons by December 1939 and her draught when fully loaded, with five double bottom tanks pressed up, was 22ft forward and 23ft 6in aft.

Protection and Concealment

Limited splinter protection was provided to critical areas of the ship: 20lb (0.5in) high tensile steel plate was used for the outboard sides and roof of the main transmitting station, and there were chest-high bulwarks surrounding fore control; the after TS remained unprotected. The steering gear compartment was protected by 20lb bulkheads and a 40lb roof, while all magazines, shell and handing rooms were also encased, most likely with additional 40lb plate. Splinter mattresses were fitted around the forward screen, wheel and chart houses, compass platform and after control. They consisted of boiler felt and asbestos sheet wrapped and quilted into 5ft x 2ft x 4in canvas squares, and provided roughly the same level of protection as mild steel.

Protection against moored mines was provided by clump and chain rigged Mk V paravanes, deployed from existing derricks on the forecastle via a system of cables and chains anchored to a clump casting welded underneath the forefoot while in dock. Twelve Type R chemical smoke floats were carried for the purposes of concealment but the white chemical smoke, lacking a steam ship's complementary oil-enriched funnel smoke, was not very effective.[15]

Plans to paint AMCs to resemble warships were reversed by September 1939 advice explaining that camouflage rendered civilian shipping conspicuous. Commissioned auxiliaries, including AMCs, would employ their peacetime schemes to create doubt and either induce the enemy to close an AMC or hesitate to attack an unarmed vessel. However, light colours were toned down and matt paint replaced gloss, as in the remainder of the fleet. Repainted prior to sailing, *Kanimbla* served with her McIlwraith McEacharn black hull, red funnel with black top and dulled white superstructure. The white was replaced by a more discrete light grey in early 1940, and the ship also adopted various funnel schemes and false flags to disguise her British origins. She adopted a hull scheme described as 'lavender' with buff topsides during an April 1941 refit in Bombay before an overall dark grey scheme in the same port during July 1941 as Admiralty policy changed.

Personnel

When committing to convert three Imperial AMCs in 1921, the Australian Government also offered to provide their crews, who would be paid at Admiralty expense. British merchant crew in *Arawa* and *Moreton Bay* were signed on under specific conditions for naval service and bolstered by Australian reservists, but Australians opting to stay in *Kanimbla* were directly enrolled in the Royal Australian Naval Reserve (RANR) as 'Hostilities only' men. The response amongst *Kanimbla's* crew was less than enthusiastic, but luckily most of the engineering department volunteered, which was just as well for a navy devoid of diesel propulsion experience.

Kanimbla deployed with a complement of 32 officers and 259 ratings. Cdr Getting and the commissioned gunner were the only permanent service officers in the ship, and Lt-Cdr Branson resumed his RN status. Five part-time RANR officers served with 24 RANR (Seagoing) merchant service officers with prior naval training or temporary appointments. 184 partially-trained RANR ratings served on the lower deck, with the backbone of the ship's complement being provided by one RN and 32 RAN ratings as well as 42 from the fleet reserve (time-expired regular sailors liable for call-up in an emergency).[16]

Trials, Workup and Deployment

Kanimbla's conversion was completed on 23 November 1939. In addition to the areas discussed above, a range of other work had been completed including the installation of secondary lighting, additional wireless and visual signalling equipment such as Aldis lamps, projectors and mechanical semaphores, flag lockers, darkening ship screens, towing gear, and an after steering position. A mainmast gaff for the ship's colours was also fitted. *Kanimbla* had sailed on 13 and 20 November for sea and gunnery trials in which a number of defects, particularly with the 6in mountings and the blast damage they created, were identified and subsequently rectified.

The new warship thereafter sailed on 27 November to provide her inexperienced crew with an intense 12-day period of workup both at sea and in Melbourne's Port Phillip Bay, approximately twice the allotted period for a cruiser being brought forward from reserve with a regular crew. With many initially suffering sea sickness in heavy weather, the ship's company were drilled in general seamanship, boat lowering, handling and hoisting as well as streaming and recovering paravanes. Considerable time was spent simulating day and night encounters in which guns crews and control personnel practised loading, training, laying, ammunition supply and fire control procedures, culminating in sub- and full-calibre main armament firings. Cdr Getting considered the period highly successful and the gunnery results quite satisfactory by the time the ship returned to Sydney on 8 December. A large number of small defects had shown up, but these were easily rectified before *Kanimbla* finally sailed for her war station, Hong Kong, on 13 December 1939.

Kanimbla's initial China Station service saw her conducting patrols with the aim of preventing German trade through the eastern Soviet Union. She intercepted eleven vessels belonging to nations in the process of being overrun by Hitler's armies, as well as a suspected Soviet blockade runner, and forced them into Allied ports under guard. She was transferred to the East Indies Station in mid-1940 to escort the valuable Mediterranean and Indian convoys transiting around the Cape of Good Hope. Her finest hour was undoubtedly training, leading and largely manning a flotilla of small craft in the amphibious assault on the Iranian port of Bandar Shahpur on 26 August 1941 as part of the Anglo-Russian occupation of that country. At a time when any addition to Allied shipping capacity was exceedingly welcome, *Kanimbla* captured 41,638 tons of Axis merchantmen sheltering in the port while putting a company of Indian infantry ashore. Three German and three Italian ships were taken before they could scuttle, along with two Iranian gunboats and a 6,000-ton floating dock. The Australians thereafter performed one of the greatest salvage feats of the war in raising one of the two German freighters that did manage to sink themselves.[17]

Analysis

The conversion of *Kanimbla* and two other Imperial AMCs in Australia was the result of two decades of deliberate contingency planning and far from the hastily-arranged emergency measure that might be assumed. It also reflected a significant Australian commitment to Imperial defence far from her own shores. As the RAN had advised the Admiralty earlier in the year, the five-week time frame for three semi-complete conversions was unachievable due to a lack of experience, finite engineering capacity and reliance upon the UK for specific specialist advice. *Kanimbla*, largely due to the delays in arranging her buoyant cargo, took 11 weeks and a day to convert and finally sailed for her war station 14 weeks after being taken in hand. In comparison, *Arawa* and *Moreton Bay* had taken just under seven weeks to convert and ten weeks to deploy. Thankfully there was less of an operational imperative to get the Australian ships to sea than in the European war zone, and no untoward ramifications ensued. *Kanimbla*'s experience was not an isolated one amongst overseas conversions.

Kanimbla's fighting attributes should undoubtedly be the main focus of any technical analysis of her transformation into an auxiliary warship. In common with, at a minimum, those AMCs not fortunate enough to receive a full director system and longer-ranged guns, she suffered serious shortcomings in relation to the threat environment in which she was expected to operate. It was not

HMS *Kanimbla* manoeuvring alongside the freighter *Vladimir Mayakovsky* on 20 March 1940. The Russian vessel was intercepted and detained off Japan during patrols to interdict contraband bound for the German war effort. Extensive splinter matting around the navigating and flying bridges is visible as well as the paravane working arrangements of chains at the bow and two 5-ton derricks. (Author's collection)

anticipated that AMC's would fight regular warships or operate in waters exposed to a significant aerial threat. However, actions against enemy merchant raiders were very much in mind, and in this regard the gunnery and direction systems *Kanimbla* was provided with were not up to the task. German auxiliary cruiser conversions took far longer but were undertaken with a care and thoroughness born of a very different operational imperative from that of the British, that of maximum offensive capability while operating in enemy-dominated seas. They were smaller and often faster than their British counterparts, the latter being selected for their superior seakeeping qualities.

A significant disparity in battleworthiness was brought home to the RN in July and October 1940 when HM Ships *Alcantara* and *Carnarvon Castle* were outfought and heavily damaged by the German raider *Thor*, before HMS *Voltaire* was lost to the same opponent in April 1941. Despite mounting 5.9in (15cm) guns of similar antique vintage, the German weapons were modified with increased elevation to achieve a range of approximately 18,000 yards and *Kanimbla* would have found herself outranged by almost 4,000 yards, as had been the case in the aforementioned actions. Despite her decent turn of speed as a liner, post-conversion she was only capable of a disappointing 16 knots, one knot in excess of the minimum Admiralty requirement, and the ship struggled to keep up with the fast trooping convoys she was employed to protect in the Indian Ocean. In action with a faster opponent she would have been outranged, and the enemy would have had little trouble in dictating the range of the engagement and breaking off at will.

The light type director system, which could supply neither integrated elevation transmission nor coordinated electric firing, was derided as entirely unsuitable for long range engagements. Wiring arrangements were also flawed. Guns crews were forced into inefficient local control as electrical systems, utilising undivided and exposed wiring runs, were quickly shorted by shrapnel. Men were also cut down due to a lack of gun shields and inadequate splinter protection in control positions. All of these shortcomings represented lessons learned during the First World War and remedied in the regular fleet. However, contingency planning, affected by questions of finance and resources, awaited the availability of more sophisticated equipment from regular cruisers, and arguably stands as the biggest failing of British preparations for war. *Kanimbla's* wiring runs, while still undivided, were at least run below decks and therefore better protected, and she received shields for her three after guns in Hong Kong. These, as well as the later fitting of improved splinter protection for ammunition parties, control personnel and HA crews, might have seen her fare better in action. However, it should also be noted that in the trade protection role, aside from single-ship encounters, the presence of AMCs capable of inflicting career-ending damage was sufficient to deter merchant raiders from approaching escorted convoys.

Action experience prompted a number of improvements across the AMC fleet. Ships with Mk VII guns were slowly issued with 28lb 2oz, 12 dram cordite SC140 'super charge' propellant and 6crh projectiles to increase their range. *Kanimbla* was issued with supercharges in August 1940, increased to 277 in November

Kanimbla as she appeared with overall dark grey scheme applied alongside in Bombay, India 26–31 July 1941. (Hiram Ristrom)

Kanimbla at War

War Stations

Australian Station:	6 September 1939 – 28 December 1939
China Station:	28 December 1939 – 9 August 1940
East Indies Station:	9 August 1940 – December 1941
Australian Station:	December 1941 – 31 May 1943

Commanding Officers

6 September 1939 – 3 October 1939:	Lieutenant Commander Geoffrey CF Branson RN
4 October 1939 – 9 November 1939:	Commander Frank Edmund Getting RAN
10 November 1939 – 24 March 1941:	Acting Captain Frank Edmund Getting RAN
25 March 1941 – 2 February 1943:	Acting Captain William Leslie Graham Adams RN, OBE
3 February 1943 – 31 May 1943:	Acting Commander Frederick Ross James RAN

Total Steaming

Total Steaming Distance 6 October 1939 – 2 April 1943:	213,044 miles
Total Steaming Time 6 October 1939 – 2 April 1943:	16,164 hours
Average Speed:	13.08 knots

1941, improving her range with the existing 4crh shells to at least 16,175 yards; 6crh shells were received in April 1941, increasing maximum range to 17,870 yards. Small improvements were also made to fire control communications, and a simple AMC fire control table was issued. This was at least an improvement on the initial system, which relied exclusively on human input and output. Remote-controlled searchlights and depth charge throwers were fitted in April 1941.

Kanimbla's HA armament and its control system were inadequate against virtually any form of attack by modern aircraft. AMCs suffered from the enormity of the problem facing the late 1930s Admiralty in adequately improving anti-aircraft fire power across the fleet, and their needs were never addressed. *Kanimbla* did not receive any form of director control while in service, rendering her heavy HA armament virtually useless aside from boosting morale. Her close-range anti-aircraft armament was not augmented until the arrival of four single 20mm Oerlikons in April 1942 and a further two that August.

Thankfully, buoyant cargo was a success in that numerous AMCs sustained heavy underwater damage during the war but only *Rawalpindi* capsized, and then only after shellfire from the battleship *Scharnhorst* detonated a magazine and broke her in two. Ten ships were lost to submarine torpedoes but another five survived. *U-99* expended no fewer than six torpedoes to despatch the shattered, but still floating wreck of HMS *Patroclus* on 4 November 1940. HMS *Jervis Bay*, sister to *Arawa* and *Moreton Bay*, was knocked out of the fight with the heavy cruiser *Admiral Scheer* in 15 minutes, yet remained afloat and drawing German fire for a further two hours while the majority of her convoy escaped. Compartments filled with steel drums were also found to provide a level of protection from bursting shells to vital machinery in their vicinity. However, some sinkings from single hits highlighted the need for even more drums, and *Kanimbla* received a further 4,600 during an April 1941 refit.

Despite raider hunting in the Indian Ocean, including searching for the *Admiral Scheer*(!), *Kanimbla's* ship's company never faced the trials to which some of her less fortunate sisters were exposed. Aside from warning shots fired ahead of recalcitrant merchantmen, she only ever used her armament in action once. A rail locomotive escaping from Banda Shahpur was engaged as *Kanimbla* also fired at aircraft flying over the anchorage. Only two 6in guns crews were available due to boarding party requirements, and the locomotive escaped unscathed. Fortunately the HA system proved incapable of hitting the three misidentified RAF Hurricane fighters being targeted.[18] In Singapore at the outbreak of the Pacific War, *Kanimbla* found herself escorting the first convoy to leave the area for the Australian Station, her crew having just witnessed HM Ships *Prince of Wales* and *Repulse* sailing towards their eventual destruction at the hands of the Japanese naval air force.[19]

Conclusion

Kanimbla's war fighting capabilities left much to be desired and would have seen her at a considerable disadvantage against a comparable German auxiliary cruiser, especially as first converted in 1939. However, she was modern and her sea-keeping abilities, engineering reliability, flexibility and endurance, as well as a well-trained and resourceful ship's company ensured her success in the remote, non-combat, economic warfare operations on which she was employed. AMCs were either returned to trade or reconverted for other naval use as the war at sea progressed in the Allies' favour.

Kanimbla found herself recommissioned into the RAN on 1 June 1943 as an infantry landing ship to join HMA Ships *Manoora* and *Westralia* spearheading an Australian amphibious assault force. She excelled in this role during the latter part of the Pacific War, finally possessing a powerful HA armament that succeeded in beating off kamikaze attacks during the Philippines landings.[20] She was not returned to her owners until December 1948. Sold in 1961 to Japanese interests and renamed *Oriental Queen*, she was eventually broken up in Kaohsuing, Taiwan, during 1974. Until the end of their days, all of *Kanimbla*'s RAN ship's companies were proud of the ship's claim to having captured more shipping (neutral included) than any other vessel flying the White Ensign during the Second World War.

Acknowledgments:

This article is dedicated to the author's grandfather, former Leading Seaman Con Cannon, RANR (1919–2015). Serving as a merchant seaman in *Kanimbla* from 1937, he decided in September 1939 that if the Navy was taking *his* ship, they would have to take him too. Many thanks also to ex-Leading Seaman Hiram Ristrom, RANR's *Kanimbla* Association (disbanded 2014) for his support over the preceding decade as well as Chief Petty Officer Imagery Specialist Oliver Garside for photographic assistance.

Principal Sources:

National Archives of Australia, HMAS *Kanimbla*'s Ship's Book; The National Archives (UK); RAN Historical Collection; RAN Sea Power Centre – Australia, HMAS *Kanimbla* Ship's History File; HMS *Excellent* collection; Australian National Maritime Museum, McIlwraith McEacharn Collection; Australian War Memorial, Reports of Proceedings and War Diaries; interviews – ex-Leading Seaman Con Cannon, RANR (2005–2008), ex-Lieutenant Frank Newman, RANR (2006), ex-Petty Officer Harry Stirk, RANR (2007), ex-Lieutenant Commander Don Dykes, RANR(S) (2008) and ex-Leading Seaman Hiram Ristrom, RANR (2018).

Endnotes:

[1] Decisions taken in the early months of the war resulted in only 56 AMC conversions across the Empire with 51 serving in the RN, two in the RAN and three in the Royal Canadian Navy.

[2] These two ships were never considered part of Admiralty calculations.

[3] Many ships requisitioned as AMCs during the First World War were found unsuitable and returned. The giant trans-Atlantic liners were fast but consumed fuel at a prodigious rate, while some smaller ships struggled in the wild seas north of the British Isles. The loss of the 4,985 GRT *Clan MacNaughton* with her entire crew in 1915 was attributed to a severe gale.

[4] The 14,200-ton *Moreton Bay* and 14,500-ton *Arawa*, both 'Bay' class steamers owned by the Aberdeen and Commonwealth and Shaw Saville & Albion lines respectively, had originally been built by the Australian Government and stiffened during their construction in 1921–22 before being sold with their three sisters in 1928.

[5] The RAN authorised the requisitioning of the passenger ships *Manoora* and *Duntroon*, with the latter vessel swapped for *Westralia* following an unsatisfactory engine survey, upon a perceived German raider threat and a desire to deploy regular cruisers on active overseas service with the RN.

[6] The ship was assessed to be in an excellent, well-maintained condition and classed +100 A1.

[7] AMCs with Mk VII guns were fitted with the Pattern 710 range clock with a maximum range prediction of 14,000 yards. The modern 6in Mk XXIII gun fired 112lb shells to 24,500 yards at 6–8 rounds per minute.

[8] The results of initial gunnery trials necessitated significant additional stiffening work for both P3 and S3 mounting supports.

[9] The PIII was designed with main and auxiliary electrical circuits for the purpose, but the only guns prepared for electric firing in AMCs were the 6in Mk XII on PVII and PVII* mountings for the 11 ships eventually fitted with full director firing systems in UK yards.

[10] The modern 4in Mk XVI gun fired 35lb shells to 39,000 feet at 15–20 rounds per minute.

[11] The Admiralty had endorsed 2pdr Mk VII quadruple pom-pom and 0.5in machine guns, but none were available for any AMC at that time.

[12] A sonar set was considered desirable in 1926 but was never available due to higher priorities.

[13] The swimming pool was a large semi-fixed, salt water basin residing within No 4 hatch on B deck. It was drained before the ship sailed from Sydney, as well as in colder climates, but was filled using fire hoses when required in hot weather. The ship's company fully appreciated this asset during the ship's Indian Ocean and Persian Gulf service.

[14] As a possible explanation of the confusion, early Admiralty advice saw the two 'Bay' class ships ballasted but not embarking buoyancy drums during their conversions.

[15] The Admiralty subsequently decided to fit CSA smoke producing apparatus to all twelve Imperial diesel AMCs, including *Kanimbla*, in late 1940.

[16] As the men were serving on Admiralty account, they were incidentally entitled to a rum ration in lieu of the beer provided in Australian vessels. The arrival of a lighter loaded with barrels containing approximately 240 gallons of Bundaberg rum, to be stowed in the newly renamed spirit room, was greeted with universal approval.

[17] The German *Marienfells*, *Sturmfells* and *Wildenfels* along with the Italian *Barbara*, *Bronte*, and *Caboto* were captured, despite a number being sabotaged and set alight, along with the Iranian *Chahbaaz* and *Karcas*. *Weissenfels* sank in deep water but the valuable modern *Hohenfels* was dragged onto a sand bank while sinking and subsequently salvaged. The floating dock became Admiralty Floating Dock XXVII in 1942.

[18] The RAF had advised that any aircraft sighted would be enemy and at that stage of war, in the Middle East at least, the chances of aircraft being friendly were decidedly low.

[19] The well-known photos of the capital ships of Force Z leaving port to confront the Japanese were taken by *Kanimbla*'s ship's photographer as she lay at anchor in the Johore Strait.

[20] January 1945: 1 x 4in Mk XVI* on Mk XX mounting, 2 x single Mk VIII* pom-poms on Mk VIII* mountings, 12 x single 20mm Oerlikons (3 x Mk II guns on Mk IV mountings, 9 x Mk IV guns on 3 x Mk IV and 6 x Mk V mountings), 4 x .303in Vickers, as well as the original 3in guns. By the Borneo landings of June 1945 she carried the 4in, pom-poms and 6 x single 40mm Mk I Bofors as well as 8 x 20mm Oerlikons. The long-serving 3in had finally been removed.

THE FRENCH BATTLESHIP
BRENNUS

The battleship *Brennus* was conceived during the early 1880s, suspended by Admiral Aube, and completed to a radically modified design in 1893. She incorporated a number of striking new features, but poor stability meant that she had to be completely rebuilt before she was fit for service. **Philippe Caresse** looks at the tortured history of this powerful vessel, which never fired a gun in anger.

To follow the battleships *Hoche*, *Marceau*, *Neptune* and *Magenta*, the *Conseil des Travaux* on 10 March 1882 requested that the distinguished naval architect Louis de Bussy draw up plans for two new battleships, to be named *Brennus* and *Charles-Martel*. The first was to be laid down at Lorient naval dockyard, the second at Toulon.

On 3 May of the same year, de Bussy submitted his proposals to the Council for its consideration; they were discussed at a meeting twenty days later. De Bussy submitted an unconventional design for a ship with a length of 105 metres between perpendiculars, a beam of 19.36 metres, and a displacement of under 10,000 tonnes. Two twin enclosed turrets mounting 34cm (13.4in) guns were disposed *en echelon* amidships, as in the contemporary British *Inflexible* and the Italian *Dandolo*. The secondary armament comprised eight 14cm (5.5in) guns of which four were in twin turrets fore and aft. Four widely-spaced funnels combined with minimal superstructures added to their revolutionary appearance. A further innovation was a three-shaft propulsion system, which was designed for a speed of 17 knots and an endurance of 5,000nm. Protection for the hull comprised a 44cm (17in) belt and an armoured deck with a maximum thickness of 75mm (3in).

The Council debated at length the proposal for an enclosed twin turret; concerns were expressed regarding poor visibility and the inability of a twin turret to match the rate of fire of two single guns. Nevertheless, on 6 September the plans were submitted to the Minister of Marine, Vice Admiral Jauréguiberry, and the final project was approved in December.

Despite the relatively smooth progress thus far, on 18 September 1883 the new Minister of Marine,[1] Vice Admiral Peyron, asked the *Conseil des Travaux* to reject de Bussy's plans and to turn instead to engineer Charles-Ernest Huin with a view to securing a modification of the *Marceau* design. The reason for this *volte-face* was concern over the concept of the 'central redoubt', with the ends of the ship unprotected but tightly subdivided as in the British and Italian ships, which the Minister considered inferior to the complete belt of the latest French battleships. The ram of the de Bussy design was to be suppressed and a straight prow adopted similar to

The rudder of *Brennus*, weighing 17.10 tonnes, is transported from the Guérigny works to Lorient by rail. (DR)

that of the latest ocean liners. The after part of the ship was to be lengthened by ten metres, and the iron plating of the outer hull replaced by steel.

As far as the ram was concerned, it was the Inspector General of the Corps of Constructors, Marielle, who was categorically opposed to it. The Council also wanted the wing turrets and the turrets fore and aft relocated. The secondary armament was to comprise seventeen 14cm guns, of which one was to be mounted close to the bow. During 1883 the materiel necessary for the construction of the two ships was ordered from the various industrial contractors. The *Conseil des Travaux* would hold meeting after meeting to discuss changes in the technical characteristics.

Unfortunately for these ships, on 7 January 1886 one of the most fervent adherents of the *Jeane Ecole*, Vice Admiral Aube, was appointed Minister of Marine. The *Jeune Ecole* favoured the construction of large numbers of small units (*la poussière navale*) armed with mobile torpedoes over the battleship, which it considered costly and obsolete. Thus, on 27 January a ministerial directive ordered the suspension of work on *Brennus* and *Charles-Martel*. The latter ship would ultimately be cancelled, and the material assembled at Toulon re-used for the construction of the battleships *Carnot* and *Bouvet* of the 1890 programme. For *Brennus*, the project would be revived only after the departure of Aube from the Ministry.

On 17 October 1887 a note announced the revised characteristics of the new battleships: three 34cm guns in barbette mountings or enclosed turrets, ten 16cm guns, four 65mm guns, and eight 47mm and eight 37mm anti-torpedo boat guns. Protection would comprise an armoured belt 45cm thick, reducing to 25cm at the ends, a 60mm protective deck, and 45cm of armour on the barbettes and turrets. Maximum speed was to be 18 knots, and endurance 4,000nm at 10 knots. No mention was made of a ram bow and Huin, who saw little value in a ram, drew up plans for a vessel with a straight stem.

De Bussy's original hull-form was retained, as it did not preclude the required modifications. In January 1888 Huin submitted the fruits of his labour to the *Conseil des Travaux*. The new plans featured the adoption of the Belleville water-tube boiler, a revised arrangement of the magazines, and enclosed turrets that provided full protection for the training, elevation and reloading mechanisms without any increase in weight.

Part of the 164.7mm secondary armament would likewise be mounted in enclosed turrets. This was a new departure for the *Marine Nationale*: the turrets were trained by hydraulic rams that enabled the gunlayer to keep his gun on a moving target. On 16 May the *Conseil des Travaux* and de Bussy, now with the rank of *Ingénieur général*, once again examined the project and requested minor modifications.

On 12 November 1891 Senator Barbey, a former Minister of Marine, proposed to fit *Brennus* with a ram. However, twelve days later the *Conseil des Travaux*

Characteristics

Length	114.46m oa; 110.30m pp
Beam (wl)	20.62m
Depth of keel	7.79m
Draught	7.56m fwd / 8.18m aft
Freeboard forward	4.32m
Normal displacement	10,983 tonnes
Full load displacement	11,370 tonnes
Surface of underside	2,985m^2
Complement	667 officers & men

responded to the effect that as the ship had now been launched, this was no longer possible – a response confirmed by Lorient dockyard, where the ship was being built. Fitting out was delayed by the late delivery of the two propeller shafts ordered from the Navy's propulsion establishment at Indret, which in turn meant that the trials date had to be put back six months.

Despite her protracted building time – eleven years from laying down to completion – *Brennus* introduced a number of new features that included enclosed turrets, a powerful secondary battery, water-tube boilers, and a system of hull subdivision that was a precursor to Bertin's cellular layer. Cost of construction was 25,083,675 French francs (equivalent to £1 million sterling).

Hull and Protection

The hull was of steel with a cellular type of construction. There were eighteen watertight bulkheads below the armoured deck, and the double hull extended for the full length except for the bow compartments. As with the later battleships of the *Flotte d'échantillons* the sides of the ship had a pronounced tumblehome.

The armoured belt, which had a total height of 2.10 metres, was 400mm thick amidships, tapering to 250mm at its lower edge; these thicknesses were reduced to 300/200mm at the bow and 300/180mm at the stern. It was manufactured by Schneider, Saint-Chamond, and Forges & Aciéries de Châtillon. The armoured deck was 60mm thick; the plates were supplied by the Forges Nationales de la Chaussade at Guérigny. The sides of the conning tower were protected by 120mm plates, and the communications tube was formed by hoops with a thickness of 150mm.

Armament

Brennus had a twin 340mm turret forward and a single turret aft. The turrets were built by the Farcot company, whose factories were located at Saint-Ouen, in the northern outskirts of Paris. The two guns in the forward turret were mounted in the same cradle and elevated together. Training of the turrets was by four crossing chains on a drum fixed to the pivot tube and operated by hydraulic rams. Only 39 rounds per gun were provided. Although in theory the firing cycle should have been one round every two minutes, in practice it was twice as slow;

Armament 1895

Three 340mm 42-cal Mle 1887 in one twin, one single turret
Ten 164.7mm 45-cal Mle 1893 QF in four single turrets, six casemates
Four 65mm 50-cal Mle 1891 QF in single open mountings
Fourteen 47mm 40-cal Mle 1885 QF
Eight 37mm 40-cal Mle 1885 QF
Six 37mm Mle 1885 revolver cannon
Four 450mm torpedo tubes

Calibre	Shell weight	Muzzle velocity	Firing Cycle
340mm	420kg (CI)	640m/s	1.5rpm*
164.7mm	45kg (CI)	800m/s	2/3rpm
65mm	4.10kg	715m/s	8rpm
47mm	1.5kg	610m/s	12rpm
37mm	0.5kg	435m/s	20–25rpm

Calibre	Angle of elevation	Range
340mm	+10°/-4°	10,900m
164.7mm	+20°/-6°	10,450m
65mm	+18°/-20°	5,500m
47mm	+24°/-21°	4,000m

* Never achieved – see text.

this was attributed to the operation of the hydraulic mechanism of the breech and the loading tray.

The forward turret and its barbette were armoured with plates 455mm thick, the after turret with 405mm plates. There were 80mmm plates on the roof and the armoured hood had a similar thickness. The armour for the turrets was supplied by the Marrel steelworks.

The secondary armament comprised four 164.7mm guns in single turrets and six in casemates. The turrets were protected by armour plates 100mm thick, and the central redoubt for the casemate guns had the same level of protection.

Brennus was fitted with four above-water trainable tubes for 450mm torpedoes; two torpedoes were provided for each tube. The Mle 1889 torpedo had a length of 5.01 metres and an overall weight of 502kg that included the 75kg warhead. There was a dual range/speed setting for the torpedoes: 600 metres at 29.5 knots and 800 metres at 28 knots.

Machinery

Brennus was the first French battleship to be fitted with the new water-tube boilers, which were built by the Delaunay–Belleville company at their works in Saint-Denis (Paris). The boilers were rated at 17kg/cm^2 and dimensions were 2.8m long by 3.47m high; grate area was 102.24m^2 and heated surface area 3,254.56m^2. There were 32 boiler bodies in six paired boiler rooms.

The boilers supplied steam for two vertical triple expansion (VTE) reciprocating engines built by the

Propulsion Machinery

Boilers	32 Belleville
Engines	two 4-cylinder VTE
Propellers	two 4-bladed; diameter 5.40m
Rudder	single; surface area 24.33m^2
Designed power	12,500CV
Max power on trials	14,133CV
Speed	17 knots (des); 17.10 knots (trials)
Coal (full load)	706 tonnes
Endurance	2,805nm at 11 knots
	757nm at 17 knots

Navy's in-house propulsion establishment at Indret. Each of the engines had a theoretical power rating of 6,250CV, and shaft revolutions at maximum speed were 90rpm. The two steel propellers were likewise designed and manufactured by Indret.

Onboard electricity was supplied by four H-type dynamos each rated at 400A/80V and powered by piston engines. Lighting for the working spaces and mess decks was provided by 345 electric lamps below the armoured deck and 182 above. There were also 31 lamps for the signal projectors and navigation lights. Total current draw for lighting was 330A.

Other equipment powered by electricity was as follows:

– two Thirion no 5 bilge pumps each rated at 30 tonnes per hour
– seven ventilators for general purposes rated between 10,920m^3 and 6,084m^3 per hour
– two Indret ventilators for the engine rooms each rated at 50,000m^3
– six ventilators for the boiler rooms each rated at 80,000m^3
– six ash hoists manufactured by the Maupéou works each rated at 1,080kg per hour

Equipment

When *Brennus* was completed, she was equipped with two large sighting tables for fire control located in the conning tower. A ministerial directive dated 10 April 1908 decreed the installation of two Barr & Stroud 2-metre rangefinders, to be mounted on pedestals on either side of the navigation bridge.

The military masts, of which there were two when the ship was first completed, were constructed of hardened steel. The foremast had a height of 22.35 metres above the deck and a diameter of 1.80 metres.

The 60cm searchlight projectors were of the Mangin type. They were disposed on the upper platforms of the military masts and on the upper deck (*pont des gaillards*). When the military mainmast was landed, the after projector was relocated to a circular platform atop the after bridge deck.

The capstan motor was a Bossière model and was rated at 35 tonnes. There were also five Bossière single- or

Equipment

Searchlights	six 60cm Mangin projectors
Boats	two 10-metre steam pinnaces
	one 7.65-metre White launch
	one 11-metre pulling pinnace
	one 10-metre pulling cutter
	two 9-metre pulling cutters
	four 8.5-metre whalers
	two 5-metre dinghies
	two 3.5-metre punts
	two 5.6-metre Berthon canvas boats
Anchors	two Marrel 13.72-tonne anchors
	two 3.61-tonne kedge anchors

double-ended steam winches. The complement of boats is detailed in the accompanying table. The two 10-metre steam pinnaces and the 9-metre cutters could mount a 47mm gun and a 30cm searchlight projector.

Launch and Trials

After many false starts, *Brennus* was officially laid down at Lorient naval dockyard on 2 January 1889. The protective plating for the armoured deck was delivered in October of the same year, and the boilers installed during 1891. The engines were mounted from 1 December 1891, supervised by engineers Jouasse and Touze of *Construction Navales*.

Brennus was launched on 17 October 1891; the Minister of Marine, Senator Edouard Barbey, and numerous civilian and military personalities attended. In order to make an impression on the general public who had flocked to see the event, the main turrets were in place but with wooden guns. After launch the hull was towed directly to the fitting-out quay.

In April 1892 at Le Havre, when one of the 164.7mm turrets was being transferred from a barge to the cargo ship *Caravane*, which was due to transport it to Lorient, the cables of the crane gave way and the turret, which weighed 55 tonnes, crashed down onto the barge. The latter sank, carrying to his death one of the workers and leaving two others badly injured in the water. The turret was subsequently recovered and duly delivered to *Brennus*.

On 12 January 1893, the boilers were lit for the first time and static trials conducted with the starboard engine, followed on 8 May by the port engine.

On 17 June the stability trial was undertaken. All the guns were trained to starboard and the entire crew was assembled on the same side. The ship took on a list of 28 degrees, and the barrels of the 164.7mm casemate guns were half-submerged in the water. The metacentric height was calculated to be 0.749 metres at a displacement of 10,815 tonnes with a maximum draught of 7.78 metres. The cause of the instability was principally due to the

The launch of *Brennus* at Lorient on 17 October 1891. Note the absence of a ram, which was unusual for a battleship of the period. (Musée de la Marine)

The imposing mass of the superstructure of *Brennus* is particularly in evidence in this photo taken during fitting out. Note the Bullivant anti-torpedo nets, which would subsequently be removed. (Musée de la Marine)

massive superstructures, funnels and military masts. It appears that the thickness of the upper edge of the armoured belt, which was 20cm below the waterline, prevented the ship from righting herself.

The day after this catastrophic inclination trial, the workshop responsible for the manufacture of the propellers notified Lorient that there would be a seven-month delay in delivery. It was therefore proposed to use the propellers of the battleship *Neptune* that, after small adjustments, could be used in the interim.

Naturally the partisans of the *Jeune École* took full advantage of the problems with the new battleship to initiate a negative campaign in the press. The senior officers of the Navy were taken to task over the affair, and the Minister requested an enquiry to establish who was responsible for the failings.

On 2 August, *Brennus* left Lorient for Brest. Forty-eight hours later, her commanding officer received authorisation to disembark 70 tonnes of inessential materiel in order to lighten the ship. On 5 September a preliminary trial of the machinery took place with 16 boilers lit. On 17 October, the port of Brest submitted a first proposal to lighten the ship; this was discussed by the *Conseil des Travaux* on the 30th. The report concluded that there was 412 tonnes of excess weight, and that most of this was high in the ship. As the ship had been more than twelve years in gestation and build, the designers had lost control of the weights in their attempt to incorporate the latest equipment in an established hull. The other cause of the lack of stability was the pronounced tumblehome of the sides of the hull, adopted in order to save weight, free up arcs for the guns mounted on the beam, and to create a steady gunnery platform with moderate roll.

In order to remedy the principal flaws in the design, the following modifications were implemented:

– The bridge extensions above the main turrets and to the sides were reduced.
– The height of the superstructures was reduced by 1.9 metres by suppressing the deck linking the fore and after bridges and lowering the admiral's chart house
– The Bullivant torpedo nets and the boat rails were removed; the boats were now carried one deck lower when stowed.
– The military mainmast was reduced to a stump, and a simple pole mast stepped in its place.
– The reserve anchor was disembarked.
– The tops of the funnels were modified.
– Various inessential items of equipment were landed.

At virtually the same time, advantage was taken of the ship's docking for modifications to fit the correct propellers and to replace the original 164.7mm guns by the latest Mle 1893 QF model. Finally, in order to improve stability a light caisson structure with a timber frame and steel plating (*soufflage*) was fitted at the waterline over the armour belt.

Brennus in Lorient roads before the major reconstruction which took place at Brest. The boat booms are fully extended. (DR)

A fine view of *Brennus* in her original configuration. (Bougault)

Brennus during her power trials on 9 January 1896; she attained a maximum speed of 17.1 knots with natural draught. (DR)

The ship was officially manned for trials on 1 August 1895, and conducted preliminary sea trials on 21 and 30 August. The fuel consumption trials were run on 19 September. Engine problems resulted in a spell in the dockyard at Brest 12–30 October. On 22 December a second stability trial was judged to be satisfactory: metacentric height was 1.047 metres. On 9 January 1896 *Brennus* ran her full power trials, during which she attained a speed of 17.10 knots with 14,133CV. Gunnery trials began on 14 January and were entirely satisfactory.

The ship's first commanding officer, CV Besson, made the following assessment of *Brennus*:

> The ship has good qualities but also major defects. The most serious problem is that when engaging on the broadside the ship is effectively unarmoured. The 340mm turrets are too heavy and are not balanced, and when both are trained on the same beam the heel of the ship is such that the upper edge of the belt is level with the water. This means that the only side protection against enemy shell is provided by the 10cm upper belt. The thickness of the armour belt at its lower edge – only 25cm – and the large turning circle are further weaknesses.
>
> The positives are the ship's high speed and the disposition of her guns. The 164.7mm turret guns are exceptionally well placed, although it is a pity that they cannot be trained manually. In terms of habitability the ship is superb; the officers quarters are particularly well-appointed, and the messdecks for the crew more than satisfactory.
>
> All in all, despite her faults, *Brennus* is a magnificent fighting ship, as powerful as any other battleship afloat.

On 11 January 1896, a dispatch from the Minister announced the entry into service of the battleship and her assignment to the Mediterranean Squadron. *Brennus* left Brest on the 20th of the same month. Having passed the Strait of Gibraltar, the ship began her 24-hour trial off Barcelona that took her to the Cape of Antibes. The trial, conducted at a constant speed of 16 knots, revealed the poor quality of the condenser tubes, which subsequently had to be replaced during a refit carried out between 1 February and 15 March. On 1 April the C-in-C of the Mediterranean Squadron, Vice Admiral Gervais, hoisted his flag in *Brennus*.

Showing the Flag

On 14 April 1896, the fleet embarked on a cruise to Corsica and North Africa. *Brennus* followed in the wake of the battleship *Redoutable* and was followed by *Marceau*, *Magenta*, *Neptune*, *Amiral Baudin*, *Courbet* and *Dévastation*; numerous cruisers and torpedo boats accompanied the battleships. There were port calls to Bastia, Saint-Florent, Calvi, Propriano and Porto Vecchio. Afterwards, the ships sailed for La Goulette, the port of Tunis, and *Brennus* would be the second battleship to enter the lake at Bizerta after *Redoutable*. This was a major event for the period, as it demonstrated to the British and the Italians that the French

Navy now had an important base on the Tunisian coast.

Brennus would return to Toulon on 12 June. The summer grand manoeuvres began on 9 July with visits to Hyères and Algeria. It was during a port call in Corsica that Admiral Gervais almost disappeared, drowned. The admiral, intending to stretch his legs on the stern walk, forgot that the duckboards had been removed and fell onto the upper edge of the armour belt directly beneath. Badly bruised and shaken, he was bed-ridden for several days. The Mediterranean Squadron returned to Toulon

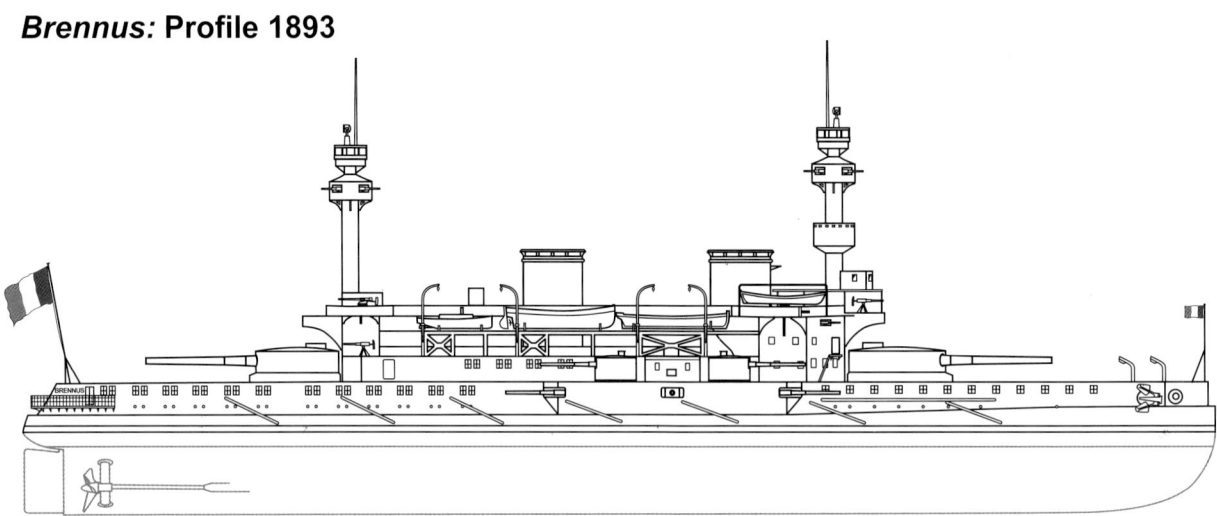

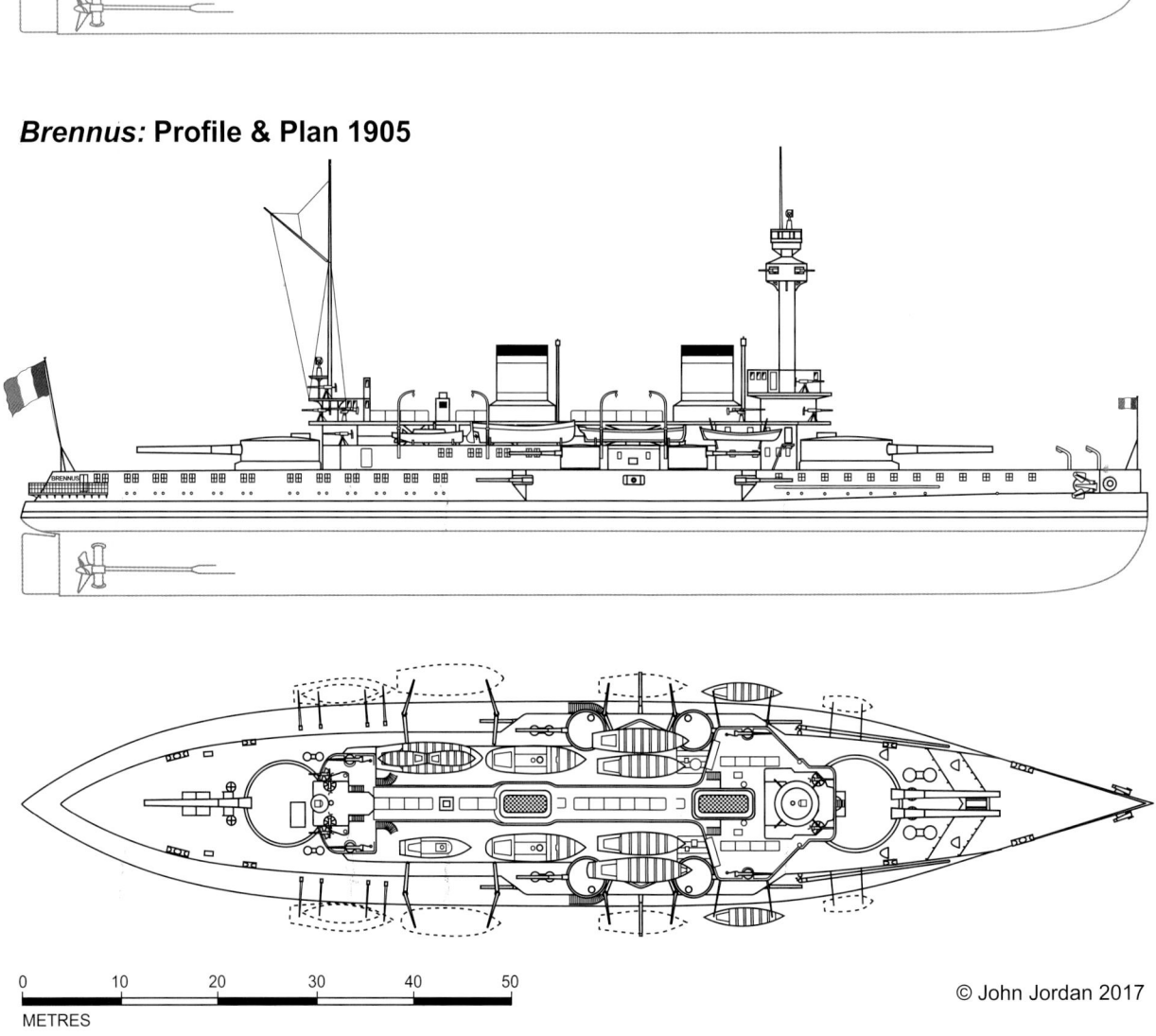

***Brennus:* Profile 1893**

***Brennus:* Profile & Plan 1905**

© John Jordan 2017

on 1 August. On the 24th of the same month *Brennus* was towing a target for the squadron during a gunnery exercise. Following a pointing error, a revolver cannon on the destroyer *Vautour* fired on the bridge of the battleship; luckily there was only one casualty, who was struck in the foot. On 15 October Admiral Gervais was replaced by Vice Admiral Cavalier de Cuverville.

The year 1897 was unremarkable; there were the customary visits to Villefranche, Les Salins d'Hyères and Golfe-Juan. During August the new battleship *Charles-Martel* joined the Squadron. On 15 October, Vice Admiral Humann hoisted his flag in *Brennus*.

On 16 Apil 1898, President Félix Faure was received on board *Brennus* when the battleship was moored at Villefranche. Following various manoeuvres, President Faure suggested a race between the battleships. After two hours at full power, *Brennus* was ahead of the three ships of the *Marceau* class and *Formidable*, but trailed in the wake of the new battleships *Charles-Martel* and *Carnot*.

The summer manoeuvres occupied the crews from 8 to 30 July, and during September Minister of Marine Édouard Lockroy embarked in *Brennus* to observe gunnery exercises that ended with the destruction off Cape Cépet (Saint-Mandrier) of the old gunboat *Arrogante*.

Vice Admiral Fournier became C-in-C on 1 October, and shortly afterwards the Fashoda crisis broke. Tensions with Great Britain prompted the recall of those on leave on 18 October. The two battle divisions, which comprised *Brennus* (flagship, 1st Div), *Masséna* and *Bouvet*, *Charles-Martel* (flagship, 2nd Div), *Carnot* and

Commanding Officers

CV Besson	25 Jul 1894 – 1 Apr 1896
CV Puech	1 Apr 1896 – 15 Oct 1896
CV Gadaud	15 Oct 1896 – 15 Oct 1897
CV Boisse	15 Oct 1897 – 1 Oct 1898
CV Boué de Lapeyrère	1 Oct 1898 – 15 Oct 1900
CV Le Léon	15 Oct 1900 – 18 Aug 1901
CV Donin de Rosières	18 Aug 1901 – 20 Aug 1903
CV Sauvan	20 Aug 1903 – 10 Aug 1905
CV Guillou	10 Aug 1905 – 25 Aug 1907
CV Lecuve	25 Aug 1907 – 6 May 1908
CV Duvauroux	06 May 1908 – 15 Nov 1909
CV Maudet	15 Nov 1909 – 15 Nov 1911
CV de la Riviere	15 Nov 1911 – 1 Jan 1912
CV Vincent	1 Jan 1912 – 11 Oct 1912
CV Jeanselme	11 Oct 1912 – 21 Apr 1913
CV Jezequel	21 Apr 1913 – 18 Sep 1913
CV Durcy	18 Sep 1913 – 26 Jun 1914

Jauréguiberry, embarked combat munitions. The international crisis ended on 5 November, and the ships returned to their customary activities.

At the time, Captain Boué de Lapeyrère, the future C-in-C of the *Armée Navale* during the Great War, was the commanding officer of *Brennus*. An unusual anecdote has been related by a number of authors concerning the figurehead of the battleship – *Brennus* was the last Navy vessel to have this feature. When Boué de Lapeyrère found that his ship lacked a figurehead, he salvaged a bust of the

The new silhouette of *Brennus* following the replacement of the military mainmast by a simple pole. She is seen here in Toulon roads; on the left can be seen the new battleship *Charles-Martel*. (DR)

The twin 340mm turret of *Brennus*, with the conning tower and the navigation bridge behind it. (DR)

Brennus in dock. The rudder and one of the two four-bladed steel propellers, which had a diameter of 5.40m, are in the foreground. (DR)

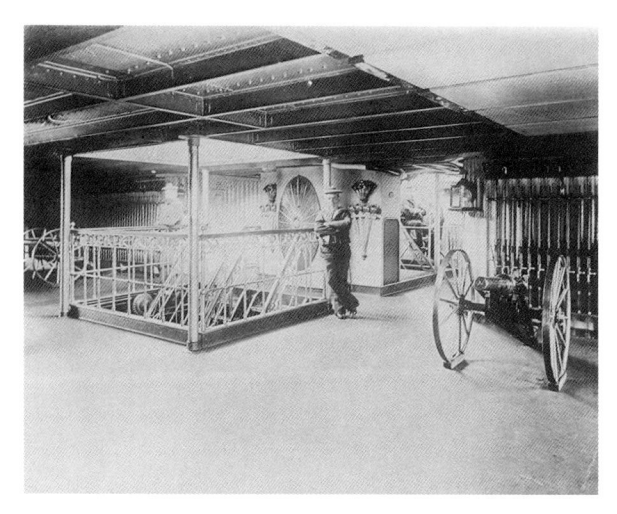

The *salle d'armes* on *Brennus*, with the two 65mm field guns for the landing party. (DR)

goddess Ceres that he found in one of the warehouses on the naval base. He then summoned the master carpenter and the following conversation ensued:
– You see this woman?
– Yes, Sir.
– I want you to make her into a man.
– As you wish, Sir.
– I want you to plane off the breasts and carve a chest suitable for a Gaul.[2]
– Yes, Sir.
– Finally, once this woman has become a man, you will cut off the top of the head and carve a rebate around it to allow a helmet to be fitted.
– Yes, Sir.
– Carry on!

Then, turning to the chief engineer, Boué de Lapeyrère said: 'And you, you will make me a Gaulish red and gold copper helmet with wings.'

The improvised figurehead was duly installed on the bow, and was carefully removed and stowed below whenever the battleship went to sea.

On 11 October the Mediterranean Squadron sailed for the Levant. Port calls were made at Piraeus, where a dinner was organised for the Queen of Greece and the Prince Regent, then Beirut, Jaffa, Haifa, Tripoli, Rhodes, Smyrna and Salonika. The ships returned to Toulon on 21 December.

Towards the end of 1899, the CO of *Brennus* wrote the following note regarding the military mast:

> ... its presence is of limited use if one takes into account the dubious advantages to be gained from the four 47mm guns mounted on a platform subject to vibration and with poor arcs of fire, and the problems that would result from this mast being toppled onto the conning tower and the forward turret, which would be completely immobilised.

However, these observations were never acted upon.

The year 1900 saw the customary sorties with visits to

Officers from other ships of the squadron are received on board the flagship. The after 164.7mm turret and casemate guns are prominent on the right of the photo. (Philippe Caresse collection)

Brennus moored to a buoy in the anchorage at Villefranche. On the right of the photo is the battleship *Jauréguiberry*. (Philippe Caresse collection)

Flag Officers

VA Gervais	1 Apr – 14 Oct 1896
VA Cavalier De Cuverville	15 Oct 1896 – 14 Oct 1897
VA Humann	15 Oct 1897 – 30 Sep 1898
VA Fournier	1 Oct 1898 – 30 Sep 1900
CA Besson	18 Aug 1901 – 9 Sep 1903
CA Jauréguiberry	10 Sep 1903 – 2 Jul 1905
VA Fournier	3 Jul 1905 – 3 Aug 1905
CA Jauréguiberry	3 Aug 1905 – 9 Aug 1905
CA Germinet	10 Aug 1905 – 25 Jun 1906
VA Fournier	26 Jun 1906 – 4 Aug 1906
CA Germinet	04 Aug 1906 – 16 Aug 1907
CA Guillou	15 Nov 1909 – 14 Nov 1911
CA Sourrieu	14 Nov 1911 – 31 Nov 1911

Notes:
VA	*Vice-Amiral*	Vice Admiral
CA	*Contre-Amiral*	Rear Admiral
CV	*Capitaine de Vaisseau*	Captain

Looking forward onto the after 340mm turret. The lighter guns mounted on the after superstructure platforms are 47mm Hotchkiss anti-torpedo boat guns and, outboard of them, two of the four 65mm QF guns. (DR)

Villefranche, Ténès (Algeria) and Mers el-Kebir, and on 30 June the Mediterranean Squadron made its way to the Atlantic for combined manoeuvres with the Northern Squadron. There was a port visit to Brest 9–12 July, and to Cherbourg on the 13th, where a naval review was organised in the presence of President Émile Loubet; the fleet then returned Brest on 23 July.

The Heroism of LV de Mauduit-Duplessix.

On 21 October 1899 the destroyer *Framée* was launched at the Chantiers de la Loire shipyard of Nantes. The ship had a displacement of 347 tonnes at full load and a

The destroyer *Framée* a few days before her sinking. (Philippe Caresse collection)

Brennus at her moorings off Toulon. (DR)

maximum speed of 26 knots. Armed with a single 65mm QF gun, six 47mm and two 381mm torpedo tubes, she belonged to a type of vessel favoured by the adherents of the *Jeune Ecole*, who saw them ushering in the end of the battleship.

Commanded by *lieutenant de vaisseau* de Mauduit-Duplessix, *Framée* ran her trials successfully before arriving in Brest on 26 July 1900. She then joined the Mediterranean Squadron, which at that time comprised the battleships *Brennus* (VA Fournier), *Charlemagne*, *Gaulois*, *Charles-Martel* (CA Roustan) and *Jauréguiberry*, with VA Gervais in overall command in *Bouvet*. The squadron also conprised three armoured cruisers, six protected cruisers, four coast defence battleships, the torpedo cruiser *Foudre*, two destroyers and four torpedo boats.

The fleet sailed at 1030 on 1 August for Charente Maritime (SW France) in company with *Framée*, which replaced the destroyer *Chevalier*. On 7 August, following a port visit to Arcachon, the recently-completed destroyer joined the bulk of the squadron at sea that was now headed back to Toulon.

During the night of 10/11 August, off Cape Saint Vincent, the CO of *Brennus* summoned *Framée* and sent the following signal, which was communicated visually using a luminous semaphore: 'Instruct *Foudre* to take up position at the rear of the line and the torpedo boats to position themselves on the flanks.'

Having passed *Charles-Martel*, *Framée* was making 16 knots and crossing ahead of the flagship. While the signals were being transmitted, the crew of *Brennus* realised that she was approaching dangerously close to her starboard bow. The officer of the watch, LV Dumesnil, conscious of the danger, gave the order: 'Hard starboard and reverse engines'.

On *Framée*, the captain hurried to his post on the bridge and ordered 20 degrees port, but the destroyer continued to turn and, seeing the danger, de Mauduit ordered full speed ahead, aiming to cross the bow of the battleship. Despite this manoeuvre the stem of *Brennus* sliced into *Framée* abreast her third funnel. The little ship immediately heeled over onto her port side and capsized, keel up, while her propellers continued to turn. The collision took place at 0007 precisely.

The shock of the impact brought a large number of officers and men on deck, and a rescue was organised. Buoys, lifebelts and lengths of timber were thrown into the sea. Despite these efforts many were drowned in the wake of the battleship. Part of the destroyer's crew had been asleep at the time of the collision and had no chance of reaching the upper deck. Nevertheless, the chief engineer managed to get two men from the engine rooms to safety before he himself disappeared beneath the waves. Captain de Mauduit-Duplessix himself stood on the side of his ship and urged his crew to save themselves. Quarter-master Rio, who was on the deck of *Brennus* to port, tossed him a lifebelt and encouraged him to grasp it, but the CO refused his help and went down with his ship. Only fourteen men were saved out of a crew of 61.

The fleet remained on the spot for two and a half hours. The ships illuminated the waters with their searchlights while their boats continued to search for survivors. At about 0300 Admiral Fournier regretfully gave the order to resume their passage. The cruiser *Galilée* and the torpedo gunboat *Dunois* remained behind until dawn without finding any more survivors.

The squadron arrived back in Toulon on the evening of Tuesday 14 August, then sortied a week later for gunnery practice against the Mèdes rock on the island of Porquerolles and a short visit to Golfe-Juan.

The End Approaches

When he left his command, Boué de Lapeyrère wrote in his report:

> *Brennus* is a mediocre sea-boat due to the lack of freeboard forward, but in terms of the disposition and power of her guns she should be taken as a model. Her superstructures are not excessive, and the parts of the ship above the waterline are arranged in a rational fashion. Unfortunately the same cannot be said of the compartments below the armoured deck, which constitute an unfathomable labyrinth, cramped and airless.

Brennus was placed in 2nd category reserve on 17 October 1900, and underwent a short docking period; she was recommissioned on 1 June 1901. Machinery problems delayed her re-entry into service, but she made a port visit to Algiers later that month. Following the summer grand manoeuvres, the battleship was assigned to the Reserve Division on 10 August, alongside the battleships *Hoche*, *Carnot* and *Amiral Baudin*, the torpedo cruiser *Foudre* and the torpedo gunboat *La Hire*. With reduced complements, these ships would make only brief sorties to the anchorage at Les Salins d'Hyères. *Brennus* flew the flag of Rear Admiral Besson, who participated in the sorties of July 1902 off the Tunisian and Algerian coasts. That same year, it was proposed to installed wireless transmission on board, as well as a manual loading system for the fore 340mm turret in an attempt to improve the rate of fire. The latter modification was not carried out, despite the firing cycle being in excess of four minutes per round.

Following a refit that took place 15–31 January 1903, *Brennus* rejoined the 1st Battle Division, which also comprised the battleships *Saint-Louis* and *Hoche*, for a visit to Cartagena in honour of King Alfonso XIII of Spain.

During the years 1904 and 1905, Minister of Marine Pelletan, an unconditional supporter of the theories of the *Jeune École*, judged the grand manoeuvres of the fleet to be too onerous and cancelled them outright.

Brennus was in refit 11–28 March 1904, then made brief sorties to the coasts of Provence. Following another spell in dockyard hands from 15 May to 16 June 1905, the battleship hoisted the flag of Admiral Fournier on 3 July and moored at Les Salins two days later. The Squadron was at Bizerta from 13 July and saluted the coffins of the crew members lost on board the submarine *Farfadet*, which were laid out on board the tug *Cyclope* (*Farfadet* had sunk on a routine patrol from Sidi-Abdallah to Bizerta on 6 July). Following the customary summer exercises, Minister of Marine Gaston Thomson, who was due to observe manoeuvres

The stern of *Brennus*; the upper edge of the armour belt can be clearly seen. (DR)

Brennus at her moorings in Toulon roads. (DR)

at Hyères, was embarked at Marseille. There followed sorties to Corsica, Golfe-Juan and Les Salins from 17 to 28 October.

At that time one of the ship's commanding officers, CV Servan, reported:

Beneath the armoured deck the ship is a real labyrinth; this hinders the movement of personnel and makes the compartments other than the engine and boiler rooms uninhabitable. Ventilation is totally inadequate, and the temperature in these compartments and in the Poste Central excessive.

Brennus in Toulon roads. The overall grey livery tells us that this photo was taken after January 1908. (DR)

Shikishima at Sasebo 13 December 1915. The first of four battleships authorised under the Second Period Expansion Programme of 1895, she was ordered from Thames Ironworks, Blackwall, and completed in 1901. (Courtesy of Fulvio Petronio)

Yamamoto Gonnohyôe (Gombei), often called the 'father of the Imperial Japanese Navy'. It was he who overcame the dominance of the Army, securing equality with the sister military service, and 'cleansed' the IJN prior to the Sino-Japanese War of 1894–95. However, one of the less happy consequences of equal status was the dual command system that deepened the rivalry between Navy and Army to an extent that Yamamoto failed to anticipate. (Courtesy of Lars Ahlberg)

Navy Minister Admiral Saigô Tsugumichi. He promoted Yamamoto Gonnohyôe and allowed him a free hand. Courtesy of Lars Ahlberg)

Brennus at her moorings in Toulon roads. (DR)

at Hyères, was embarked at Marseille. There followed sorties to Corsica, Golfe-Juan and Les Salins from 17 to 28 October.

At that time one of the ship's commanding officers, CV Servan, reported:

Beneath the armoured deck the ship is a real labyrinth; this hinders the movement of personnel and makes the compartments other than the engine and boiler rooms uninhabitable. Ventilation is totally inadequate, and the temperature in these compartments and in the Poste Central excessive.

Brennus in Toulon roads. The overall grey livery tells us that this photo was taken after January 1908. (DR)

Admiral Fournier would again embark on *Brennus* for the grand manoeuvres of 1906 – they were reinstated by the new Minister of Marine, Gaston Thomson – and would be at Marseille for the laying of the first stone of the Rove Canal. On 4 December the transport *Dives*, which was moored in Toulon roads, broke her anchor chain and collided with the battleship on the port side close to the stern. The *Dives* ended up grounded at Mourillon, on the east side of the dockyard, while *Brennus* suffered only superficial damage.

During 1907 Toulon Dockyard submitted to the *Comité Technique* a proposal to lighten the upperworks of the battleship by removing the military mast. Given the age of the ship, the response was negative: '… it would be better for the Navy if *Brennus* were to be left in her current state than to remove superstructures.'

From the beginning of 1907 *Brennus*, *Charles-Martel* and *Hoche* made up the 1st Division of the Reserve Squadron. On 15 February the squadron was dissolved, becoming the Training Division (*Division d'instruction*) of the Mediterranean Squadron.

On 12 June, in the forward 340mm turret, a master gunner carelessly set fire to a rag soaked in petrol. The resulting fire was quickly mastered, and the badly-burned petty officer was dispatched to the naval hospital at Saint-Mandrier.

The battleships *Patrie* and *République* entered service during 1907, and from 25 August *Brennus* was placed in 'normal reserve'.

On 15 November 1909 *Brennus* was assigned to the Torpedo School as a replacement for *Charles-Martel*, which joined the 2nd Squadron at Brest. Despite her age and reduced status, it was during 1909 that work was begun to improve the rate of fire of her main guns. The modifications were not completed until 1911, and it is not known whether gunnery trials followed.

Brennus would be present at the Great Naval Review of 4 September 1911 off Toulon, when President Armand Fallières embarked on the old battleship *Masséna*. At 0500 on the following day *Brennus* was leaving the anchorage when she collided with the steamer *Magali* from Marseiile, which had a large number of spectators

Despite her faults *Brennus* had an unusually 'modern' silhouette for a battleship of the period. Note the two Barr & Stroud 2-metre rangefinders on either side of the bridge structure. Beyond *Brennus*, on the right of the photo, can be seen a battleship of the *Danton* class and the armoured cruiser *Jules Michelet*. (DR)

Brennus during the Great Naval Review held off Toulon on 4 September 1911. Shortly afterwards, in January 1912, she was placed in reserve. (DR)

on board. Happily there were no casualties, but *Magali* was compelled to run aground at Saint-Mandrier to avoid sinking.

On 1 January 1912 *Brennus* was again placed in normal reserve, then in 'special reserve' from the 24th. She paid off on 1 April 1914.

During the Great War the battleship was hulked, becoming an annex to the 5th Depot of ships' crews. In 1915 the three 340mm guns were disembarked and sent to Ruelle to be re-bored as 400mm howitzers. *Brennus* was also cannibalised to provide parts and equipment for other units of the *Armée Navale*. She was finally stricken on 22 August 1919 and put up for sale on 30 October.

On 13 January 1921 she was handed over to the Boursier, Borrely and Sidensner Company as payment for the refloating of the wreck of the battleship *Liberté*. Having served briefly as a store, she was broken up for scrap at Toulon in 1922.

Editor's Note: This article was translated from the French by the Editor. The official plans of *Brennus*, with the sole exception of a plan view of the upper decks dating from about 1905, do not appear to have survived. The two profile drawings therefore had to be carefully assembled using the deck plans, line drawings (often crude and inaccurate) published in secondary sources, and photographs of the ship.

Endnotes:
[1] Between 1880 and 1890 there were no fewer than fifteen different Ministers of Marine.
[2] Brennus was a renowned Gaulish chieftain, He defeated the Romans at the Battle of the Allia (18 July 390 BC). In 387 BC he led an army of Cisalpine Gauls in an attack on Rome and captured most of the city, holding it for several months. Brennus' sack of Rome was the only time in 800 years the city was occupied by a non-Roman army before the fall of the city to the Goths in 410 AD.

THE GENESIS OF THE SIX-SIX FLEET

It is widely acknowledged that Japan won the Russo-Japanese War of 1904–05 with what has come to be known as the Six-Six Fleet.[1] **Hans Lengerer** looks at the process by which this fleet came into being and acknowledges the contributions of Yamamoto Gonnohyôe and Arichi Shinanojô.

The Chinese War Indemnity

Following China's defeat in the Sino-Japanese War (SJW) of 1894–95 she was compelled, according to Article IV of the Treaty of Shimonoseki, to pay reparations of 200,000,000 taels in Kuping silver, a figure which corresponded to Yen 311,072,865. A further 1,500,000 taels (¥2,380,102) had to be paid for the temporary occupation of Weihaiwei by the Imperial Japanese Army for three years as a guarantee that the terms of this treaty would be observed (Article VIII). Article II of the treaty signed on 8 November 1895 concerned the handing back of the Liaodung peninsula, for which China had to pay 30 million taels (¥44,907,499) as compensation.

The total of Yen 364,868,587[2] exerted a greater stimulus to economic, industrial and financial activities than in any period since the Meiji Restoration. Arguably the most significant aspect of this economic and industrial boom was an extensive armaments programme which aimed to place Japan among the leading naval powers of the world, and to raise the force strength and capability of the IJN to a point at which Japan could withstand the pressure of another 'Triple Intervention' and also defeat the primary hypothetical enemy, Russia.[3]

While the expenditures for the expansion of the Imperial Japanese Army (IJA) amounted to Yen 56,781,708, those for the expansion of the IJN were ¥139,157,097 plus ¥30,000,000 as a supplementary fund for warships and torpedo boats and ¥579,762 for establishing an iron foundry (the Yawata steel works).[4] In other words, the IJN received roughly 47 per cent of the Chinese indemnity.[5] This considerable sum was used to quadruple the force strength of the IJN by building the so-called Six-Six Fleet (mostly in British shipyards), to expand and equip the naval dockyards and repair facilities, and to improve training. Recognising that a well-developed shipbuilding industry was the prerequisite for a large navy, the government supported the private shipbuilding industry by 'encouragement' laws and made great efforts to build up related industries. This aggressive expansion policy enabled the IJN to attain within eight years the standard of 200,000 tons promoted by Navy Minister Kabayama in September 1890 as a 'minimum necessary force' which would permit Japan to defeat the Imperial Russian Navy in the war of 1904–05.

Lessons of the Sino-Japanese War

While China operated its Peiyang Fleet as an escort force to enable army forces to be transported by sea, Japan considered the purpose of the Navy to be the annihilation of the enemy fleet, thereby securing and maintaining control of the seas. The IJN fought the Battle of the Yellow Sea with a fleet composed of cruisers and coast defence ships, defeated a Chinese fleet built around two big armoured ships,[6] and attained maritime supremacy. The Chinese Northern Fleet was all but annihilated in Weihaiwei by the night attacks of IJN torpedo boats.

The composition of the Japanese fleet which fought against the Chinese represented the thinking of the French *Jeune Ecole* and of the Russian Stepan Makarov, while the Chinese fleet had the large armoured ship at its centre and was therefore closer to the force structure favoured by the British Royal Navy and the contemporary US naval historian Alfred Thayer Mahan. The theories of the *Jeune Ecole* were particularly attractive to small and weak countries and had been adopted by Japan early in the 1880s for financial reasons. Superficially considered, the sea battles of the Sino-Japanese War appeared to bear out these theories, and one might have expected the IJN to continue on this path. However, the strategic goal of the Navy – that of securing and maintaining sea control – was now seen as attainable only with the acquisition of large armoured ships as the core of a battle fleet, and the Chinese indemnity enabled Japan to fund the necessary programmes and to create a balanced naval force.

The Origins of the Six-Six Fleet

In the spring of 1895, when the end of the Sino-Japanese War was in sight, the head of the Bureau of Naval Affairs (*Gunmukyokuchô*)[7] in the Ministry, Yamamoto Gonnohyôe, was ordered by Navy Minister Saigô Tsugumichi to establish the key principles upon which future planning should be based. These studies were to focus on i) the general political situation in Asia and ii) the strategic planning of the world powers in Asia. The rearmament of the IJN was to be based upon these findings, with the primary aims of maintaining the independence of Japan by supremacy at sea and building the naval forces and facilities necessary for the defence of the country.

Fuji at Kure between 1917 and 1919. *Fuji* was the first of the six 1st class battleships ordered from Britain under the 1892 and 1895 naval programmes. She was built by Thames Ironworks, Blackwall. (Courtesy of Lars Ahlberg)

Immediately after the 'Triple Intervention',[8] Yamamoto made the following policy submission to the Minister:

> If our current naval forces are evaluated, one gets the impression of a naked man armed with a sword confronting an enemy protected with a helmet and armour. The traditional mental attitude of the IJN is a factor, but the shortage of funding for naval construction also plays a part. The cruisers *Naniwa* and *Takachiho* are good examples: they are armed with main guns whose calibre is excessive for their size and build; protection is somewhat neglected compared with offensive power. We must make every effort within the financial possibilities to obtain well-balanced ships in the future. In particular, we need to ensure that defensive capabilities match offensive capabilities.
>
> As for naval operations, there is one country[9] that Japan will have to confront in East Asia in the future … otherwise there are few countries which might oppose Japan alone. During the present discussion about the forces of the IJN, it is necessary to consider the case that one country or allied countries might advance into East Asia. A fleet capable of matching these forces should therefore be prepared.
>
> Japan should possess a main fleet composed of armoured battleships and first class cruisers. The light forces, particularly those below the small cruiser, should be operated as an auxiliary force. Depending on the result of the comparison between Japanese forces and those of the aforementioned countries, the main fleet should comprise six armoured battleships and six first class cruisers.

Naval Expansion Planning after the Sino-Japanese War

In July 1895 Navy Minister Saigô submitted to the Cabinet the New Warship Building Programme (*Shinkan seizô keikaku*), which was based upon Yamamoto's assessment. Saigô explained the rationale for the programme as follows:

> – England and Russia have different interests in Eastern Europe and Asia; it is therefore unlikely that they will form an alliance against Japan. In order to determine the size and power of the fleet, either Britain, or Russia allied with France (or with one or two other minor powers) should be the yardstick. It will be sufficient to build a fleet superior to that which can be dispatched to Asia by one or other of these powers.
>
> – Judging from the present situation, Britain might be able rapidly to dispatch to Asia between one third and one half of the modern armoured ships stationed in the Mediterranean Sea. The situation of Russia is similar. It is necessary to meet either eventuality.
>
> – If the above assessment is correct, Great Britain or Russia with her allies would be able to dispatch five or six modern armoured ships to Asia. In order to oppose them, new construction of four first class large armoured ships is required; with the two battleships already under construction this will provide a force of six modern large armoured ships.

This would be the first IJN programme to meet a specific threat. In accordance with the lessons learned during the

Table 1: The 1892 Naval Programme

Type	1892 Naval Programme	To be built under FY 1893	To be completed 1896–97
Battleships	4 x 11,400 tons	2 x 11,750 tons	*Fuji* (12,649 tons)
			Yashima (12,517 tons)
1st class Cruisers	4 x 5,200 tons		
2nd class Cruisers	2 x 3,800 tons		
3rd class Cruisers	1 x 2,700 tons	1 x 2,700 tons	*Akashi* (2,800 tons)
4th class Cruisers	3 x 1,630 tons		
Despatch Vessel	2 x 1,800 tons	1 x 1,800 tons	*Miyako* (1,800 tons)
Torpedo-Gunboat	3 x 870 tons		
Total	19/87,800 tons	4/28,000 tons	4/29,766 tons

Notes:
1 In the autumn of 1892 Navy Minister Vice Admiral Nire Kagenori (8 August 1892 – 10 March 1893), the predecessor of Saigô Tsugumichi, submitted a plan to the Diet for the expansion of the IJN to 120,000 tons in 16 years at a cost of ¥59,197,419. The core of this plan was the construction of four armoured battleships of 11,400 tons and four 1st class cruisers of 5,200 tons. Two battleships, one cruiser and one despatch vessel were to be funded under a ¥19,558,543 budget submitted to the fourth Diet session in December 1892. The Upper House reduced the budget but the government persisted with its request, whereupon an Imperial Statement was issued on 10 February 1893 by which the Emperor Meiji would donate ¥300,000 per year for a period of six years from the budget of the Imperial Household and all officials were ordered to gift 10 percent of their salaries for the same period. The plan was again submitted to the Diet and the budget, amended to ¥18,082,526, authorised. It was subsequently decided to reduce the building period of the two battleships from seven to five years, which meant that the budget had to be revised upwards to ¥23,834,955. (The process is described in detail in *Kaigun Gunbi Enkaku*, Chapter 14.)
2 *Fuji* and *Yashima* were the first battleships built for the IJN and were to oppose their Chinese counterparts *Ting Yuen* and *Chen Yuen* (7.400 tons, four 30.5cm main guns in two twin turrets), which outgunned any IJN warship in the ten-year period before the Sino-Japanese War. It was the beginning of the construction of capital ships equivalent to those of the large naval powers.

Sino-Japanese War, the core of the fleet was to be a force of armoured battleships. *Fuji* and *Yashima*, currently building in Britain, were included in this force structure, and there were to be four new ships. This main force would be supported by six first class cruisers, together with a number of minor vessels.

The original construction programme dating from 1892 is outlined in Table 1. Following the Sino-Japanese War the situation in the Far East was much changed, and with the large Chinese war indemnity available the programme of 1892 was revised in 1895 – see Table 2. The 1895 programme totalled 17 major warships, plus 75 destroyers and torpedo boats.

Saigô proceeded to give a lengthy exposition to explain the mission of the IJN and the forces required to fulfil it. The principle points are summarised below:

– The aim of the IJN is to attain naval supremacy. To achieve this the IJN has to have a strong main battle fleet to oppose any enemy.

Table 2: The Revised Naval Programme of 1895

Type	No	Displacement/Type
Battleships	4	15,000 tons
1st class Cruisers	4	7,300 tons
2nd class Cruisers	3	*Yoshino* class
3rd class Cruisers	4	3,000 tons
Despatch Vessel	2	1,900 tons
Torpedo Gunboat	5	*Tatsuta* class
TB Tender & Repair Ship	2	*Balkan* class

– The main force of the main battle fleet is to comprise armoured battleships.
– Besides the battleships the main battle fleet needs a support force composed of cruisers and minor warships.
– The strength of the IJN should be calculated in the future on the basis of the strength of the fleet capable of being dispatched to East Asia by a single power or that of the combined forces of two powers. The IJN should be sufficiently strong to oppose these forces.
– To meet the above criteria a fleet composed of six battleships, 18 cruisers, six torpedo gunboats and two TB tenders needs to be established. This programme proposes the construction of four armoured battleships to supplement the two battleships already authorised under a previous programme. In addition, it is proposed to build nine cruisers, three torpedo gunboats and one TB tender under this programme. The force strength defined in this paper will be completed by ships already built or building, or with ships to be proposed under a later programme.[10]
– This programme will extend over ten fiscal years (FYs). It should not be considered as set in stone. Instead, the world situation and the force strengths of other powers will be constantly monitored, and it may be necessary to revise current intentions and modify the programme accordingly.
– In addition to the proposed high-speed cruiser force, a 'protection' force will be formed with the ships already completed, because the latter are unable to operate together with the main battle fleet. The protection force will be able to carry out the duties which have to be performed in peacetime.

Yashima, sister-ship of *Fuji*, passing through the swing bridge at Newcastle-upon-Tyne shortly after her completion; she was built by Armstrong, Elswick. (Courtesy of Fulvio Petronio)

The proposal was considered at a meeting of the Cabinet, which decided the programme would be divided into two periods to facilitate funding.

On 26 December 1895 the first tranche was submitted to the 9th session of the Diet and approved with one small modification: ¥203,440 were subtracted from the design funding. It became the First Period Expansion Programme, with ¥94,776,245 authorised, with continuation funding for the seven fiscal years from 1896.

The second tranche, to which two 1st class cruisers had been added, was submitted to the 10th session of the Diet in December 1896, and was approved unamended as the Second Period Expansion Programme with ¥118,324,718. The budget was expanded to total ¥213,100,964 to fund the entire programme from FY 1896 to FY 1905.[11]

Table 3: Warships Built Under the First & Second Period Expansion Programmes

Type	Planned	Built	Names
Battleships	4	4	Shikishima, Asahi, Hatsuse, Mikasa
1st class Cruisers	6	6	Yakumo, Asama, Azuma, Tokiwa, Izumo, Iwate
2nd class Cruisers	3	3	Kasagi, Chitose, Takasago
3rd class Cruisers	2	3	Niitaka, Tsushima, Otowa
Torpedo Gunboat	3	1	Chihaya (subsequently reclassified as Despatch Vessel)
Shallow-draft Gunboat	–	3	Uji, Fushima, Sumida
Destroyers	12	23	Ikazuchi, Inazuma, Murakumo, Shinonome, Akebono, Sazanami, Yugiri, Shiranui, Usugumo, Kagero, Oboro, Akatsuki, Kasumi, Shirakumo, Asashio, Harusame, Murasame, Hayatori, Asagiri, Niji, Ariake, Fubuki, Arare
1st class Torpedo Boats	16	16	Hayabusa, Manazuru, Kasasagi, Chidori, Kori, Aotaka, Shirataka, Hato, Tsubame, Hiburi, Kiji, Sugi, Uzuura, Kamome, Hashitaka, Otori
2nd class Torpedo Boats	37	37	Nos 29 – 49 (= 21 boats), Nos 60 – 75 (=16 boats)
3rd class Torpedo Boats	10	10	Nos 50 – 59
Zatsuekisen (Zassen)	584	584	

Note:
The revisions were necessary for organisational reasons and for trade protection. The process is described in detail in *Kaigun Gunbi*

Shikishima at Sasebo 13 December 1915. The first of four battleships authorised under the Second Period Expansion Programme of 1895, she was ordered from Thames Ironworks, Blackwall, and completed in 1901. (Courtesy of Fulvio Petronio)

Yamamoto Gonnohyôe (Gombei), often called the 'father of the Imperial Japanese Navy'. It was he who overcame the dominance of the Army, securing equality with the sister military service, and 'cleansed' the IJN prior to the Sino-Japanese War of 1894–95. However, one of the less happy consequences of equal status was the dual command system that deepened the rivalry between Navy and Army to an extent that Yamamoto failed to anticipate. (Courtesy of Lars Ahlberg)

Navy Minister Admiral Saigô Tsugumichi. He promoted Yamamoto Gonnohyôe and allowed him a free hand. Courtesy of Lars Ahlberg)

The First and Second Period Naval Expansion Programmes provided for the realisation of the Six-Six Fleet as conceived by Yamamoto Gonnohyôe. Table 3 lists the warships built under these programmes, which were as usual subject to revision. These ships were generally completed before the Russo-Japanese War and provided the foundation for the IJN's defeat of the Russian fleet.

The Man Behind the Six-Six Fleet Programme

After a more or less unexceptional career as a naval officer[12] Lt-Cdr Yamamoto Gonnohyôe (1852–1933) became adjutant to Navy Minister Saigô Tsugumichi in July 1887. In June 1891 he was transferred to the Navy Ministry as Secretary (*Kambo shugi*) and entered the political stage. Promoted Rear Admiral, as head of the Naval Affairs Bureau he was at the centre of the Navy's administration and was not afraid to carry out unpopular measures such as the enforced retirement of 97 naval officers, among them eight of flag rank, before the Sino-Japanese War, and the promotion of young, well-educated officers in their place.[13] In 1893, as the *éminence grise* behind the Navy Minister, he embarked on a struggle to place the staff organisation of the IJN on a par with that of the IJA.[14]

Yamamoto was promoted Vice Admiral in May 1998 and became Navy Minister on 8 November of the same year. He remained in this post until his resignation on 6 January 1906, on which date he was succeeded by Vice Admiral Baron Saitô Makoto.[15] Taking into account Yamamoto's time as Secretary, he was a key figure in the ministry for 14½ years, during which time he was at the centre of naval policy.

The IJN required some ten years to prepare for the Russo-Japanese War despite the most favourable conditions, namely good cooperation at the highest levels between the political body and the military, particularly with regard to the basis of diplomacy and the defence of the country. Yamamoto's personal decisions deserve the highest praise. He was responsible for the selection of Vice Admiral Tôgô, whose career thus far had been unexceptional, to command the fleet. The appointment of Tôgô surprised many high-ranking officers, but it proved to be a fortunate choice, and he is still remembered today as one of the outstanding admirals of the IJN.

On 7 January 1906 the Saijoni Kinmochi Cabinet was formed and Saitô Makoto was appointed Navy Minister; Saitô held this position for seven years until 1913 and his successor, Admiral Katô Tomosaburô, filled this post for eight years despite some cabinet changes. This was advantageous in terms of continuity: despite regular changes in the composition of the cabinet, these three prominent admirals each held their posts for long periods of time, and were able to concentrate their abilities upon the organisation and expansion of the IJN and the selection of its leaders.

Hatsuse at Shimuzu on 3 February 1902, shortly after her arrival in Japan. Built by Armstrong at Elswick, she had a similar hull-form to the Armstrong-built *Yashima*. (Courtesy of Lars Ahlberg)

Asama, the first of a pair of armoured cruisers ordered from Armstrong, Elswick, under the First Period Expansion Programme of 1895. (US Naval History and Heritage Command, NH 58986)

Asama's sister *Tokiwa* early in her career. Despite being ordered from three different European countries, all six armoured cruisers had a similar displacement and a common British-pattern armament of four 8in guns and twelve/fourteen 6in QF guns. (US Naval History and Heritage Command, NH 58682)

Why A Six-Six Fleet?

Saigô's justification for the size and power of the new ships deserves attention. He was keen to point out that the 15,000-ton displacement of the battleships in the new programme[16] matched that of the latest British construction. The advantages were stated to be as follows:

- Battleships of that size cannot pass through the Suez Canal. An enemy fleet with comparable ships would therefore have to be sent around the Cape of Good Hope, which would be time-consuming and costly. A considerable quantity of coal would have to be embarked, as with Britain neutral the necessary coaling stations would not be available.
- Should the enemy send a battle fleet to the Far East, most of our forces would be tasked with the defence of the naval ports, shipyards and locations along our coastline well-suited to the establishment of a base. Therefore we must expand our navy ... by means of the proposed large construction programme. The enemy has no shipyard or dock capable of repairing his big ships in the Far East and the construction ... or expansion of his facilities would require considerable expenditure ...
- The warships the enemy is able to dispatch to the Far East via the Suez Canal would necessarily be limited to 2nd class battleships and cruisers. If we have a force composed of six battleships and four[17] powerful 1st class cruisers of 7,000 tons our country can be adequately defended ...
- The cruisers in this programme are all big ships we have never seen in this country, and they are equivalent in size and power to any armoured cruiser or older-type battleship that the enemy can send to the Far East.
- The 2nd and 3rd class cruisers are all superior to those currently in service; they have high speed and are armed with many QF guns.
- The battleships are comparable in fighting power to the British *Centurion*, which is currently the largest ship in the Far East, and displacement is 5,000 tons greater. The 1st class cruisers are similar to the British *Edgar* currently deployed to the Far East, and the 2nd class cruiser is an improved *Yoshino* class ...

The above reasoning underpinned the Six-Six programme. The Japanese concluded that, given the constraints on the deployment of 1st class battleships

The armoured cruiser *Yakumo*, ordered from the German AG Vulcan shipyard, at Stettin following her completion in 1900. (US Naval History and Heritage Command, NH 58993)

from European waters to the Far East, the IJN battle fleet proposed was sufficient both to defend Japan and to defeat any fleet that one or two allied powers might dispatch.

The Man in the Shadows: Vice Admiral Arichi Shinanojô

The genesis of the Six-Six Fleet has rightly been credited to Yamamoto Gonnohyôe. However, he was not the first to propose a fleet based around six battleships; this idea predates the Sino-Japanese War.

On 11April 1891 Vice Admiral Arichi Shinanojô[18] submitted a memorandum entitled 'Proposal for Coastal Defence' (*Kaibô Ikensho*) to Navy Minister Vice Admiral Kabayama Sukenori (17 May 1890 – 7 August 1892). He likewise concluded that, taking into account the current strength of the principal naval powers and the probable strength of an attacking force, a force of six battleships should form the core of Japan's defensive fleet. His arguments were backed by data on the dispatch of probable ship types and took into account the constraints of the Suez Canal, the alternative long voyage around the Cape of Good Hope, coal consumption and supply. These arguments clearly underpinned those used by Navy Minister Saigô in 1895 to justify the proposed expansion of the IJN after the Sino-Japanese War to the Cabinet and the Diet. The following statements are taken directly from Arichi's memorandum:

- From a national standpoint the IJN needs to be at the forefront and coastal defences put in order.
- In order to determine naval force strengths and strategy it is first necessary to study the enemy's attack strategy. Compared to the strength of the enemy the IJN is very weak. The enemy will occupy an undefended bay on the coast of the mainland or an offshore island, establish his base and carry out attack operations from there. If our navy is sufficiently strong in comparison to the enemy, he cannot attack us directly but would be forced to destroy our navy before occupying a part of the mainland. The targets of the enemy fleet are naval ports and our fleet. Our naval ports need to be able to defend themselves without the support of our fleet. Our fleet should be stronger than that of the enemy, otherwise there is no chance for victory. We must, therefore, have a strong fleet as the principal component of our national defence. Decisions on the warship types required and their numbers should be made on the basis of the geographical position of our country and the strength of the attack force an enemy country may be capable of dispatching to Japan. Consequently, studies of the warship types and numbers an enemy can dispatch and the replenishment capabilities available to them are of supreme importance and need to be undertaken urgently.
- If the enemy approaches our homeland with the intention of destroying our fleet, he might be able to defeat us; however, fighting ships (*Sentôkan*) cannot be dispatched without supply bases in the vicinity of our homeland. Therefore, Japan should possess a strong fleet for national defence and aim to annihilate an attacking enemy. For this reason the construction of six battleships with a displacement of 10–12,000 tons should be approved as soon as possible. These ships are equivalent in power to 18 enemy cruisers (*Junkôkan*, later *Junyôkan*), which may appear at important military strategic locations.
- Because the depth of the Suez Canal is 28 *shaku* (8.53m) it cannot be passed by warships whose draft exceeds 25 *shaku* (7.62 metres). If the enemy intends to dispatch from Europe warships superior to ours they would have to go around the Cape of Good Hope, but this long distance is very disadvantageous.
- During the grand manoeuvres of the French Navy in 1890 the coal bunkerage of one battleship was sufficient for five days. According to naval adviser Ingles coal stowage for a battleship is sufficient for three days. If the average consumption for two speeds is calculated, coal has to be supplied every four days. For instance, a 10,000-ton battleship consumes 200 tons every 24

Vice Admiral Arichi Shinanojô. His essay on coast defence contained most of the arguments found in the later rationale for the Six-Six Fleet. However, at the time when the budget for this fleet was approved Arichi had retired from the IJN, and his name is seldom linked to Japan's (largely) British-built Six-Six Fleet which defeated the Russians in the Russo–Japanese War of 1904–05. (Courtesy of Lars Ahlberg)

hours. If four ships were dispatched their coal consumption per week would be 5,600 tons. East Asian countries depend on Japan for most of their coal; in the principal ports such as Shanghai and Hong Kong, the ratio of the coal supplied by foreign countries [to the coal supplied by Japan] is between 1:7 and 1:9. If Japan stops her exports it will no longer be possible to provide coal supplies in the vicinity of Japan. Given this situation, no enemy country can dispatch large battleships to attack Japan in the event of war and thereby exercise sea power, even if it has a very strong navy at its disposal in peacetime. It will therefore be forced to dispatch a force of cruisers for the long transit to Japan. To oppose the enemy force the IJN should establish a fleet composed of battleships, together with supporting cruisers. The construction of six new-build battleships would permit the IJN to form a fleet with the already-completed warships which could easily undertake coast defence battles or a blockade. The addition of these six battleships would increase IJN force strength by 60–72,000 tons ..., enabling it not only to ensure national defence but to attain sea supremacy in Asia.

- The expense of such a construction programme is the major concern. Each ship would cost between ¥6 million and ¥7 million; for six ships the total cost would be ¥36–42.[19]
- The coasts of the Japanese mainland have a length of 15,200 miles; Japan has many good ports and important bays, and coal production is well developed. The building of the 'Nicaragua' or 'Panama' Canal has begun. Following its completion Japan will no longer be a country in the Far East at the far end of the trade routes, but a trade centre for Asia; the further development of commerce is certain. Japan's influence will rise and change the situation in Asia. Negotiations with foreign countries will be more frequent; it will be a time when Japan can be ambitious, and much will be achieved. If the expansion of the Navy is considered only then, it will be too late. Those in charge of national defence have to plan today.
- If the aforementioned planning is carried out, a revision of the existing treaties appears to be assured.[20] Trade will be protected in peacetime, and trading rights in Asia expanded. In wartime the duties of the fleet – fighting against a strong enemy and blockade – will be made easier.

The battleship *Asahi* at Sasebo 8 June 1901, shortly after her arrival in Japan. She was built by John Brown & Co, Clydebank. (Kure Maritime Museum)

The battleship *Mikasa* during the early 1900s. The last of the six battleships to be built under the First Period Expansion Programme, she would be Admiral Tôgô's flagship at the Battle of Tsushima in 1905. *Mikasa* was built by Vickers at Barrow. (US Naval History and Heritage Command, NH 58973)

The Defensive Character of the Six-Six Fleet

The defence of Japan's approaches and coastline against an attacking fleet was the key element in all these deliberations. Offensive warfare in distant waters was never mentioned, and the characteristics of the six battleships and six armoured cruisers reflected this strategic approach.

The final pair of armoured cruisers, *Izumo* and *Iwate*, were built in Britain and funded under the Second Period Expansion Programme. They were completed by Armstrong, Elswick, in 1902. This is *Izumo*; the photo was probably taken in 1902, shortly after the ship's arrival in Japan. (US Naval History and Heritage Command, NH 45847)

All six battleships and four of the armoured cruisers were designed by leading British naval architects to Japanese requirements,[21] and the plans had to be approved by Japan. Due to the rapid progress in technology, the guns and protection systems were subject to incremental improvements in the course of the building programme. However, even though the details of the individual ships were different, armament and speed were virtually the same to permit their operation as a single unit in a single formation.

The battleships had two more secondary 6in guns than their British counterparts. This was made possible by a reduction in coal bunkerage and the acceptance of relatively spartan living quarters. As the ships were intended to operate in Japanese waters, being tasked with the interception and defeat in combat of a fleet approaching the Japanese mainland after a very long voyage, these features were considered less important;[22] deployment to distant seas to fight against an enemy fleet stationed there was out of the question.[23] This emphasis on fighting power and protection at the expense of range and habitability is testimony to the essentially defensive character of Japan's first modern battle fleet, which was built in preparation for the clash with Russia. This tendency can also be seen in later warships.

Finally, it should be noted that the concept of a Six-Six Fleet as a definition of naval force structure was the result of a new process of reasoning, and would subsequently be further developed with the so-called Eight-Eight Fleet.[24]

Conclusion

The victory in the Sino-Japanese War brought Japan a huge war indemnity which made possible enormous modernisation and expansion programmes which, in turn, enabled the IJN to win the naval war against Russia. Without this money and public support for a growth in naval force strengths, the efforts of Yamamoto would have failed, the Six-Six Fleet concept could not have been realised, and Japan would either have been compelled to back down from a military confrontation with Russia, or waged war and lost. The subsequent history of the Imperial Japanese Navy and of Japan would then have been very different.

Endnotes:

1. According to the late Rear Admiral Takasu Kôichi, who conducted extensive research into the warship building programmes of the Meiji era (1868–1912), this term was not used at that time but introduced later when the 'fleet' had been built.
2. The expense of the occupation force at Weihaiwei was separated and placed under 'General Accounts'
3. Russia had been seen as Japan's major potential opponent until 1882, when China assumed this role. Following the victory over China in 1895, Russia was again seen as the number one threat.
4. Between 1896 and 1903 a total of ¥20,771,581 was spent, but the majority of this money was raised from bonds.
5. Before the victory in the Sino-Japanese War public opinion towards the IJN was unfavourable, but this situation changed thereafter and the IJN was supported by the whole nation and the subject of high expectations.
6. These ships nevertheless withstood all the attacks of the Japanese warships and remained operational even after more than 200 hits from small- and medium-calibre guns. This convinced the Japanese that armoured ships were more or less immune against such calibres, and strengthened their belief that armoured ships had to be opposed by armoured warships mounting heavy guns.
7. This bureau was the liaison organ between the naval policy organ (*Gunsei*), represented by the Navy Ministry, and the naval command organ (*Gunrei*), represented by the Naval General Staff
8. The Triple Intervention was a diplomatic intervention by Russia, Germany, and France on 23 April 1895 over the terms of the Treaty of Shimonoseki signed between Japan and Qing Dynasty China that ended the First Sino-Japanese War.
9. That 'one country' was, of course, Russia.
10. This is a very interesting passage for Yamamoto's Fleet Building Programme (*Kantai kensetsu keikaku*). Originally the Six-Six Fleet was to have been a Six-Six-Six Fleet. However, budgetary constraints resulted in the suppression of one group of armoured cruisers; in their place six older vessels (the central battery ship *Fusô*, the ex-Chinese turret ship *Chin'en*, and the coast defence cruisers of the *Matsushima* class) were added as a second line fleet.
11. The actual cost amounted to ¥219,097,962.
12. The only event which stands out is his embarkation on the German warships *Vineta* and then *Leipzig* for a journey around the world; he was very interested in the world situation and had considerable practical experience.
13. Among them Ijuin Gorô, Takarabe Takeshi, Takeshita Isamu, Saitô Makotô, Shimamura Hayao, Katô Tomosaburô, and Okada Keisuke, who subsequently contributed much to the development of the IJN.
14. On 28 December 1903, immediately before the RJW began, the Wartime IGHQ Regulations (*Senji Daihonei Jôrei*) were revised and the IJN obtained equal command rights with the IJA following a bitter ten-year campaign.
15. During his time as Navy Minister Baron Yamamoto was promoted to full admiral in 1904 and twice became Premier: 20 February 1913 – 24 March 1914, when his Cabinet fell in the course of the Siemens bribery affair, and then from 2 September 1923 – 19 December 1923.
16. *Fuji* and *Yashima*, the design of which was derived from the *Royal Sovereign*, displaced only 12,500 tons; the new ships were based on the British *Majestic*.
17. Later increased to six by the addition of two armoured cruisers; final displacements were around 9,500 tons.
18. Arichi was promoted Rear Admiral in June 1886 and CO of Yokosuka naval port, was chief of the NGS from 15 May 1889 to 16 January 1891, and succeeded Vice Admiral Inoue Yoshika as C-in-C of the Standing Fleet on 17 June 1891; he was promoted Vice Admiral on 12 December 1892. On the same day he was transferred to Kure Naval Station as C-in-C before again taking over command of the Standing Fleet (11 May – 15 November 1895). His second command was cut short by the *Thales* Incident (the stop and search of the British merchant vessel *Thales*, thought to be carrying a ringleader of the resistance movement against the Japanese on Taiwan, by the despatch vessel *Yaeyama* on the high seas on 20 October 1895). Britain insisted that the C-in-C be disciplined and deprived of his post for life. There was no corresponding article in the Japanese penal code, but the IJN was forced to reprimand Arichi in order to avoid the risk of conflict with Britain. With the sanction of the Emperor, Arichi and the CO of *Yaeyama*, Captain Hirayama Toshirô, were transferred to the reserve. Hirayama became director of the High Merchant Ship School in Tôkyô; Arichi was ennobled in June 1896, became a member of the Upper House in July 1897, and Privy Councillor in April 1901. He died in April 1919, age 74.
19. A lengthy explanation of possible budgetary savings and tax increases followed. Written in 1891, three years before the beginning of the Sino-Japanese War and four years before the war indemnity, Arichi's analysis demonstrates that the Six-Six Fleet could never have been built within the current national budget.
20. Vice Admiral Arichi is referring to the inequitable treaties forced upon the Tokugawa Bakufu by the USA and the European powers soon after the opening of Japan by force. The struggle to free Japan from these treaties was one of the most important tasks of the Meiji government. Britain made concessions in 1893 (effective after five years), and other nations then followed.
21. The other two armoured cruisers were ordered from France and Germany; all six are described in *Warship 2014*, 70-92.
22. The same arguments were made later in the shipbuilding race between Japan and the USA with regard to the passage of the Panama Canal.
23. During the First World War, when the IJN was tasked with the protection of sea lanes and convoy escort, the COs acknowledged that their ships were shorter-ranged than their RN counterparts.
24. There was also an Eight-Eight-Eight Fleet before 1918, the third 'eight' being constituted by older ships, just as with Yamamoto's original Six-Six-Six Fleet concept.

THE RISE OF THE BROWN CURTIS TURBINE

While the early marine turbine is generally associated with Charles Parsons, the competitive development of the Curtis turbine has received less attention. **Ian Johnston** assesses its acquisition under licence by John Brown & Co Ltd, and its adoption for many of the Royal Navy's major warships during the First World War.

When the Parsons turbine was introduced to the Royal Navy in the early years of the 20th century, it quickly replaced reciprocating machinery in most classes of warship. And so it remained for almost ten years: the world's largest navy propelled by Charles Parsons and his turbine. However, in the years that followed, that would change as the Curtis turbine, known in the UK in its slightly modified form as the Brown Curtis turbine, began to be fitted in RN warships, leading to its selection to power the largest and fastest warships then built, the battlecruisers of the *Renown* and *Hood* classes. This turbine was invented in the United States by Charles Curtis, although the version that went to sea with the Royal Navy was developed by John Brown & Co Ltd at Clydebank on the River Clyde. By the end of the First World War, almost half of the major warships in the Royal Navy were being fitted with Brown Curtis turbines.

This non-technical article considers some of the events that led to the adoption of the Brown Curtis turbine in the UK before and during the First World War, and is based on primary source documents from the records of John Brown & Co Ltd. These include Company minutes, letter books, tender documents and a series of letters written by Charles Curtis in New York to Stephen Pigott and Thomas Bell, both of whom were in Clydebank.

Principal Participants

Charles Gordon Curtis (1860–1953) was a civil engineer and an attorney, and the inventor of the turbine to which he gave his name. He sold the rights to his turbine to the General Electric Co. He was Vice Chairman, later Chairman, of the International Curtis Marine Turbine Company of New York.

Thomas Bell (1865–1952) was a marine engineer of repute and Director of the Clydebank shipyard of John Brown & Co Ltd from 1909 until 1935, having been Engine Works manager previously. In 1903, Bell served on Cunard's Turbine Committee. He went to General Electric in 1907 to appraise the Curtis turbine and subsequently negotiated the UK rights to build and license the turbine. Bell retired in 1935, although he remained on the Board of Directors until 1946. He was knighted in 1917.

Stephen Pigott (1880–1955) was a mechanical engineer in the employment of the General Electric Co in Schenectady, working on the development of the Curtis turbine. He moved from there to Clydebank in 1908 to continue development work, and rose to become Director in charge of John Brown's Clydebank shipyard after succeeding Sir Thomas Bell. He was knighted in 1939.

The Early Development of the Marine Turbine

The introduction of the marine turbine in the early 1900s was of major importance in ship propulsion, conferring clear advantages over then-standard reciprocating machinery. Within just a few years of its appearance in the UK, Parsons' turbine was selected in 1903, after much deliberation, to power the two largest ships in the world, the Cunard liners *Lusitania* and *Mauretania*. This was a bold if risky decision for such an emerging technology. In 1905 turbines were chosen to power the battleship *Dreadnought* and this decision, together with a uniform main armament, is often cited as the reason why this ship was considered revolutionary.

The Admiralty's appraisal of the turbine cited the following advantages:

– great savings in weight
– reduced number of working parts
– smooth operation
– ability to be started at short notice
– reduced coal consumption
– reduced engine room complement

Although undoubtedly a significant improvement overall, the turbine had a few disadvantages, among them its high rotational speed which, when coupled directly to the propeller shaft (direct drive), turned the propeller more quickly than was hydrodynamically efficient. At cruising speeds turbines were less efficient than reciprocating engines; this meant they consumed more coal, a not-insignificant issue for ocean-going navies such as the Royal Navy. While these problems were recognised from the outset, it took the best part of a decade before a

Charles Gordon Curtis. (Author's collection)   Sir Thomas Bell. (Author's collection)   Sir Stephen Pigott. (Author's collection)

successful solution was in place, namely the introduction of gearing whereby a gearbox placed between the turbine and propeller shaft reduced the rotation of the shaft, thus turning the propeller at slower and more efficient speeds.

Although Parsons was first to develop and exploit his turbine through licensing in the UK, there were other interesting variations on turbine design, notably Rateau in France, De Laval and Ljungstrom, both in Sweden, and Curtis in the USA. Although Parsons had invented his turbine in 1884, it wasn't until 1897 that it finally captured the imagination of the Royal Navy when the launch *Turbinia* steamed down the ranks of warships assembled at the Spithead Fleet Review of that year at speeds in excess of 30 knots. From then on developments moved rapidly in both mercantile and naval spheres, and in November 1905 John Brown & Co signed a licence agreement with the Parsons Marine Steam Turbine Company backdated to 1 January 1905 and valid for 15 years. It specified a payment of £2,500 per turbine at the outset, and a royalty for each horsepower developed on full-power contractors trial.

In 1901, the US electrical generating company Westinghouse acquired a Parsons licence to adapt this turbine for use in power generation in land power stations. It was for this purpose that Charles Curtis initially developed his turbine. It differed from the Parsons type in principle: the Curtis was an impulse turbine while the Parsons was a reaction type. The Curtis turbine required less exacting tolerances and was more compact, with a shorter rotor than the Parsons, both highly desirable characteristics. In 1896 Curtis patented his turbine, and in 1901 he sold the rights to the General Electric Company. Early versions of the turbine were horizontal, but it was in the vertical position that it proved most successful in generating power on land. The first vertical Curtis turbine was supplied to the Newport and Fall River Co in Newport, Rhode Island in 1903.

In 1903, the German electrical equipment manufacturer AEG acquired a licence from General Electric to manufacture Curtis turbines for power stations in Berlin. However, AEG was quick to develop a marine version to power the small steamer *Kaiser* built by AG Vulkan at Stettin in 1905, a vessel that was subsequently used by the German Navy to test turbine propulsion. AEG Curtis turbines would later go on to propel a large number of German warships.

In the United States, interest in the turbine resulted in a decision by the Navy in 1905 to build three light cruisers, one with reciprocating engines (*Birmingham*), one with Parsons turbines (*Chester*) and one with Curtis turbines (*Salem*). The outcome of these trials resulted in the Curtis turbine being selected for installation in a new battleship, although unlike the position in the UK where every battleship after *Dreadnought* would be turbine powered, the US Navy took a more cautious approach by constructing a sister vessel with reciprocating machinery to further assess the merits of both methods of propulsion. Thus General Electric Curtis direct drive turbines were fitted in *North Dakota*, built at the Fore River shipyard, while her sister *Delaware* was fitted by the Newport News Shipbuilding Co with triple expansion machinery. Both ships were laid down late in 1907.[1]

The Fore River shipyard in Massachusetts became the main licence holder for the Curtis turbine in the United States, and supplied the Imperial Japanese Navy with Curtis turbines for the armoured cruiser *Ibuki* and the battleship *Settsu*. The Curtis turbine was a successful rival to the Parsons type, and was in use (or about to be in use) in the USA, Germany and Japan. In 1907, shipbuilders and marine engineers John Brown & Co Ltd sent their Engine Works Manager, Thomas Bell, to the USA to investigate the technical and commercial merits of this turbine. It may be that the British Admiralty were behind this development, although surviving John Brown company records make no mention of being prompted by the Admiralty, and it seems likely that Brown's acted independently.

Having already built a number of Parsons turbine installations to power ships, John Brown & Co was in a position to understand the technical issues and have an informed sense of where future developments would go in the then-relatively new domain of turbine design. In 1946, when Sir Thomas Bell finally retired from the John Brown Board, he made a farewell speech in which he referred to significant events in his working life. Part of this long speech is included here as his remarks set out the tone, if not the detail of the great events that shaped turbine propulsion in the UK, and in particular of the early Parsons turbine and the origins of the Brown Curtis turbine.

Sir Thomas Bell's Speech

Speaking firstly of the period when John Brown & Co acquired the Clydebank shipyard in 1899 Sir Thomas Bell says:

> This started in 1899, and it was at a very difficult time for engineering was making tremendous strides and advances: Sir Charles Parsons' wonderful marine steam turbine had just come in, and I remember well how very unwillingly Clydebank came to take a prominent part in assisting the development. It was like this: In 1901 we got the contract for two 18-knot, 23,000-ton vessels *Caronia* and *Carmania* and everything was going well with the vessels when Cunard decided to investigate what would be involved in building two great 25-knot steamers which, if ultimately to be built, were to be named *Lusitania* and *Mauretania*, and when investigating it they found they would need to get the great power of 70,000hp to power it, almost double the power of any vessel yet constructed. So the question came up with Cunard as to whether, in view of this new invention of Sir Charles Parsons, there would be any greater risk in building turbines of that power than in balancing and generally running such ponderous engines as these reciprocators had to be.
>
> So they adopted the very wise arrangement of 'passing it on to someone else', and they instituted a Turbine Committee with scientists and representatives of the Admiralty and of the big societies of Lloyd's and the Board of Trade: also a few marine engineers, among them being Andrew Laing and myself, and we had all sorts of channel trips in turbine- and reciprocating-engined ships and altogether running quite exciting races.
>
> The final result of the Turbine Committee's findings was that they quite thought the risks were in favour of the turbine – at any rate it would be easier to balance it than the other one, and Cunard, after a lot of thinking, started

This postcard photograph taken in 1903 at Bay Ridge, Brooklyn, shows the team responsible for the early development of the Curtis turbine. Stephen Pigott is at centre in the white shirt. It is not clear who 'yours truly' is at left but this might conceivably be Charles Curtis. (National Records of Scotland, UCS 1-118-GEN-523-6)

arranging to build ships with turbine machinery but, to our horror, they thought a very good way of further testing this was to stop the reciprocating engines they had commenced for the *Carmania* and to fit turbines instead. They thought it would make excellent experience for them. We could not well refuse or say anything about it seeing we had agreed to put turbines in the two much larger vessels so, with the best grace we could muster, we set about designing these turbines – at least Parsons did that working in conjunction with us – and finally, I may tell you, after some hairsbreadth escapes and troubles, we finished the turbines and got them aboard, and on the sea voyages she actually kept level with the *Caronia*, the reciprocating ship, and most of the trouble was due to the number of ill-designed auxiliaries *Carmania* had and how serious a matter their consumption of steam was, which really did away with any savings the turbines themselves had over the reciprocators. Well anyway, the ultimate success of the *Carmania* turbines, while we lost a lot of money doing it, was such that we were recouped the loss by the Admiralty giving us the orders for several fairly large vessels and of our managing to get a price that made up for *Carmania*. The next problem was the building of the turbines for *Lusitania* and *Mauretania*, and there it was a different matter. These turbines were really like huge Frankensteins – even Andrew Laing was not free from fears, but in the end we, John Brown, happened to be the first ship done whereas we were both manoeuvring not to be the first ship. These ships, as you know, 'filled the bill' for many years. They had their troubles also, but they were queens of the Atlantic.

After that, things went very quietly excepting for a little trouble that most of us found, which was in connection with the fine clearances of the blading required for economical steam consumption. Of course, most of you gentlemen know that, measuring in thousandths, it is always good to remember that a 64th is 15/1000 [of an inch], and lo and behold, in some of the smaller turbines in which the designed blade clearances were 30/1000 of an inch, Mr Parsons said 'we will make the blade tips a little thinner'. He did not tell how many turbines in those days came back with no blades in them.

Well, that being so, I heard of the Curtis turbine in America and that it had impulse blading which permitted fairly good clearances, so, with the permission of our Board, I went over to America. I am telling you all this because it is the genesis of the Brown Curtis turbine of which our friend Sir Stephen Pigott was the High Priest. I saw Mr Curtis and found him anything but a businessman – a delightful person to meet and a genius, but not a very great businessman, and I naturally went then to Schenectady to see the General Electric Works there, where Curtis turbines were being manufactured. I was greatly impressed by them and with the big turbines they were making for land purposes. On the evening of my visit to Schenectady, a pleasant young fellow came up to me in the hotel in the historic and quaint old town of Albany and introduced himself as Charles Curtis' assistant and also that he had seen me that afternoon in the Works, and further that he was going down the Hudson River to New York by the late-night boat. I never was more interested in my life than on that evening chatting with my new acquaintance, whose name was Stephen Pigott. Till very late, or perhaps it was the following morning, we chatted first about the turbines and later about life in Colombia University where he had just completed his studies and obtained his degree. I never saw him again for a year and a half because my negotiations with the Syndicate that had Curtis in tow failed and I came back with empty hands. But a year afterwards, the Syndicate got in touch with us again and this time were reasonable and we took out the Rights but made a proviso that we only did it on condition that Curtis' Assistant, Mr Stephen Pigott, should come over and remain with us [at Clydebank for] at least 12 months. This they finally assented to and during that year we carried out very ambitious tests and trials, and at the end of the time, mainly due to Mr Pigott we evolved the Brown Curtis turbine. I am telling you this because it is due to this new turbine that we really acquired a special standing with the Admiralty who at that time, from 1910 onwards, had so many important naval contracts to give out including the battlecruiser *Tiger* whose turbines were the greatest power yet constructed, and led to us obtaining the order for *Repulse* and finally the contract for *Hood* because none of the other companies had tendered with Brown Curtis turbines and they were the ones which the Admiralty wished.[2]

Checking Sir Thomas Bell's recollections above with the archived directors' minutes of John Brown & Co, it is recorded that negotiations with Curtis were underway during the early part of 1908, and that on the Board meeting of 20 May 1908 an agreement had been signed. At the same meeting it was noted that Thomas Bell had met with Charles Parsons. Parsons was undoubtedly concerned that his turbine was being challenged in the home market.[3]

The New Turbine Works at Clydebank

In June 1908, the Clydebank Board passed expenditure of £14,000 to set up 'Curtis demonstration plant', which was in fact a shed set up like a ship's boiler and engine room combined where experimental Curtis turbines made in the engine shops could be steamed at full power. It was also reported in the same month that a letter had been received from Charles Parsons. Although the contents are not described, it was agreed that a letter should be returned stating that John Brown & Co 'will carry out its obligations to Parsons in full'. This meant that for every Curtis turbine built at Clydebank, Parsons would receive the same royalty payment as if it had been a Parsons turbine. This was clearly a contentious issue and one which Curtis understandably objected to, not least as in contributing to Parsons profits it made Curtis turbines more expensive. John Brown & Co nevertheless sought legal opinion on the subject from Sir Alfred Cripps QC in April 1909, who advised that they had no option but to pay Parsons, as the agreement with his company was for the manufacture of any kind of turbine.[4]

In July 1908 it was agreed that all British applicants for Curtis rights should be referred to John Brown & Co. This would be a profitable arrangement for John Brown. In the same month, 28-year-old Stephen Pigott travelled to the UK with his young family on the *Lusitania* to begin a one-year appointment at Clydebank, an appointment that eventually would last for the rest of his working life. Pigott, in conjunction with the engineering staff at Clydebank, began making modifications to the Curtis 'demonstration' turbine, the objectives of which were described as follows:[5]

- to gain experience with a turbine type that could use superheated steam as in land turbines
- to achieve economy at low powers without the disadvantage of close-fitting parts and the fine adjustment of parts that this required
- simplification of connection and better engine room arrangements
- higher propeller efficiency through an increase in propeller size.

At the beginning of 1909 the Admiralty, evidently impressed with developments at Clydebank, placed an order with John Brown & Co for the 'Town' class cruiser *Bristol* to be fitted with Curtis turbines. This would be followed at the end of 1909 with an order for a second 'Town', *Yarmouth*, to be built by the London & Glasgow Shipbuilding Co, with Curtis turbines constructed by John Brown. Throughout this period a number of modifications were made to the Curtis turbine, and by March 1909 the facility built to steam the experimental turbines was in operation, allowing Thomas Bell to make an initial report to his Board. No fewer than three experimental turbines were built before the turbine was considered to have met expectations. The turbine that resulted as a consequence of these experiments was given the name Brown Curtis.

Later that year, the Fairfield, Yarrow and Thornycroft companies made application to John Brown & Co for

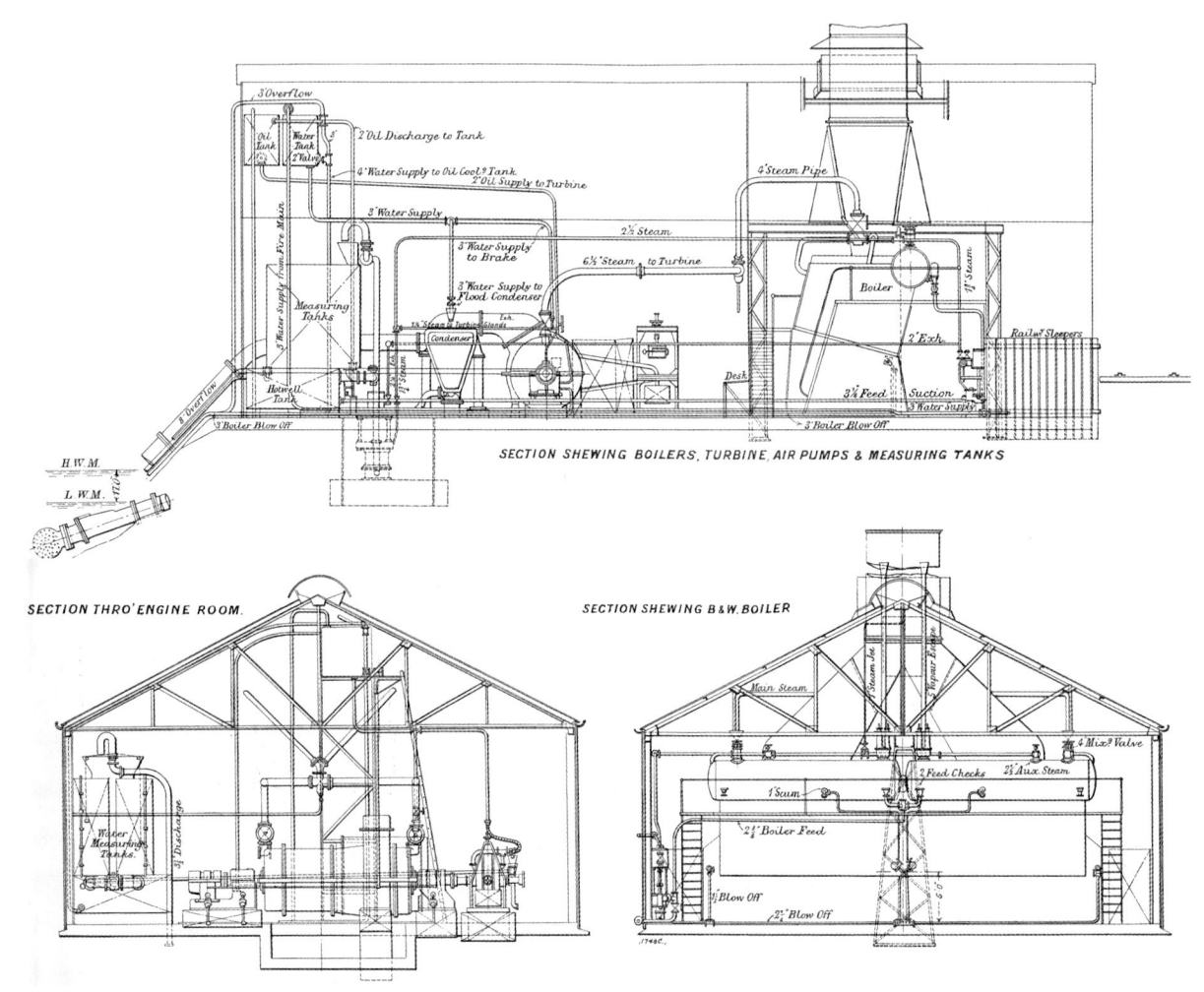

Longitudinal section and end elevations through the shed erected at Clydebank where experimental Curtis turbines were run at various powers (including full power) and modified for marine purposes. In the longitudinal section view two two Babcock & Wilcox boilers are at right, the Curtis turbine (end on) and condenser are left of centre, and measuring tanks for determining the amount of water consumed by the turbine at left. (*Engineering*, 30 September 1910, page 465)

THE RISE OF THE BROWN CURTIS TURBINE

One of three experimental turbines constructed at Clydebank used to develop the Brown Curtis type. This one was fitted in the 'Town' class cruiser HMS *Bristol*. The top half of the casing has been removed and is lying in the foreground. (National Records of Scotland UCS 1-118-390-1)

Curtis licences, and this would be followed soon after by most of the major marine engine builders in the UK. In September the Admiralty placed an order at Clydebank for three destroyers of the *Acorn* class, two (*Acorn* and *Alarm*) to be powered by Parsons turbines, and one (*Brisk*) with Curtis.

Clydebank minutes for 16 November 1909 make reference to a Brown Curtis turbine to be 'modified for the

HMS *Brisk*, the first British destroyer to be fitted with Brown Curtis turbines, running successful trials on the Clyde probably in May 1911. (National Records of Scotland, UCS 1-118-395-3)

purpose of carrying out a test on HMS *Neptune* and HMS *Indefatigable*'.[6] The detail of what this refers to and whether or not the tests were carried out is unclear, although the Parsons machinery for both ships (under construction at Portsmouth and Devonport respectively) was manufactured at Clydebank.

On 23 February 1910 *Bristol* was launched, and after fitting-out began a series of trials which proved successful. One internal John Brown report says that the Admiralty was so satisfied with the Brown Curtis turbine that other contractors had been invited to inspect them.[7]

On 18 August 1910 Thomas Bell wrote to Captain Fuji of the Imperial Japanese Navy, who had expressed great interest in the Brown Curtis turbine. After saying that he had sent drawings and information to Armstrong at Elswick, Bell continued:

> What I would like to explain to you is that this information represents the whole of the results of and experience acquired by us at great cost, not only from the Curtis Co, but as the result of over twelve months of experiments with turbine buckets and nozzles in the 2,500hp experimental turbine which we designed, constructed and erected in our yard at the request of the British Admiralty, who in return for same have awarded us the contract for the machinery of the cruisers *Bristol*, *Yarmouth* and of four TBD's representing a total of 98,000shp.
>
> Should we be fortunate enough to be given the contract for this machinery we should be delighted to place the whole of our experience at the service of the Japanese Admiralty. After carefully considering the risks we run in withholding the information at the present time, we have come to the decision and I trust you will be able to place it before your Government in a favourable light. You will notice that in the drawings which we have sent to Elswick, we have embodied our very latest investigations and designs, ie we adopt a compound Curtis steam turbine in which both the ahead and the astern turbine are compounded; thereby we obtain an astern turbine on each shaft but still are able to keep the weight and size of each

Taken in December 1913, this image shows work proceeding on one of the two IP turbine rotors for *Tiger* during the fitting of blades in the central LP ahead stage. At left is the HP stage of the IP turbine with the LP stage at centre, and the HP astern turbine fully bladed at right. Main bearing shells for *Tiger* lie in the foreground. (National Records of Scotland, UCS 1-118-418-62)

unit considerably below that of the Curtis turbines which you have at present working satisfactorily in the *Ibuki*, and in addition are able to promise a steam consumption nearly 20% below that obtainable in *Ibuki* had she been run with a satisfactory vacuum.[8] [It is not clear what machinery Bell is referring to, but it could be for Armstrong Whitworth's tender for *Kongo*.]

On 23 August 1910 a further letter from Bell to Fuji included the following descriptive statement:

> The compound Curtis turbine can hardly be termed new; all we do in it is to divide the stages into two turbines instead of one. By this means whilst increasing steam economy we greatly reduce the overall length of the turbine. Further, owing to the construction of the Curtis turbine we are able to combine a high pressure ahead and high pressure astern turbine in the same casing with the consequent great saving in floor space.[9]

Throughout 1910, Brown's began to offer Brown Curtis as an alternative to Parsons turbines in their tenders. While success was achieved with destroyers and cruisers, the next achievement, more prestigious as well as commercially attractive, would be to fit Brown Curtis turbines in a capital ship. It was not until January 1912 when the tenders were submitted for the battlecruiser *Tiger* that John Brown, who offered Parsons as well as their own Brown Curtis turbines, were successful in this.

Correspondence Between Curtis and John Brown

It was at this time in 1912 that correspondence between Charles Curtis in New York and Stephen Pigott, and occasionally Thomas Bell in Clydebank, begins. Pigott had by now become an indispensable member of Clydebank's engineering staff. The correspondence is voluminous, and what follows is a selection part-quoted from the originals. It is also one-sided, as Stephen Pigott's replies have not survived. This correspondence shows that technical details of British and US warship machinery was being exchanged and discussed in detail and that the commercial interests of the International Curtis Marine Turbine Company were uppermost in Curtis' mind. (The letters quoted in this section are from UCS1/11/6–11, Charles Curtis correspondence.)

7 May 1912
Curtis wrote to Pigott saying that the International Curtis Marine Turbine Company had awarded Pigott $900 (£184 4s 2d). This was in recognition that John Brown & Co had been successful in their tender for the battlecruiser *Tiger*, to be fitted with Brown Curtis turbines:

> I am requested by the Directors of our Company to say that they wish to congratulate you most heartily for the admirable work you have been doing for the John Brown Company but which has resulted largely for the benefit of our company, and particularly to congratulate you upon getting the award for the large battleship cruiser [*sic*] the reference for which we saw in the newspapers here some weeks ago.

14 June 1912
E Norton (International Curtis Marine Turbine Company) wrote to Pigott forwarding technical information regarding the turbines of the destroyer USS *Henley* under construction at the Fore River shipyard, Quincy.

23 October 1912
Curtis wrote to Pigott on the contentious subject of royalty payments, which his letter explains as follows:

> The real underlying trouble with the situation is that John Brown (and possibly the other shipbuilders – I do not know) have to pay the Parsons Company a royalty which is very large, comparatively speaking, which they feel bound for certain reasons other than legal or moral to go on paying for the full life of the Parsons Agreement, that is until 1920. In other words they have got to go on paying Parsons 1s 10.5d per HP on large ships and 1s 3d on small ships for every turbine that they build, apparently no matter of what type, and without any reduction until January 1st 1920.
>
> In the assumption that 1000 HP is done each year 500 of which is in large ships and 500 of which is in small ships, the average rate of royalty is 1s 5.6d shillings and the total amount of royalty to be paid to the Parsons Company would be 10,920 shillings. The total royalty to be paid to our company would only be 2,567 shillings or just one quarter of the amount which would be paid the Parsons Company before their Contract expires. And yet the facts are that we have contributed an important improvement in the shape of the Balanced Drum, having such advantages in the way of economy and elimination of the Dummy rings that we have been able to get it introduced in some of the largest and most advanced warships of the most progressive nations, whereas the Parsons Company has furnished no substantial improvement and so, as far as we know, in the case of your Brown-Curtis turbines you are not using any of the features that were covered in Parsons' earlier patents, and which were the foundation and consideration for making the Agreement with Parsons. In other words, the Parsons Company has contributed nothing in recent years, while we have contributed an important improvement, and yet we are expected to content ourselves with an average royalty during the next few years which is about one quarter of the average royalty which John Brown & Company has been paying.

Although Pigott's reply has not survived, Bell picked this issue up and responded on 2 November 1912:

> Dear Mr Curtis,
> As regards paying royalty to Parsons, I would tell you, for your private information, that people in very high posi-

tions in the British Navy were so impressed with the wonderful additional tactical advantages obtained in ships fitted with turbines and oil-fired boilers that they felt a big debt of gratitude to Sir Charles Parsons, and I was given to understand that, if we agreed to pay Parsons' turbine royalty during the continuance of our license [sic] whether the turbines were his or other types, all opposition would be withdrawn to the introduction of your turbine into the British Navy.

In view of this, and also in view of the fact that our Company was not prepared to carry out a wearisome and expensive law suit with the Parsons Company over the matter, we thought it well to agree, and I am sure you will not be the losers by this.

Personally I think very highly of your reversed drum proposal, but, of course, we are running very big risks should there be the slightest hitch in the 'TIGER' proposal, as, with her enormous power, a single failure on trial, and the time lost in examining any such part of the turbine, could easily mean a loss to us of from £8,000 to £10,000.

1 November 1912

With the decision of who would build and engine the five battleships of the *Queen Elizabeth* class still to be decided, Curtis wrote to Pigott regarding the comparative cost of Parsons and Brown Curtis machinery:

> I should think that you can build Brown Curtis turbines, particularly of the Balanced Drum type, with an equivalent economy at not over 80% of the cost of the Parsons.

6 December 1912

Curtis wrote to Pigott on the subject of forthcoming US battleships:

> Dear Mr Pigott,
> Although the navy department seemed to have decided some time ago to adopt the reciprocating engines in their next Battleship, I have succeeded, I think, in reopening the question. At any rate the Bureau of Steam Engineering has asked me to prepare a special design for this ship, preferably with four shafts, and involving our latest improvements, and also to make recommendations as to the propeller revolutions. They have assured me that if we can come within ten per cent of what they anticipate from the reciprocating engine at 10 knots and ditto 10 per cent better at full power (21 knots), they will favour the turbine, but they must be convinced that we can do this. We are now laying out a four shaft arrangement consisting of a high pressure turbine on one shaft comprising of about six wheel stages, an intermediate balanced drum turbine on the second shaft, and a low pressure turbine on the first shaft. The horse power of the ship will be about 30,000 at 21 knots. We have laid out propellers turning at 250 revolutions, another set turning at 230 revolutions, and another set turning at 265 revolutions, and we are now designing the turbine. At low power all the wheel stages will be operating in series. At full power the second, third and fourth will probably be by-passed.

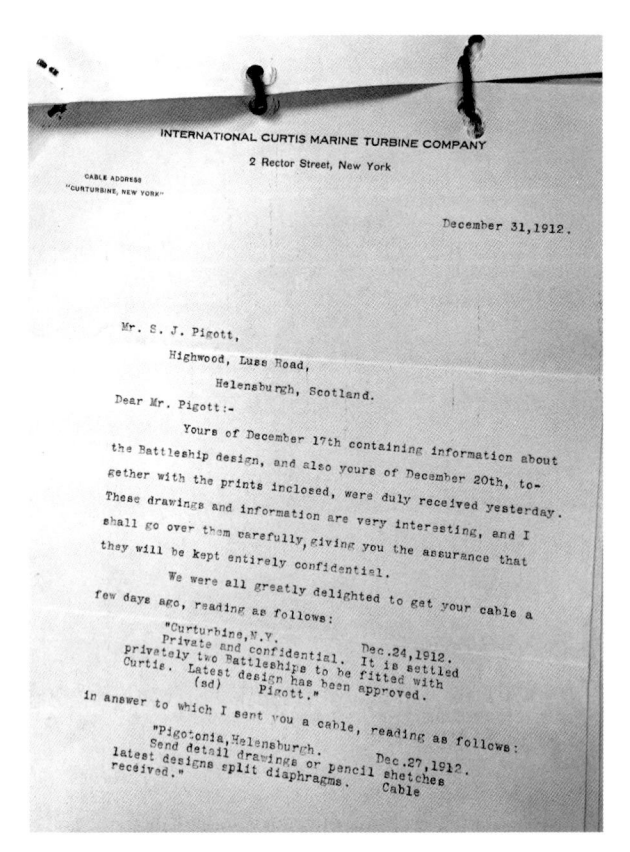

A typical letter from Curtis to Pigott, this one dated 31 December 1912. It is acknowledging the decision to fit Brown Curtis turbines to *Barham* and *Valiant*. Curtis assures Pigott that drawings and information he sent will be kept confidential. (University of Glasgow, Archive Services, Charles Curtis letterbooks, UCS 1/11/5–7)

> I wish very much you would make any suggestions you can on this line, and I wish particularly you would send me the estimated water rate curve of your latest designs for Battleships at different powers, giving me the assumed steam conditions at each power, and also the shaft revolutions, diameters of the different parts, peripheral speed, etc, so that we can see how our calculations work out compared with yours. I wish you would also let me know just how the turbines are designed, that is, how many velocity stages, number of rows of buckets, in the drums etc. I will not show these designs or figures to anybody, but will keep them entirely to myself. Of course the best way to get economy at low power is with a properly designed geared cruising turbine, but the Bureau of Steam Engineering will not consider putting anything of this kind in the next Battleship because they regard it at the present stage as experimental.

On 31 December 1912 Curtis gave Piggott a further $900 'with hearty congratulations and appreciations' along the lines of the earlier award.

On the same day but in a separate letter, Curtis thanked Piggott for sending him details and drawings of the battleship design. This is a reference to the *Queen*

THE RISE OF THE BROWN CURTIS TURBINE

One of *Barham*'s turbine sets in the erecting shop in January 1915. At right is the HP ahead turbine, and permanently coupled at left is the LP ahead and astern turbine. Two identical sets drove the inner shafts, and an IP ahead and HP astern drove the outer. (National Records of Scotland, UCS 1-118-424-281)

Elizabeth class, of which *Barham* and *Valiant* were to be powered by Brown Curtis turbines.

> These drawings and information are very interesting, and I shall go over them carefully, giving you the assurance that they will be kept entirely confidential. We are also delighted to get your cable a few days ago reading as follows: '24 December 1912. It is settled privately two battleships to be fitted with Curtis. Latest design has been approved.'

21 January 1913
Curtis wrote to Pigott congratulating him and John Brown & Co on having built Brown Curtis turbines either directly or through sub-licensees of over 1,000,000hp.

22 January 1913
Curtis wrote to Pigott regarding a successful suit that had been taken out against the Cramp shipyard in Philadelphia by the Fore River Co (who held the marine rights in the US) together with the International Curtis Marine Turbine Co, as Cramp had infringed 'our main fundamental turbine patent'.

> In addition to the two destroyers which the Cramps began to build and on which the suit was brought, they have taken contracts for seven additional boats, making in all about 150,000hp. I do not know that there is anything else to be said, except that the Cramps have come round, and express a desire to get together and settle the matter on the best terms they can, without taking the question before a referee.

4 March 1913
Curtis wrote advising Pigott that the chief engineer of Newport News Shipbuilding & Dry Dock Co, Mr Bailey, 'is going abroad to investigate the latest practice there in the development of our turbines'. Newport News had just won the contract for USS *Pennsylvania* with four-shaft Curtis turbines.

I shall appreciate it very much if you will talk to Mr Bailey as frankly as you would us and show him and explain to him all your latest developments, so far as is permissible under the John Brown Company practice. I have no doubt that Mr Bailey will go into the matter very carefully and make a good job of the *Pennsylvania*, and we are doing everything in our power to help. I am sure that you will do the same.

13 March 1913
Curtis wrote to Pigott:

I enclose blueprint showing the layout which is to go in the United States battleship *Pennsylvania* just ordered from Newport News Shipbuilding Company. These, by the way, should be kept entirely confidential, so far as the Chief Engineer of the Newport News Company is concerned. He will visit you at Clydebank in the near future, and while I am not under obligations to keep the above confidential, I would not want him to think that I am sending you the layout of the machinery just as they are going to build it.

There is not the slightest question in my mind but that you ought to bring down your revolutions far below what you have been working with in England. The revolutions you have been using, and the Parsons people have been using, are far too high both for propeller efficiency and for holding power.

28 Nov 1913
Curtis wrote to Pigott:

I wish you would drop me a line at once giving the actual results obtained on some of the latest Parsons Battleship turbines in England. I want this for my own private information, and will not let it get out.

What I want is to form as accurate an idea as possible of what the Parsons people will be able to offer in the way of economy for the next United States Battleships in order that we can design something which will be better but not too expensive.

19 March 1914
Curtis wrote to Pigott. He was anxious to know if Brown Curtis turbines would be selected for any of the four battleships which he noted had been passed by Parliament. [Reference to the *Royal Sovereign* or *Revenge* class battleship; there were five in total and all were fitted with Parsons turbines.] Curtis went on to say that two battleships would be built in the US and that he hoped to get one or both of them, and that both Newport News and the New York Shipbuilding Co were 'very friendly.' He stated that 'Fore River have discredited themselves so thoroughly by their bad work and the troubles had with their turbines that I am afraid they will not be in the running'.

At this point the correspondence stops and thus there is no further insight into the discussion between Curtis and Pigott regarding the powering of subsequent British and American capital ships.

Conclusion

Brown Curtis turbines were subsequently fitted in *Renown*, *Repulse*, *Furious*, *Hood*, *Rodney* and *Nelson*. From *Furious* onwards these were of the single reduction-geared type. John Brown supplied Brown Curtis machinery for two Russian battleships, and for the Chilean battleship *Almirante Cochrane* which was under construction by Armstrong Whitworth.

Paradoxically, the Curtis turbine did not enjoy anything like this success in the US Navy, where for a combination of reasons turbines got off to a bad start, and then issues to do with improving mechanical efficiency led the US Navy down the path of turbo-electric propulsion. Two battleships, *Florida* and *Utah*, were laid down in March 1909 both fitted with Parsons direct drive turbines. These were followed in 1910 by *Wyoming* and *Arkansas* also with Parsons turbines; then came the reversion to triple expansion engines for *New York* and *Texas* in 1911. In 1912 *Nevada* was laid down fitted with Curtis turbines, although sister *Oklahoma* had triple expansion machinery. In the following class, *Pennsylvania* had Curtis and *Arizona* Parsons. Of the three battleships laid down in 1915, *New Mexico* had turbo-electric propulsion, *Idaho* Parsons and *Mississippi* Curtis direct drive turbines. Thereafter succeeding classes employed exclusively turbo-electric drive driven by either General Electric (Curtis) turbines or Westinghouse (Parsons).

Sources:
The archives of John Brown & Co Ltd, held by the University of Glasgow.
UCS 1/11/6 - 11 Charles Curtis correspondence.
UCS 1/1/13 Minutes of John Brown & Co. Clydebank
Transcript of Sir Thomas Bell's retirement speech in possession of the author.

Endnotes:
[1] The merchant ship *Creole* built by Fore River in 1907 was fitted with Curtis turbines that proved to be unsuccessful.
[2] Quoted from a transcript of Sir Thomas Bell's speech in possession of the author.
[3] UCS 1/1/13 Minutes of John Brown & Co Ltd.
[4] UCS 1/1/13 Minutes of John Brown & Co Ltd.
[5] *Engineering*, 30 Sept 1910, page 465.
[6] UCS 1/1/13 Minutes of John Brown & Co Ltd.
[7] John Brown & Co Ltd., Manager's Reports for 21 September 1909, held by Sheffield City Archives (not catalogued).
[8] UCS 1/11/4 letterbooks of John Brown & Co Ltd.
[9] UCS 1/11/4 letterbooks of John Brown & Co Ltd.

BATTLECRUISER *TIGER*

The Arrangement of the Main Engines: Correcting a Century of Erroneous Consensus

Working from contemporary photographs taken in the engine works of the Clydebank shipbuilder John Brown & Co Ltd, **Dr Brian Newman** critically examines published descriptions of the Brown Curtis turbine machinery installation in HMS *Tiger*, and attempts to establish the actual configuration.

The battleships and battlecruisers of the Royal Navy have been discussed widely in the century since *Dreadnought* made her dramatic *début*. The debate has largely focused on specifications, operational histories and on how external appearance was modified by the imperatives of war.[1] There has, on the other hand, been little discussion of the internal arrangement of these vessels, and this aspect has been poorly recorded by comparison.[2] This article on the arrangement of the main engines of the battlecruiser *Tiger* makes but a small contribution to redressing this imbalance, but in the process exposes more than a century of widely accepted but inaccurate data about the vessel.

Clydebank's association with the Curtis turbine began

Many of the Clydebank photographs in the care of the National Records of Scotland are of stunning quality and this one is no exception. *Tiger* is shown alongside the west quay of the fitting-out basin at Clydebank just before she departed for her sea trials. (National Records of Scotland, UCS 1-118-418-138)

with a visit by engine works manager Thomas Bell (1865–1952) to Charles G Curtis (1860–1953) at the Schenectady works of The International Curtis Marine Turbine Co in 1907. At this time the engine works under Bell were unsurpassed in their accumulated experience in the design and manufacture of high-power marine steam turbines, as evidenced by its construction of the machinery for the Cunard liners *Carmania* and *Lusitania*. By mid-1908 Clydebank had acquired a licence to manufacture Curtis turbines and provided a new facility (see the accompanying article by Ian Johnston) in which three 'experimental' turbines were tested under full load conditions[3] with a view to persuading the Admiralty of their efficiency and reliability.

The association with Curtis brought to Clydebank a brilliant American engineer, Stephen J Pigott (1880–1955), whose outstanding ability was crucial to the success of the Brown Curtis marine turbine. Initially appointed for one year to oversee the construction of the experimental turbines, he remained at Clydebank for the rest of his working life, retiring as Managing Director in 1948. Clydebank prepared tenders offering Brown Curtis or Parsons machinery for *Tiger* by January 1912, and by April 1913 Curtis turbines had been specified. The Admiralty had already selected them for the cruiser *Bristol*, completed in 1910, and *Yarmouth* and *Southampton* in 1912; *Tiger* was the first British capital ship so fitted.

Tiger fitting out under the cantilever crane at John Brown's on the Clyde. The starfish of the tripod foremast, topped by the fire control director for the main guns, is prominent. (NRS UCS 1-118-418-148)

Because their differences were in detail it is generally accepted that the *Lion* class consisted of *Lion*, *Princess Royal* and *Queen Mary*, but that *Tiger* constituted a distinct class because of the modified position of 'Q' turret, the adoption of a secondary armament of 6in guns (replacing the 4in guns of the earlier ships), and the adoption of Brown Curtis instead of Parsons turbines. *Tiger* was authorised under the Naval Estimates for 1911 and the contract was awarded to the Clydebank shipyard of John Brown & Co Ltd, which allocated yard number 418 to both hull and machinery.

Research into the machinery of *Tiger* was prompted by a photograph (see Fig 2) of her machinery in the erecting shop. The arrangement of the turbines posed some questions, leading the author to consult the leading sources of information. It was immediately apparent that there was a significant discrepancy between the photograph and the descriptions in those sources.

The Current Consensus

Contemporary sources such as The Institution of Naval Architects and The Institute of Marine Engineers, which in normal circumstances debated such matters at the most expert level, were silent on the arrangement of *Tiger's* machinery, probably because of the impending war; the subject was not discussed by the Naval Architects until 1919 and not at all by the Marine Engineers or the other leading professional marine technology institutions.[4]

The most widely disseminated reference sources, Brassey's *Naval Annual* and Fred T Jane's *Fighting Ships*, did not discuss the arrangement of the machinery in detail, giving only the type, number of shafts and combined horsepower. The 1912 edition of *The Naval Annual* stated that *Tiger* was to have Parsons turbines, although no other data was provided.[5] This indicates that it was assumed that, as Parsons turbines were the established choice for such vessels, this convention would continue to be observed. Similarly, in the edition of *Fighting Ships* for that year Parsons turbines were claimed for *Tiger*;[6] *The Naval Annual* repeated the statement that *Tiger* was to be fitted with Parsons turbines in its 1913 edition.[7]

The 1913 edition of *Fighting Ships* was more accurate, though not completely so, listing Brown Curtis turbines and a designed shaft horsepower (shp) of 100,000[8] – the designed power was actually 85,000. The following year it listed both *Queen Mary* and *Tiger* with Brown Curtis machinery.[9] This introduced a new error: *Queen Mary* was in fact powered by Parsons turbines constructed at Clydebank. These publications did not usually provide detailed descriptions of machinery or other arrangements; they dealt in general particulars such as tonnage, dimensions, armament, armour, speed and complement, and so their content should be judged on that basis. Nevertheless, they were widely consulted and respected sources enjoying quasi-official status, and so must bear some responsibility for the subsequent dissemination of misleading information. In addition, the name *Tiger* perhaps encouraged an assumption by the respective editors that she was to be merely a lightly modified *Lion*, and that she would show few differences in her principal particulars.

It is reasonable to assume that the attribution of Parsons turbines to *Tiger* was based on the monopoly status of these designs in British capital ships to that date – in fact in all British turbine-powered warships until the completion of *Bristol* – and that when Curtis turbines were first specified it was for vessels which did not attract the same degree of public interest that a 'dreadnought' would have elicited. It was also reasonable for such publications to assume that, even though the turbines were not of the Parsons type, they would be arranged in a similar fashion.

One of the earliest descriptions of the vessel stated that 'The British Admiralty have, as usual, issued no particulars of the ship, but the broad features of the design which are given here may be taken as fairly accurate.'[10] It continued:

> The propelling machinery is of the Parsons steam turbine type, arranged for four screws. The turbines are designed to give 99,000 horsepower, and are placed in two watertight compartments, so that in each engine room there is one high-pressure ahead, and one high-pressure astern turbine mounted separately on the outer shaft, and a low-pressure ahead and astern turbine within one casing on the inner shaft.[11]

HMS *Tiger*: Plan of Hold with Parsons Turbines

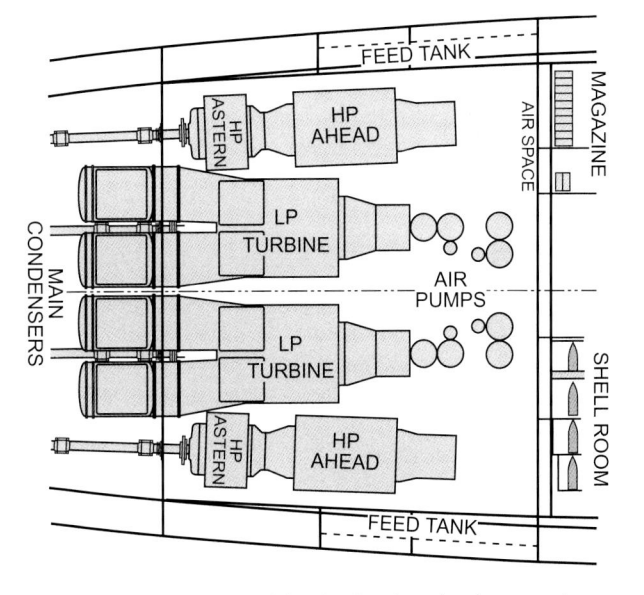

Fig 1: A detail from the Hold drawing for *Tiger* showing an early design option for the vessel with Parsons turbines, as is evidenced by the HP ahead and astern turbines on the outer shafts with their great length by comparison with the Brown Curtis turbines actually fitted. The drawing is titled 'A Battlecruiser to be Named …' but it is clearly for *Tiger*, as the magazine (at right) is contiguous with the engine room. (© John Jordan 2018)

Fig 2: The photograph which initiated this investigation, showing the starboard inner shaft turbines erected in the engine works at Clydebank on 2 April 1914. At left is the starboard HP ahead turbine permanently connected to the starboard combined LP ahead and astern turbine at centre – an arrangement at odds with all but one of the published descriptions of this machinery over almost 100 years. At right is the temporarily-connected port HP ahead turbine. (Courtesy of Ian Johnston)

The assumption that the machinery in the preceding units of the 'class' (ie *Lion*, *Princess Royal* and *Queen Mary*) would be repeated in the new vessel was a reasonable one; the writer was after all sufficiently well informed as to mention the increased power of the projected new vessel over her earlier 'sisters', although less so when describing the '… new 14-inch [sic] main battery …'.[12]

A crucial influence on the subsequent inaccuracy of descriptions of *Tiger*'s machinery is the 'Plan of Hold' (Fig 1)[13] in the National Maritime Museum collection at Woolwich. The full title of this drawing is: 'Battle Cruiser to be named'. However, the magazine (for 'Q' turret) immediately forward of the engine room bulkhead at frame 194 shows that this is not a class drawing but specifically for *Tiger*, although clearly at an early stage in the design process and assuming propulsion by Parsons turbines. The evidence for this is absolute: in the *Lion* class the aft boiler room was situated immediately forward of the engine room. However, the proportions of the HP turbines, with LP turbines on separate shafts, clearly show that these are of the Parsons type, in the standard Parsons arrangement. Unfortunately, no 'as fitted' machinery arrangement drawings for *Tiger* are known to have survived in archives or private collections.

It was only after the war that detailed information on the design of British warships was disseminated more widely, one of the first contributions coming from Sir Philip Watts (1848–1926), Director of Naval Construction (DNC) from 1902 until 1912. This, a source which would normally be considered authoritative, nevertheless proved to be otherwise:

The machinery of the *Tiger* is capable of developing an aggregate shaft horse-power of 108,000, giving her a speed of 30 knots. It consists of a four-shaft installation of Brown Curtis marine steam turbines, arranged in two sets, each set consisting of one HP ahead turbine and one HP astern turbine on the wing shaft, one LP ahead turbine and one LP astern turbine in the same casing on the outer shaft, which work in series and are directly coupled to the two shafts working together.[14]

Watts' description of 'wing' and 'outer' shafts is somewhat contradictory and one must assume that he confused 'outer shaft' with inner shaft, which would bring the description into line with others cited below. The wording '… two shafts working together …' confirms this. It is also probable that he consulted the Plan of Hold (redrawn as Fig 1) and assumed from it the arrangement cited.

Another account one could reasonably assume to be authoritative was published under the auspices of the DNC which described the machinery arrangement thus: 'Each engine room has: Wing: 1 HP Ahead, 1 HP Astern. Inner: 1 LP Ahead and Astern in same casing'.[15] DK Brown described this source as follows: 'Originally confidential, this two-volume work forms an excellent factual account of warship designs of the period'.[16]

The detailed design of British battleships was barely discussed after the Watts/d'Eyncourt[17] papers of 1919 until the publication of Oscar Parkes' landmark work in 1957. Parkes conformed to the consensus with his description of the machinery in *Tiger*: '… Each of the four shafts was available for ahead and astern working, with the turbines in two sets each with HP ahead and HP astern on the wing shafts and one LP ahead and one LP astern in the same casing on the inner shaft …'.[18]

A new generation of researchers emerged in the 1970s, producing works of greater technical content at a more detailed level of analysis than previously. One of this new generation, John Roberts, wrote:

> The Brown-Curtis turbines were arranged in two engine rooms divided by a centreline bulkhead, each having a separate condenser room abaft of it. A high-pressure ahead and a high-pressure astern turbine was connected to each wing shaft and a low pressure ahead and astern turbine, in the same casing, to each inner shaft.[19]

Another of this new generation of writers was RA Burt, whose description broke with the consensus: '… Four shafts were fitted, with the HP ahead, and LP ahead turbines being arranged on the inner shafts, and the impulse turbines located on the outer shafts.'[20] This description is confused: the Curtis turbine was essentially an impulse machine.[21] However, Burt does accurately describe HP and LP turbines on a single shaft. It is possible that he was confusing IP (intermediate pressure) with impulse, but if this is the case it would be of interest to know the source,[22] as none of those cited in the published canon described such an arrangement.

A series of research monographs on the development of the marine steam turbine was produced between 1982 and 1986. In the first, *Tiger's* machinery was described thus:

> High pressure ahead and astern cylinders worked on the wing shafts and the low pressure ahead and astern turbines

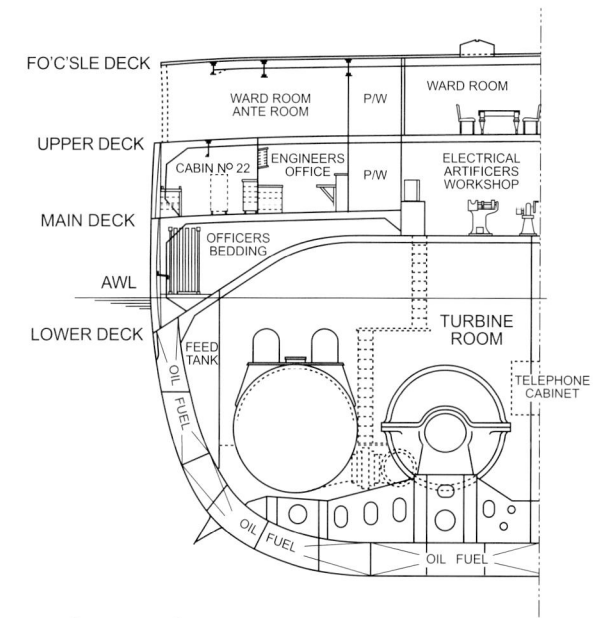

Fig 3: A detail from the 'as fitted' frame plan showing the starboard engine room of *Tiger* looking aft at frame 208. On first analysis, the drawing seemed to support a four-independent-turbines per shaft hypothesis, but other evidence eventually discredited this; it confirmed an HP on the inner and revealed an IP on the outer shafts. (© John Jordan 2018)

in a common casing worked on the inner shafts. Brown-Curtis also won the orders for the 'Renown' class with 92,500kw (126,000hp) using the same arrangement.[23]

Jung compounds his erroneous description of *Tiger's* machinery in referring to the 'Renown' class: his ascription of the arrangement of the turbines in *Renown* and her sister was also incorrect. DK Brown, on the other hand, was correct in stating that: 'To speed production, they (*Renown* and *Repulse*) had the same machinery as *Tiger*…'[24] Brown is probably citing d'Eyncourt's 1919 paper[25] although this is not stated.[26]

Misleading Evidence

The initial photograph of *Tiger's* machinery was puzzling, as the arrangement did not accord with that in any of the above descriptions. It was to be more than a year before access to other photographs of the machinery in the works from different perspectives became available, and in the interim it was assumed that the photograph showed a set of machinery for what was without question *Tiger*,[27] comprising four independent turbines per shaft (HP ahead and astern and LP ahead and astern) – this arrangement is hereinafter referred to as 'four-independent-turbines'. Evidence supporting this assumption initially came from three other sources, the first of which stated: 'The turbines on each of the four shafts constitute a separate unit, each shaft having high-pressure impulse and low-pressure reaction turbines along with astern turbines….'[28] A second almost identical description followed a month later.[29] A letter from Curtis replying to Pigott provided crucial and apparently authoritative evidence confirming the above descriptions:

> I note particularly what you say in regard to the four-shaft high speed cruiser having four independent complete turbines aggregating 30,000 horse power ... Regarding the use of four low pressure turbines on Battleships or large Cruisers, I have always doubted whether it would be wise to do this on Battleships. My idea has always been that it would not probably be wise unless the horse power became very large, something like 60,000 horse power or more. I doubt if it will be advisable for a Battleship of less than 50,000 horse power, but it seems to me very desirable in the case of a large high powered Cruiser ...[30]

Curtis, in referring to '*the* four-shaft high speed cruiser' [author's italics] suggests that the subject was more than a speculative design, and it is apparent that he was responding to an arrangement which Pigott had proposed; given the date of his letter, this arrangement can only have been for *Tiger*, as Pigott was working on the design of her machinery at that time. The reference by Curtis to 'four independent complete turbines aggregating 30,000 horse power' may be explained by the addition of the output of the astern turbines into the equation, which is what it is contended Pigott did. Thus each astern set could produce 8,750shp per shaft, which when aggregated results in a total of 35,000shp, or about 41 per cent of the total for the ahead turbines at designed full power, a figure which other authorities cite for contemporary capital ships.[31] The letter reveals the thinking of those most intimately involved in the design of these turbines: it justifies the adoption of fully independent turbines for each of the four shafts in vessels with machinery above 60,000shp – exactly that specified for *Tiger*.

Taken together, the initial photograph, the descriptions in *Engineering*, *The Shipbuilder* and, crucially, the Curtis letter clearly pointed to four independent turbines on each shaft. Further support was evident in the arrange-

Fig 4: One of the intermediate pressure turbine rotors which drove the outer shafts, photographed on 1 December 1913. At left is the multi-collar thrust shaft followed by the ahead IHP stage and the ILP stage. At right behind the figure is the HP astern turbine. (NRS, UCS 00100118-0418-00064)

Fig 5: One of the HP ahead and LP ahead/astern turbine rotors for *Tiger* assembled in the erecting shop on 17 September 1913. The HP ahead rotor at left is clearly intended to be permanently connected to the combined LP ahead/astern rotor at right via the shaft coupling with integral multi-collar thrust shaft. (NRS, UCS 00100118-0418-00063)

ment of the turbines in the cruisers *Bristol*, *Yarmouth* and *Southampton*[32] and the AEG–Curtis turbines in German battleships,[33] which featured self-contained independent stages of HP and LP ahead and astern turbines per shaft – this arrangement being standard for Curtis marine turbines up to that time.[34] Providing further support for the four-independent-turbine hypothesis is a detail from a 'frame drawing',[35] part of the 'as built' record of the vessel, which must be regarded as accurate, unlike the misleading Plan of Hold discussed above. The evidence for this lies in the dated annotations in two colours which were added to the drawing when a change of use of various compartments occurred, the drawing representing the documentation of the vessel in service. The section is looking aft on the starboard side at frame 208, and clearly indicates a large turbine on the outer shaft – at odds with all of the published 'consensus' descriptions of the machinery cited above, and what is clearly, because of the similarity of its profile to those in the photographs at Clydebank, an HP on the inner – equally at odds with the consensus.

This provided further support to a four-independent-turbines hypothesis in which the inner and outer shaft turbines were off-set fore and aft so as to be accommodated within the diminishing beam of the vessel at that frame. In addition, the drawing indicates two condensers on the starboard side, suggesting four in total for the vessel, which was also compatible with the four-LP-turbines component of the hypothesis. Additional confirmation came with access to surviving photographs of *Tiger*'s machinery in the shops at Clydebank, which show four condensers awaiting transfer from the erecting shop to the fitting-out basin.[36]

Accurate Data

An explanation for the arrangement in the initial photograph was possible only when other surviving images of *Tiger*'s machinery in the shops at Clydebank became available.[37] These confirmed the permanently coupled HP and LP sets,[38] and added weight to the evidence for a four-independent-turbines arrangement with views of the

Fig 6: Assembly of the port HP and LP sets photographed on 12 March 1914. The tachometer drive at left shows that no power take-off from that end of an HP turbine was possible, and this applied to the temporarily-connected starboard HP at right, showing that this arrangement was indeed a temporary one intended to balance the axial thrust of the two HP turbines in order to distribute more equally the load on the problematic multi-collar thrust bearings then commonly specified. (NRS, UCS 00100118-0418-00078).

four condensers (assuming one for each LP turbine), which in turn ostensibly confirmed the frame drawing arrangement in that particular context. They also confirmed (by views from other perspectives) that the arrangement in the initial photograph was a temporary one for test purposes, and not viable for a marine (or any other) engine because there was no shaft coupling at either end of the assembly.[39] This to an extent undermined the four-independent-turbines hypothesis.

In addition, there were six views[40] of rotors which, given their dimensions and design, were clearly not destined for a *Tiger* with a four-independent-turbines arrangement. Initially, it was assumed that these were mis-catalogued items, and intended for one of the Russian battleships *Imperatritsa Mariya* or *Imperator Alexander III*[41] then being constructed in the engine works.

The next step in the confirmation of the four-independent-turbines hypothesis focused attention on the next set of high-power Brown Curtis turbines installed in a British warship – those intended for the battlecruiser *Repulse*.[42] The surviving official records of the construction of this vessel stated that:

The design then contemplated 110,000 SHP, oil fuel boilers and engine rooms reduced in length and width and an entirely new design of turbines, ie one HP and one LP ahead for each set ... It was afterwards decided to utilize *Tiger*'s patterns and repeat *Tiger*'s machinery in all respects43

This is evidence that the four-independent-turbines arrangement discussed two years earlier by Curtis and Pigott was more than a speculative one, for the 'contemplated' design is precisely that described in the letter cited above, but was clearly rejected because of the 15-months completion demanded by the Admiralty.[44] However, prompted by this statement, the surviving photographs of *Repulse*'s machinery[45] were accessed as were the 'as fitted' engine room drawings,[46] which in combination showed HP and LP ahead and LP astern turbines on the inner shafts, and IP ahead and HP astern[47] on the outer. Close comparison of the IP turbine rotor photographs for both vessels showed them to be identical, as were the LP rotors, thereby confirming the statement in the ships' covers and demolishing the now misleading evidence of the four-independent-turbine hypothesis advanced above.

The frame drawing discussed above is anomalous in this context, as an HP on the inner shafts and a large turbine on the outer were accurately described. However, such was the weight of evidence for a four-independent-turbine arrangement at that stage, it was assumed that it conformed to that deduction, when in fact the large turbine was an IP set and not, as first believed, an LP.

An Explanation for the Temporary Assembly

Once the actual arrangement of the turbines was established, it became necessary to explain the puzzling assembly illustrated in the initial photograph. The photographs in the erecting shop show two distinct sets. The initial photograph (Fig 2), dated 13 December 1913, shows the starboard HP turbine at left and the starboard LP turbine at centre, with the temporarily-connected port HP turbine at right. The other photographs of the assembly (Figs 6 & 7), which are dated 12 March 1914, show a mirror image of the initial image, transposing port for starboard.[48] Another photograph (Fig 5), dated 17 September 1913, shows unequivocally that an HP and an LP turbine were permanently coupled on the same shaft – an arrangement at odds with all of the descriptions in the canon of literature cited above apart from the arrangement described by Burt.

It also clearly shows (at the extreme left) that the forward section of the main shaft was hollow and could not mount a coupling flange by which power could be transmitted. This is supported by a photograph (Fig 6) showing a tachometer coupled to this section of shaft,

Fig 7: Another view of the assembly on 12 March 1914. Note that the steam pipe has been removed. It is apparent from the enlarged detail that there is no shaft coupling at right, confirming that the assembly was a temporary one which could not function as a power unit in that form. This also demolished the misleading evidence of a four-turbines-per-shaft arrangement, until that time the standard form of Brown Curtis marine turbines. The other enlarged detail shows the wording 'SHP ahd' chalked on the temporarily coupled HP turbine on the right. (NRS, UCS 00100118-0418-00077)

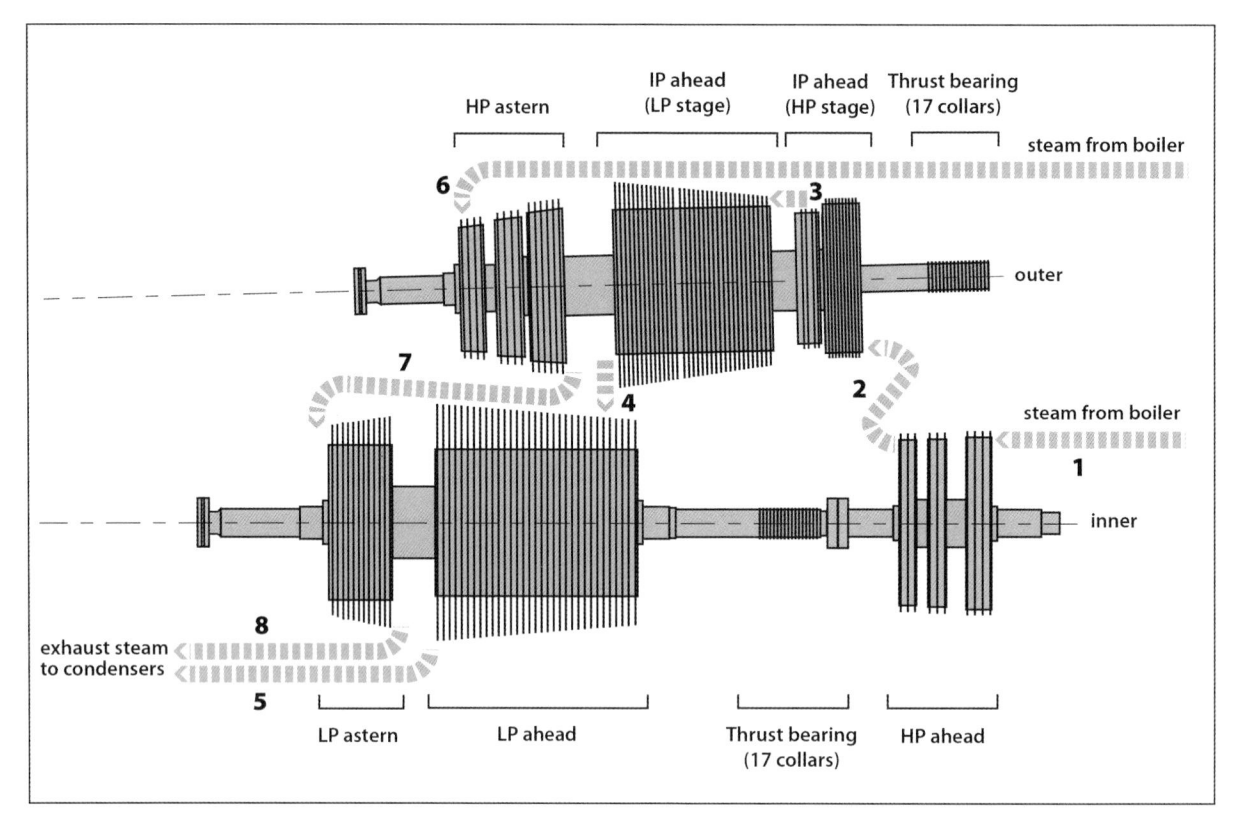

Fig 8: Schematic arrangement of the port engines in *Tiger* based on the National Records of Scotland photographs. Steam from the boilers was admitted to the HP ahead turbines (1) to the HP stage (2) of the IP ahead turbines then to the LP stage (3) of the IP turbines in the same casing before exhausting into the LP ahead turbines (4) and thence to the condensers (5). For astern running steam passed from the boilers to the HP astern turbine mounted on the IP turbine shaft (6) and exhausted to the LP astern (7) and on to the condenser (8). The starboard engines were a mirror image. (Drawing by Ian Johnston)

which was designed to facilitate the slinging of the rotor when being lifted into or from the casing.

Fig 7 shows the other end of the assembly (with enlarged detail), which matches exactly that at the forward end in having no coupling flange by which engine power could be transmitted; it is therefore clear that the assembly as photographed could not function as a propulsion unit. It also indicates that this is actually another HP ahead turbine, an arrangement completely at odds with a functioning main engine (or any other type of engine) in having no means to transmit the mechanical power developed.

The temporary HP turbine is facing the opposite way to that which would be normal in the vessel. This means that in order to revolve in the same direction as the permanent HP it would of necessity have to be from the opposite side of the vessel: with a permanent port HP, the temporary HP would have to be from the starboard engine. Therefore, the permanent as-fitted arrangement from forward to aft on the inner shafts was: HP ahead, LP ahead and LP astern, the latter integral with the LP ahead turbine and in the same casing. An IP turbine with IHP and ILP ahead stages and an HP astern stage drove the outer shafts (see Fig 8), the port and starboard engines looking forward revolving in anti-clockwise and clockwise directions respectively.

Evidence supporting this comes in two forms. Firstly, the words 'SHP ahd'[49] chalked on the casing at right (see

Fig 9: Detail from the initial photograph (Fig 2) showing the permanent bolted connection between starboard HP and LP ahead turbines.

Fig 10: Detail from the initial photograph showing that the connection between the starboard LP ahead and the port HP ahead turbine was a temporary one and was via the shaft coupling only.

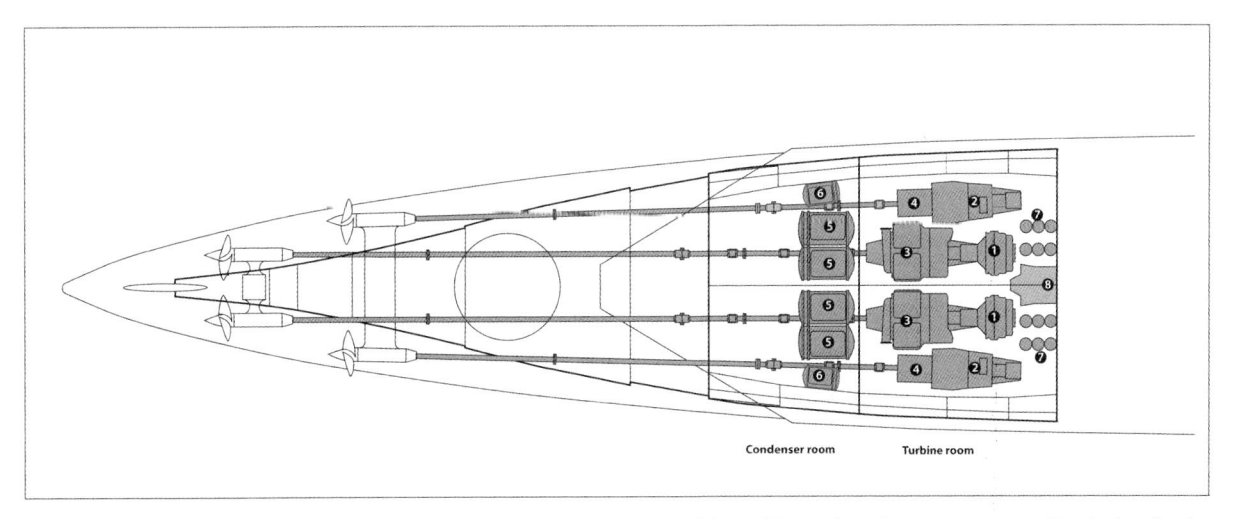

Fig 11: Plan of the machinery arrangement in *Repulse* which, in the context of the turbine and condenser rooms, was identical to that in *Tiger*. (1) HP ahead turbine, (2) IHP and ILP ahead, (3) LP ahead and astern, (4) HP astern, (5) Main condensers, (6) Auxiliary condensers, (7) Main air pumps, (8) Feed tank. (Drawing by Ian Johnston based on Archives Services, Glasgow University, UCS1/110/443/3)

enlarged detail at Fig 7). Secondly, the flange connecting the HP ahead and LP ahead/astern casings at left is illustrated in detail here (Fig 9), clearly showing the number of bolt holes and the graduated pitch thereof, decreasing from bottom to top; it will be noted that all of the bolts have been fitted. The corresponding flange on the right-hand ahead turbine casing (Fig 10) is clearly not playing any part in connecting the turbine casings, being connected to the permanent assembly by the shaft coupling alone.

A further example of casual editorial standards loosely related to *Tiger's* machinery exists in the form of a photograph of the high- and low-pressure turbines of 'HMS *Repulse*' presented in *Engineering*[50] which actually shows the turbine set for *Tiger* as illustrated above at Fig 6. This misleading article, however, contains a clue as to the reason for the temporary assembly shown in the photographs in the form of a caption which states '… High-Pressure and Low-Pressure Turbines being Prepared for Steam Balancing in the Shop …'[51] The pre-eminent expert on the design of *Tiger's* machinery, Stephen J Pigott, touched upon this aspect many years later:

> With direct-drive turbines the design was such that the steam thrust load on the turbine rotor opposed the thrust load resulting from the propeller action in driving the vessel … Even under the condition of opposed thrust forces the multi-collar thrust block had been a source of trouble and anxiety.[52]

The purpose of the temporary arrangement now seems clear: to simulate more closely actual conditions *vis-à-vis* axial loads than possible with an unopposed HP turbine. It is clear that the HP units were being lightly steamed at full speed in order to check bearings, gland integrity and balance. Recent correspondence with Geoff Horseman[53] elicited his suggestion that the LP turbine was blanked because in still air it would have generated heat and, critically, axial thrust. The implication was that a vacuum was created by an air ejector or pump in order to minimise the heat generated and to replicate more accurately the actual low pressure of the LP turbine in service conditions.[54]

Conclusion

The evidence presented above demonstrates that reliance on secondary sources alone cannot underpin serious research, and that even apparently authoritative primary sources – the Plan of Hold, the Curtis letter, and 'official' records, for example – can mislead researchers. As a result, previous commentators, both corporate and individual, each with sound claims to authority, have not been able to present an accurate description of a set of machinery which marked a watershed in the development of marine steam turbines. The actual arrangement in *Tiger* based on the machinery of the *Renown* class is shown in Figure 11.[55]

There are important mitigating circumstances: the designer (no less) of *Tiger* inaccurately described the arrangement of the machinery,[56] and his statement on the matter must have influenced the generations of writers which followed. That the 'official' record of the Director of Naval Construction[57] also did so suggests an assumption that the machinery arrangement in the *Lion* class was repeated in *Tiger* except that the turbines were Brown Curtis, not Parsons. This raises the possibility that some of those commenting on the machinery may have assumed that Brown Curtis turbines were built by a firm of that name, and differed from Parsons only in that respect.

An accurate description of the machinery in *Tiger* is now offered for interpretation. No personal criticism of the writers cited above is implied; individually and collectively they have presented a huge amount of accurate data and have made a significant contribution to the literature of the subject. In similar circumstances, without the photograph which triggered this investigation, any researcher would surely have described the arrangement of the machinery as they did.

Endnotes:

1. A notable exception is *The Battleship Builders*, an authoritative and detailed study of the construction of such vessels by Ian Buxton and Ian Johnston, Seaforth Publishing, 2013.
2. As recently as October 2015, in a detailed large-format study of about 450 pages (Norman Friedman, *The British Battleship 1906–1946*, Seaforth Publishing 2015), the propulsion machinery of these vessels was effectively ignored, thereby epitomising the general neglect of this aspect of the history and development of British battleships.
3. 'The Brown–Curtis Turbine installation in HMS Bristol', *Engineering*, 30 September 1910, 465–467 and plates 39–42.
4. The North East Coast Institution of Engineers and Shipbuilders, and the Institution of Engineers and Shipbuilders in Scotland.
5. *The Naval Annual* 1912, edited by Viscount Hythe, 185.
6. *Jane's Fighting Ships* 1912, edited by Fred T Jane, 60.
7. *The Naval Annual* 1913, 221.
8. *Jane's Fighting Ships* 1913, 45.
9. *Jane's Fighting Ships* 1914, 42.
10. 'British Battle Cruiser Tiger', *Journal of the American Society of Naval Engineers*, Vol 25, No 3, August 1913, 510.
11. *Op cit*, 511.
12. *Ibid.*
13. Hold Drawing, as designed, 1911, National Maritime Museum (Woolwich) NPC 3390.
14. Sir Philip Watts, 'Ships of the British Navy on August 4, 1914, and some matters of interest in connection with their production', *Transactions of the Institution of Naval Architects*, Vol 61, 1919, 14.
15. Director of Naval Construction, 'Records of Warship Construction During the War', Battle Cruiser Tiger, Propelling Machinery, no pagination, May 1918, I am obliged to Professor Ian Buxton for this reference.
16. DK Brown, *The Grand Fleet*, Chatham Publishing 1999, 201.
17. Sir E Tennyson d'Eyncourt, 'Naval Construction During the War', *Transactions of the Institution of Naval Architects*, Vol 61, 1919.
18. Oscar Parkes, *British Battleships*, 1973 edition, 555.
19. John A Roberts, 'The Design and Construction of the Battlecruiser *Tiger*', *Warship* No 6, April 1978, 88.
20. RA Burt, *British Battleships of World War I*, Arms & Armour Press, 1986, 218; this statement was repeated in the 2014 impression by Seaforth Publishing, 237.
21. The LP ahead turbine in the *Renown* class was of the reaction type, HMSO. Notes on Turbines with Illustrations, Fig 40, 1926. I am obliged to Professor Ian Buxton for this reference.
22. The author's contact details were forwarded to Mr Burt in October and a message in December 2017, but thus far he has received no response.
23. Ingvar Jung, 'The Marine Turbine, Part 1, 1897–1927', National Maritime Museum, Monographs and Reports, No 50, 1982, 63.
24. *The Grand Fleet*, *op cit*, 97.
25. D'Eyncourt, *op cit* 70.
26. It is possible that he referred to the Ship's Covers, National Maritime Museum, (Woolwich), ADM 463–466, letter of 19 January 1915), which also stated this.
27. The contract No 418 is legible in four places on the casings in a high-resolution copy of the photograph.
28. *Engineering*, 12 December 1913, 797.
29. *The Shipbuilder*, January 1914, 68.
30. Archives Services, University of Glasgow, UCS1/11/6, Letter Book, letter of 18 July 1912, 2–3.
31. A contemporary work comments on this aspect thus: '… Total Output of the Reversing Turbines as a Percentage of the Main Turbines, and on the Assumption of the same Total Steam Consumption – Battleships 40 to 45%, Small Cruisers 35 to 40%, Torpedo Boats 25 to 30% ….', G Bauer & O Lasche (translated by GS Swallow), *Marine Steam Turbines*, 1911, 15.
32. 'The Brown Curtis Turbine', *Engineering*, 24 May 1912, 694.
33. Bauer and Lasche, *op cit*, 186.
34. See also 'United States Destroyer Sterrett with Curtis Turbines', *Engineering*, 18 August 1911, 215, and 'USS Nevada', *Shipbuilding and Shipping Record*, 27 April 1916, 401 & 403; the turbines in both of these vessels and *Nevada*'s sister *Oklahoma* had four stages per shaft.
35. Frame Drawing as fitted (detail from) for HMS *Tiger*, National Maritime Museum, (Woolwich), NPC 3397, 1914.
36. National Records of Scotland UCS00100118-00418-00074 – 76.
37. National Records of Scotland UCS00100118-00418-00062 – 79.
38. National Records of Scotland UCS00100118-00418-00063.
39. National Records of Scotland UCS00100118-00418-00077 – 79.
40. National Records of Scotland UCS00100118-00418-00062/64/67/68/72/73.
41. Contract Nos 421 and 422 respectively.
42. There are no known photographs of the machinery for her sister ship *Renown* in the Fairfield engine works.
43. Ships Covers, National Maritime Museum (Woolwich) ADM 463-466, letter of 19 January 1915.
44. D'Eyncourt, *op cit*, 70.
45. National Records of Scotland UCS00100118-00443-00333, 336, 337, 339, 345, 347, 357, 358, 359.
46. Archives Services, Glasgow University, UCS1/110/443/3.
47. The IP turbine consisted of IHP and ILP ahead stages and an HP astern stage. See also HMSO Notes on Turbines with Illustrations, Figs 40, 41 and 141, 1926. I am obliged to Professor Ian Buxton for this reference.
48. National Records of Scotland, UCS00100118-00418-00077 – 79.
49. Starboard HP ahead.
50. 'HM Battlecruisers Repulse and Renown', *Engineering*, 11 April 1919, Fig 12 (facing page 463).
51. *Ibid.* It is assumed that this referred to the balancing of the axial thrust of the permanent HP ahead turbine.
52. Stephen J Pigott, 'Three Score Years of Development in Marine Engineering', Presidential Address, *Transactions of the Institute of Marine Engineers*, Vol XLIX, Part 9, Session 1937, 183.
53. Chief Turbine Design Engineer, Siemens Power Generation Services, C A Parsons Works, Newcastle-upon-Tyne.
54. It is also probable that the LP turbine was blanked for a static steam pressure test either before or after the balancing test.
55. Based on plan of engine room in *Repulse*, Archives Services, University of Glasgow, UCS1/110/443/3.
56. Watts, *Ships of the British Navy*, *op cit*, 14.
57. 'Records of Warship Construction During the War', *op cit*.

IN *AVRORA*'S SHADOW:
The Russian Cruisers of the *Diana* Class

Diana, *Pallada* and *Avrora* were the Russian Imperial Navy's first attempt at a true light cruiser, and the design was not a successful one. Nevertheless, the ships had active careers, and *Avrora* earned fame – and preservation as a monument – due to her association with the October Revolution of 1917. **Stephen McLaughlin** describes the design and outlines the careers of these ships.

Starting in the 1870s the Russian Navy built a series of large, long-ranged armoured cruisers intended to prey on British trade in the Pacific in the event of war. Then in 1882 Emperor Aleksandr III approved a 20-year construction programme that included not only commerce-raiding cruisers but powerful battle fleets for the Baltic and Black Seas. Although this programme was adjusted in 1885 and 1890, the underlying concept remained unaltered: battle fleets for the enclosed seas, big cruisers for the high seas.

In 1891 Admiral NM Chikhachev, director of the Naval Ministry and the professional head of the Russian Navy, sent a questionnaire to senior admirals and to the Naval Technical Committee (*Morskoi tekhnicheskii komitet* or MTK) soliciting views on a variety of issues. By this time there were no fewer than five Baltic battleships built or building, with several more at the design stage, so one of his questions concerned 'the size and characteristics of scouts needed for our battle fleet'.[1] Discussion of this question proceeded at a leisurely pace, since it soon became clear that the hoped-for battle fleet would be greatly delayed by slow construction and limited finances, postponing the need for fleet scouts; moreover, Russia's shipbuilding capacity was already stretched to the limit. By 1894 Chikhachev therefore had abandoned the idea of building fleet scouts and instead opted for more commerce raiders, but ones smaller and cheaper than the battleship-sized armoured cruisers *Riurik* (10,933 tons) and *Rossiia* (12,130 tons).

As a result, in March 1894 the MTK announced a design competition for an ocean-going, steel-hulled cruiser with the following parameters:

– displacement not more than 8,000 tons
– draft at full load not to exceed 24ft (7.3m – this would ensure that the ship could pass through the Suez Canal)

Pallada running trials in the autumn of 1901. The unshielded forecastle 6in gun is barely visible. She is still in the traditional 'Victorian' livery of black hull, white superstructures and buff funnels. (Courtesy of John Roberts)

81

- hull to be sheathed with wood and copper
- speed not less than 19 knots 'at natural draught, [with] ordinary coal and stokers'[2]
- armament to be two 8in guns (one each fore and aft) and eight 120mm quick-firing guns (QFGs) on the broadside, with maximum possible fire ahead and astern
- three torpedo tubes
- armour protection to be sufficient to resist the guns of cruisers belonging to the leading naval powers; coal bunkers to be used as part of the protection scheme
- normal coal supply to provide for a range of 9,000nm at 10 knots, with bunkerage sufficient for a maximum range of 12,000nm
- propulsion plant on two or three shafts, with Belleville boilers specified
- crew of about 500, with provisions sufficient for four months.

The results of the competition's first phase were announced in October 1894: of the nine designs submitted, three were selected for further development, with the favoured one submitted by constructors IG Bubnov and LL Koromaldi (see Table 1). However, none of the designs was considered suitable for construction, and in any case changing circumstances soon rendered the competition's criteria obsolete. Confronted by the rapidly growing German fleet, in March 1895 the Main Naval Staff (*Glavnyi morskoi shtab*, or GMSh, the Navy's chief policy and administrative organ) formed a commission to develop a new shipbuilding programme aimed at bolstering the Baltic Fleet.[3] This commission's report was then passed on to yet another commission, which was charged with working out the characteristics of the ships to be built under the new programme. On 11/23 April this second commission recommended the construction of seven cruisers for the Baltic: six of 6,000 tons and one of 4,000 tons.

However, Admiral Chikhachev was evidently growing impatient, and in March 1895, without waiting for the commission's report, he ordered the Baltic Works – a state-run shipyard that had long specialised in cruisers – to develop a cruiser design based on the British second-class cruisers of the *Astrea* class (4,360 tons, 2 x 6in, 8 x 4.7in, 10 x 6pdr, 19.5 knots). The Russian ship was to have a speed of 20 knots and the greatest possible endurance. Chikhachev saw the matter as so urgent that he ordered the work to be started immediately, 'without any sort of preliminary sketch design'.[4]

On 7/19 May 1895 Senior Constructor KK Ratnik, manager of the Baltic Works, submitted three designs to the MTK, ranging in size from 4,400 to 5,600 tons (see Table 2). However, Ratnik was not happy with the *Astrea* as a prototype; as he wrote in his cover memo:

> The Baltic Works deviates from the English cruiser *Astrea*, which was ordered [to be used] as an analogue, insofar as among the other latest cruisers of various nations it does not represent the best type.[5]

The main point at issue seems to have been sea-keeping; the *Astreas* were flush-decked ships with a relatively low freeboard. Ratnik therefore submitted a fourth design in which he took as a prototype the new British *Eclipse* (or *Talbot*) class (5,600 tons, 5 x 6in, 6 x 4.7in, 8 x 12pdr, 19.5 knots), which featured a forecastle deck. The Baltic

Table 1: Cruiser Design Competition Leaders, 1894

	Port Due	*Neuiazvimyi* ('Invulnerable')	*Trud* ('Work' or 'Labour')
Designers	IG Bubnov, LL Koromaldi	GF Shlezinger	LF Veshkurtsev
Displacement	7,960 tons	8,000 tons	7,200 tons
Dimensions	131.0m x 17.4m x 6.5m	126.2m x 17.0m x 7.0m	120.4m x 18.1m x 7.1m
Armament	3 x 8in	2 x 8in	2 x 8in
	9 x 120mm	8 x 120mm	8 x 120mm
	9 x 47mm	10 x 47mm	10 x 47mm
	11 x 37mm	12 x 37mm	10 x 37mm
	1 above-water TT	1 above-water TT	1 above-water TT
		2 submerged TT	
Protection:			
belt	3in–8in	–	–
deck (horizontal)	1in–3in	1.5in–3in	1.5in–2in
deck (slopes)	0.4in–0.66in	4in	3in
conning tower	–	6in	6in
Machinery	3 x VTE	3 x VTE	3 x VTE
	11,230ihp	11,050ihp	10,767ihp
Speed	19 knots	19 knots	19 knots
Endurance	965 tons coal	1,385 tons coal	1,244 tons coal
	9,000nm at 10kts	9,000nm at 10kts	9,000nm at 10kts
Complement	500	500	500

Table 2: Baltic Works Designs, May 1895

	4,400 tons	4,700 tons	5,600 tons	6,000 tons
Length	97.5m	106.0m	112.8m	118.26m
Beam	14.3m	14.3m	16.2m	16.92m
Draft (mean)	5.8m	5.8m	6.4m	5.9m
Armament	2 x 6in	2 x 8in	2 x 8in	2 x 8in
	8 x 120mm	8 x 120mm	4 x 6in	8 x 6in (in turrets)
	8 x 47mm	8 x 47mm	6 x 120mm	27 x 57mm*
	10 x 37mm (10 x I)	10 x 37mm (10 x I)	10 x 47mm	
	8 x 37mm (8 x V)	8 x 37mm (8 x V)	10 x 37mm (10 x I)	
		8 x 37mm (8 x V)		
	2 x 2.5in Baranovskii	2 x 2.5in Baranovskii	2 x 2.5in Baranovskii	
	4 x TT (above water)	4 x TT (above water)	4 x TT (above water)	3 x TT (above water)
	30 mines	30 mines	30 mines	
Protection:				
deck:	1.5in–2.5in	1.5in–2.5in	1.5in–2.5in	1.5in–2.5in
fwd CT	6in	6in	6in	6in
after CT	N/A	N/A	3in	?
machy hatches	5in	5in	5in	5in
hoists	1.5in	1.5in	1.5in	1.5in
funnel bases	1.5in	1.5in	1.5in	?
Machinery	7,900ihp	7,900ihp	8,740ihp	?
Speed	20 knots	20 knots	20 knots	20 knots
Endurance	450 tons coal normal	506 tons coal normal	620 tons coal normal	800 tons normal
	4,416nm at 10kts	4,652nm at 10kts	6,108nm at 10kts	?
Complement	344	374	428	?

* Ten on each of the upper and battery decks, seven in the fighting tops.

Works' design was 6,000 tons and carried an armament of 2 x 8in, 8 x 6in and 27 x 57mm. (Sir William White, the Royal Navy's Director of Naval Construction, believed that the Russian ships were 'really based upon our "Talbot" class, all the particulars of which were obtained by the Russian designers', although there is no evidence that the Russians had anything more than the published literature to guide them in their design.)[6]

On 11/23 May 1895 the four designs, having been examined by the MTK, were presented to the Grand Duke Aleksei Aleksandrovich, the emperor's uncle and, as general-admiral, the official head of the Navy. He selected the 6,000-ton design, but ordered that the 8in guns, which were mounted fore and aft, be replaced by 6in guns – probably because as quick-firers they were considered more effective for cruiser work than the slower-firing 8in breech-loaders.

By the beginning of July 1895 the Baltic Works had submitted its drawings and calculations for what was by now a 6,080-ton ship. But the MTK considered the design's stability excessive, which would have made the ship very lively and consequently a poor gun platform. The Baltic Works therefore worked out a new hull form, with the beam reduced to 53.5ft (16.3m) and the displacement to 6,006 tons. The MTK now regarded the internal arrangements as too cramped, and on 12/24 August 1895 it demanded yet another change in the hull form and authorised an increase in displacement to 6,500 tons, at the same time recommending that the hull be fuller amidships and more tapered at the ends. In reply the Baltic Works submitted two new sets of drawings, for ships of 6,500 tons and 6,540 tons, the latter being its preferred option. These designs were examined by the MTK in September 1895, when the slightly larger design was favoured. The drawings were then given to the Experimental Basin for model testing, which reported that the ship would require 12,693ihp to achieve 20 knots. The Basin also proposed its own hull form, which would require less power to achieve the same speed, but the MTK rejected it because of its reduced stability. The Baltic Works' hull form was to be used, but the MTK's review had revealed an error in hull strength calculations; correcting this led to an increase in displacement, to 6,630 tons. The new drawings and calculations were submitted to the MTK on 4/16 November 1895, and on 30 November/12 December 1895 the acting inspector of shipbuilding, EE Guliaev, recommended their approval, noting that the maximum longitudinal hull stress had been reduced from an excessive 5.89 tons/sq in to an acceptable 5.24 tons/sq in.

Despite the fact that the Baltic Works had been developing the design, on 18/30 November 1895 Chikhachev ordered that two cruisers be built at the Galernyi Island state shipyard, probably because the battleships *Petropavlovsk* and *Sevastopol* had been launched there recently, thereby freeing up a pair of slipways. The decision to build a third cruiser at the New Admiralty shipyard, where a building slip was available after the launch of the battleship *Sisoi Velikii*, was made in June 1896 (for construction dates, see Table 3).

Meanwhile, the Franco-Russian Works had agreed to design the machinery. A three-shaft plant had been adopted, apparently to extend the range – when cruising the wing engines could be shut down to save fuel, an arrangement used in the contemporary armoured cruiser *Rossiia*. However, with the ship now oriented toward Baltic operations long cruising range was less important, and on 8/20 November – only four days after the design had been submitted – Chikhachev ordered the design changed to two shafts. The chief concern was the belief that at certain rudder positions the flow of water from the centre propeller would make for a sluggish response to the helm. The Franco-Russian Works presented its new design for the two-shaft machinery plant on 3/15 January 1896. However, the naval attaché in Berlin now reported that all German large cruisers were being built with three shafts and there were no concerns about their manoeuvring qualities, so Chikhachev decided to revert to the original three-shaft plant. The Franco-Russian Works dutifully responded with yet another machinery design featuring three engines, with a total output of 12,500ihp. The MTK's engineering branch believed that it would probably be good for only 11,500ihp at natural draught. Nevertheless, the order for the machinery for the first two ships was issued on 21 September/3 October 1896.

The armament was subjected to even more changes than the propulsion plant. As approved in May 1895 it was 10 x 6in and 27 x 57mm, but the latter calibre was not in service with the Russian Navy, and in February 1896 the MTK changed the armament to 6 x 6in plus 6 x 120mm and 27 x 47mm. The 120mm gun was just entering production and was valued for its high rate of fire. Then in April 1896 news was received from the attaché in Germany that the new light cruisers M and N (*Victoria Louise* class, 6,389 tons, 19.2 knots) would be armed with 2 x 210mm, 8 x 150mm, and 10 x 88mm. This battery would give the German ships an advantage at longer ranges, then considered to be 15–20 cables (3,000–4,000 yards), where the 47mm guns of the Russian design would be ineffective. As a result, the 47mm were replaced by twenty 75mm QFGs. At the same time, due to difficulties encountered with the production of the 120mm gun, it was decided to replace these by four more 6in guns, making the armament once again ten 6in. On 20 April/2 May 1896 the MTK approved the new specifications of the design.

The design changes had been coming thick and fast, and it was hardly surprising that when *Diana*'s constructor, AI Mustafin, submitted revised weight calculations on 11/23 November 1896, they showed that the ship would be 182.45 tons overweight. To offset this, he recommended deleting two 6in guns – even with this reduction the ships would still be superior to the British 2nd and 3rd class cruisers likely to be encountered when raiding commerce. In addition, he proposed reducing coal at normal displacement to 800 tons. The MTK concurred, and also recom-

Diana, the name-ship of the class, soon after entering service in 1902. The anchor arrangements and the sponsons for the forward broadside 6in guns can be clearly seen against her white hull. (Courtesy of the late Boris Lemachko)

Table 3: **Construction Dates**

	Diana	*Pallada*	*Avrora*
Ordered	18/30 Nov 1895	18/30 Nov 1895	11/23 Jun 1896
Added to list[1]	27 Apr/9 May 1896	27 Apr/9 May 1896	6/18 Apr 1897
Construction begun[2]	Jul 1896	Jul 1896	7/19 Sep 1896
Laid down[3]	23 May/4 Jun 1897	23 May/4 Jun 1897	23 May/4 Jun 1897
Launched:	30 Sep/12 Oct 1899	14/26 Aug 1899	11/23 May 1900
Entered service	10/23 Dec 1901	May 1901	16/29 Jul 1903
Builder	Galernyi Island, St Petersburg	Galernyi Island, St Petersburg	New Admiralty, St Petersburg
Constructors	AI Mustafin	PE Andriushchenko, AI Mustafin (from October 1896)	ER De-Grofe, KM Tokarevskii, NN Pushchin (from Dec 1901)
Cost (hull+machinery)	?	?	6,400,000 rubles

Notes:
1 Added to list: the date a ship was officially added to the list of the fleet and given a name.
2 Construction begun: the date when the first iron or steel was laid on the slipway.
3 Laid down: The date of the ceremonial keel-laying, not necessarily corresponding to an important stage in the ship's construction.

mended eliminating the 1in (25mm) shields for the 6in guns and the ¾in (19mm) shields of the 75mm guns, for a further saving of 9.6 tons. This was not purely a weight-saving measure; gun shields were considered dangerous by some, since they might detonate shells that would otherwise pass by the guns harmlessly. These recommendations were approved on 12/24 December 1896.

In early 1897 concerns arose over the potential hazards of unprotected above-water torpedo tubes in a gunnery action. It was relatively simple to move the broadside tubes below the waterline, but a proposal to move the bow tube was rejected because the stemposts, with their fitting for the tubes, were already being made. A year later the gunnery armament was changed yet again as a result of a proposal by the commander assigned to *Pallada* during her construction, Captain 1st Rank AR Rodionov. On 23 May/4 June 1898 he recommended that the after broadside 6in guns be shifted aft a few frame spaces, so that the muzzles would be right at the side of the ship, allowing them to be depressed sufficiently to fire at nearby torpedo boats. Moreover, this would make room for more 75mm guns. This proposal was supported by the captains of *Diana* (MG Nevinskii) and of *Avrora* (AA Melnitskii). After review by the MTK, the final decision was to move the 6in from frame 98 to 109, and to add four 75mm guns: two on the upper deck in sponsons, and two in the admiral's dining room, between frames 125 and 126. The ammunition outfit of the 75mm guns was not increased. This change was finally approved on 12/24 January 1899.

All these changes inevitably increased the displacement: by the summer of 1898 it had risen to 6,731.28 tons. They also slowed the pace of construction, since each alteration required new drawings and calculations, which in turn had to be examined and approved by the MTK. *Diana* and *Pallada* were more than three years on the stocks, and fitting out proceeded slowly because of changes in equipment and the slow delivery of materials;

total building time amounted to about five years. *Avrora* took almost seven years to complete, despite the fact that Chikhachev's successor as director of the Naval Ministry, Vice Admiral PP Tyrtov, ordered in October 1897 that she be built without any changes from the first two. In the event, work on her was slowed by the need to correct defects identified during the construction of her sisters.

The names given to the ships reflected their commerce-raiding origins; all three were associated with sailing frigates that had distinguished themselves in Pacific waters. *Pallada*, a 52-gun frigate launched in 1852, had taken Admiral Prince Putiatin to Japan in 1853, where he successfully negotiated a treaty opening that country to Russian trade. *Diana*, a 54-gun frigate launched in 1852, had taken Putiatin back to Japan for further negotiations; while at Shimoda the ship was destroyed in an earthquake, but her crew salvaged as much material as possible and used it to build a schooner that took them to Russian territory on the Amur River. *Avrora* was a 56-gun frigate that served in the Far East during the Crimean War and assisted in the successful defence of Petropavlovsk against an Anglo-French attack in August 1854.

General Features

The characteristics of the ships are given in Table 4. The hull, of mild steel, had 135 frames spaced 3ft (914mm) apart. The framing was on the 'mixed' system, arranged longitudinally over the extent of the double bottom (frames 22 to 98) and transversely at the ends. Hull depth amidships was 38ft 8in (11.80m), freeboard was 26ft 9in (8.15m) at the bow and 17ft 9in (5.4m) aft. There were three stringers on either side of the keel, the middle stringer being watertight. The three enclosed decks were designated the upper deck, battery (or berth) deck, and the armoured deck (as the protective deck was termed in the Russian Navy). There was a forecastle deck forward and two platform decks fore and aft. Below the armoured

Avrora in Kronshtadt roadstead, 1902. In this photo the unshielded 6in gun on the forecastle is plainly visible. Note also that the fore topmast is fitted to the after side of the foremast. (Naval History and Heritage Command, NH 92877)

deck the hull was divided into fourteen main watertight compartments by thirteen transverse bulkheads, with longitudinal bulkheads outboard of the boiler rooms and magazines. Above the armoured deck there were only four watertight compartments. All three ships had wood and copper sheathing, extending 33in (840mm) above the waterline, and there were bilge keels 128ft 7in (39.2m) long. The ships had a metacentric height of 3.777ft (1.15m), which was somewhat high and may have made them lively in a seaway. (For the weight breakdown, see Table 5.)

The forecastle and upper decks had teak planking; the other decks were covered with linoleum, as were the superstructure decks. As completed the ships had two masts with an internal diameter of 30in (762mm); there was a fighting top on the foremast, and both topmasts were wooden.

The *Diana* class featured a new dewatering system, developed as a result of the loss of the battleship *Gangut* in June 1897 after striking an uncharted rock in Finnish waters. The traditional 'main drain' – a large-diameter pipe running through all the compartments and connected to all pumps – was eliminated. Although in theory this system allowed any or all pumps to be used in draining any compartment, in practice the diameter of the main drain limited its capacity, while the numerous bulkhead penetrations and valves made it overly complex. The new system featured 'autonomous' compartments, each equipped with powerful electrically-driven pumps and its own discharge pipes. In the forward and after compartments there were centrifugal pumps with a capacity of 250 tons/hour, while in each boiler room there were two pumps each with a capacity of 400 tons/hour. In addition, the three main circulating pumps in the engine rooms, used for pumping seawater through the condensers, could also be used for pumping water out of the ship; each had a capacity of 800 tons/hour. There were also three Worthington pumps for clearing the bilges, each with a capacity of 30 tons/hour.

Upon their completion *Diana* and *Pallada* were sent to join the Pacific Squadron at Port Arthur. After their arrival in the spring of 1903, the viceroy and commander-in-chief of the armed forces in the Far East, Admiral EI Alekseev, wrote a scathing report on the ships:

> The first-class cruisers *Diana* and *Pallada*, built at state shipyards in St Petersburg, lag behind their foreign-built partners [*sotovarishchei*, probably referring to the German-built *Askold* and the American-built *Variag*, which had recently joined his squadron] in all areas, with regard to speed, artillery and the finish and care taken in their design, as well as in the execution of the work. Thus, for example … there is insufficient space for carrying a full outfit of artillery shells, and some of the cartridge magazines are located next to the boilers. Their sea-going qualities are also quite poor, since the cruisers bury their bows, [and] the maximum speed does not reach 20 knots.[7]

In service the ships also proved to have poor handling characteristics, especially at low speeds and in windy

Table 4: Characteristics as Designed and Completed

Displacement:	6,731 tons normal (design)
	Diana: 6,657 tons trials
	Pallada: 6,722 tons trials, 7,081 tons normal
	Avrora: 6,897 tons trials, 7,000 tons normal
Dimensions:	406ft wl, 416ft oa x 55ft max x 21ft max (even keel)
	123.5m wl, 126.8m oa x 16.76m max x 6.4m max
Armament:	8 x 6in/45 (8 x I)
	24 x 75mm/50 (24 x I)
	8 x 37mm (8 x I)
	2 x 2.5in landing guns
	3 x 15in torpedo tubes (one above water, two submerged)
Protection:	
deck	1.5in/38mm flat, 2.0in/51mm slopes increasing to 2.5in/63.5mm at lower edge, all on 0.5in/12.7mm plating
conning tower	6in/152mm side, 3.5in/89mm tube to central post
machinery hatches	1in/25.4mm
funnel bases	1.5in/38.1mm
hoists	1.5in/38.1mm
Machinery:	three vertical triple expansion engines, 24 Belleville boilers, 11,610ihp (designed);
	Diana 12,200ihp, trials; *Pallada* 13,100ihp, trials; *Avrora* 11,971ihp, trials
Speed:	20 knots (designed); *Diana* 19.0kts, trials (measured by log); *Pallada* 19.17kts, trials; *Avrora* 19.2kts, trials
Endurance:	800 tons coal, normal,
	972 tons coal, full load
	4,000nm at 10kts (designed); 3,300nm (actual)
Complement:	20 officers, 550 men
	Avrora, 1916: 23 officers, 728 men

weather; controlling the ships in narrow waters and when mooring was difficult. This was attributed in part to the three-shaft machinery plant and to the fact that the two wing shafts diverged by 3 degrees.

These were the first Russian ships with electrical steering gear, and each ship had gear made by a different company: *Diana*'s by the Union company, *Pallada*'s by the Baltic Works, and *Avrora*'s by Siemens & Halske. There were steam and manual back-ups. The ships could be steered from the wheelhouse, conning tower, central post (hold), the after bridge and the steering-gear compartment. At 12 knots with 30 degrees of rudder the tactical diameter was 3.3 cables (660 yards), and turning a complete circle took 4 minutes 23 seconds.

Armament

As completed the ships' main armament consisted of eight 6in/45 QFGs of the Canet pattern (see Table 6 for characteristics of guns). One gun was on the centreline forward, another aft, and there were three on either broadside. The two forward broadside guns were mounted on sponsons close together under the bridge, and the foremost pair could fire directly ahead. The aft broadside guns were mounted at the after end of the superstructure, protected against following seas by a low breastwork of ⅝in (16mm) steel. They could fire directly astern, although a British report noted that they could not do so 'without danger to the after 6in gun's crew'.[8] Rate of fire was five rounds per minute, and elevation limits were +15° to –6°. There were four magazines: one

Table 5: Weights

Hull	2,671.36 tons	(39.58%)
Armour	707.45 tons	(10.48%)
Guns with hoists	386.33 tons	(5.72%)
Torpedoes/mines	114.44 tons	(1.70%)
Machinery	1,619.60 tons	(24.00%)
Fuel (normal load)	800.00 tons	(11.85%)
Equipment and rig	387.66 tons	(5.74%)
Margin	62.16 tons	(0.92%)
Total	6,749.00 tons	(100.00%)

forward, one amidships on the starboard side, and two aft. The ammunition outfit was 1,414 rounds total. There were three ammunition hoists, one forward for the forecastle gun, and two aft, one on either side.[9] Shells and cartridges were brought up on a special carrier, four at a time, and transferred to the individual guns on the upper deck by means of a monorail system.

There were twenty-four 75mm/50 QFGs, twelve each on the upper and battery decks. Rate of fire was ten rounds per minute, and elevation limits were +20° to –10°. The ammunition allowance was 6,240 rounds total (260 rounds per gun). There were four hoists on each side to supply them with ammunition. As completed, neither the 6in nor the 75mm guns were protected by shields.

The ship was equipped with eight 37mm single-barrelled Hotchkiss guns, four in the fighting top and two in each of the bridge wings fore and aft. Theoretical rate of fire was 30 rounds per minute, although this required a well-trained crew. Ammunition outfit was

4,800 rounds (600 rounds per gun). The guns in the foretop had training and elevation stops so that they could not fire into their own ship, but on the design drawings it was noted – somewhat archaically – that these stops could be removed 'in the event of boarding, [in order to] fire on her own deck'.[10]

Like all large Russian warships of the era the *Diana*s also carried two Baranov 2.5in (63mm) landing guns, provided with 1,440 rounds total.

The fire control system was provided by the St Petersburg firm of NK Geisler and Co. It consisted of electrically-driven indicators that could be controlled from the conning tower or the central post in the hold, transmitting to the guns the target bearing, range, and sight settings, as well as simple orders such as 'open fire', 'rapid fire', etc. There were also indicators in the magazines to order the type of shell to be sent up to the guns. For range-finding the ships were equipped with six Liuzhol-Miakishev micrometers, which worked by measuring the angle subtended by some vertical feature on an enemy ship, for example, the distance from the waterline to the maintop. If this height were known (reference books provided this data), the angular measure could be converted into the range to the target.

There were three 15in (381mm) torpedo tubes, one above water in the stempost and one underwater on either side mounted on the platform deck between frames 29 and 35, the tubes angled 15 degrees before the beam. Eight torpedoes were provided: two for the bow tube and three each for the broadside tubes. The torpedoes were 17ft (5.18m) long and weighed 950lbs (430kg), with an explosive charge of 140lbs (64kg); they had a range of about 660 yards (600m) at 16 knots, and a maximum range of 1,650 yards (1,500m) at a lower speed.[11] The broadside torpedoes could be launched at speeds up to 17 knots. The ships were also designed to carry 35 moored mines, to be used for protecting temporary anchorages.

There were six 75cm searchlights: one on each mast, two in the forward bridge wings and two in the after bridge wings.

Table 6: Guns of the *Diana* Class

Guns	6in/45 Canet	130mm/55	75mm/50 Canet	3in AA Lender
Calibre	6in/152.4mm	5.12in/130mm	2.95in/75mm	3in/76.2mm
Date	1892	1913	1892	1915
Weight, gun+mount	13.7 tons	1,7160kg	1.7 tons	?
Weight, barrel	5,815kg	5,290kg	879–901kg	437–440kg
Barrel length	6858mm/45 cal	7150mm/55 cal	3750mm/50 cal	2307mm/30.5 cal
Bore length	?	?	?	?
Rifled length	5349mm/35.1 cal	5862mm/45.1 cal	2943.5mm	1790mm
Rate of fire	5rpm	5–8rpm	10rpm	10–12rpm
Crew	6	10	4	?
Projectiles & Performance				
Weight	41.5kg AP	33.5kg	4.9kg AP	6.5kg Shrapnel
Charge	12.9kg	11kg	1.5–1.6kg	?
MV	790m/sec	823m/sec	820m/sec	588m/sec
Range	61 cables at 20°	18,290m at 30°	42 cables at 20°	9,500m (horizontal)

Guns	2.5in Baranov Landing Gun	2.5in AA	37mm Hotchkiss
Calibre	2.5in/63.5mm	2.5in/63.5mm	1.46in/37mm
Date	1882	1916?	1884
Weight, gun+mount	0.356 tons	1300kg	0.17 tons
Weight, barrel	106kg	?	0.13 tons
Barrel length	1210mm/19 cal	38 cal	23 cal
Bore length	1070mm/16.8 cal	2266mm/35.7 cal	?
Rifled length	778mm	?	?
Rate of fire	4rpm	?	20
Crew	4/19	?	3
Projectiles & Performance			
Weight	2.6kg	4.04kg Shrapnel	0.5kg
Charge	0.4kg	0.8kg	0.035kg
MV	370m/sec	686m/sec	440m/sec
Range	10 cables at 19°	?	15 cables at 19°

Sources: Shirokorad, *Entsiklopediia otechestvennoi artillerii*; SI Titushkin, 'Artilleriia Russkogo flota v 1877–1904 gg' (*Sudostroenie*, no 8, 1990, pp 58–64); Titushkin, 'Russkaia korabel'naia artilleriia v 1904–1917 gg' (*Sudostroeniia*, no 5, 1992, pp 50–55).

Protection

The ships were designed without belt armour on the 'protective deck' principle. The 'extra-soft' nickel steel deck plating, made by the French firm of Châtillon & Commentry, protected the ship's vitals, while the coal bunkers above it absorbed splinters and helped to preserve the ship's waterplane – hence its stability – by excluding water if the side were damaged. In the *Diana*s the protective deck extended between frames 16 and 126; over most of its length the horizontal portion was 2ft (0.6m) above the waterline, but over the cylinder heads of the main engines it rose to 3ft 6in (1.1m). The horizontal portion of the deck was 1.5in (38mm) thick, while the deck slopes gradually increased from 2in (51mm) to 2.5in (63.5mm) at the lower edge, which was 4ft 5in (1.35m) below the waterline. The protective plating was laid on 0.5in (12.7mm) deck plating. Immediately behind the outer hull plating were cofferdams 3ft 6in (1.1m) high filled with water-excluding material. Above the deck slopes were coal bunkers 10ft wide; since 2ft of coal was believed to be equivalent in protective value to 1in of iron, they were theoretically equal to about 5in (127mm) of iron.[12]

The vertical armour was nickel-chrome steel. The conning tower was protected by 6in (152mm) plates, while its roof was made of 2in (51mm) 'low-magnetic' steel in order to minimise interference with the compass located above it. The communications tube from the conning tower to the armoured deck was 3.5in (89mm) thick. The ammunition hoists were protected by 1.5in (38mm) plates, as were the funnel uptakes between the armoured and battery decks.

The ships were fitted with net defence against torpedo attack.

Machinery

The Franco-Russian Works manufactured the main machinery for all three ships. The boilers were arranged in three boiler rooms: BR1 between frames 35 and 48, BR2 between frames 48 and 62, and BR3 between frames 62 and 75. Each contained eight Belleville watertube boilers, with a working pressure of 17.2atm (253psi/17.8kg/cm^2), although on trials *Avrora* managed only 15.71atm (231psi/16.2kg/cm^2). Heating surface was 36,114sq ft (3,354.8m^2), grate surface 1,162 sq ft (107.9m^2). The steam productivity of the boilers proved greater than expected, and on her trials *Pallada* was able to maintain her speed during the last 25 minutes of the run despite four boilers going out of service.

Each ship had three vertical triple-expansion engines, in three compartments: the outboard engines were in the forward engine rooms (frames 75 to 87) and were separated by a centreline bulkhead; the engine for the centre shaft was in a compartment immediately abaft them (frames 87 to 98). The engines had the following characteristics:[13]

Cylinder diameters:
   High pressure   31⅝in (80.3cm)
   Medium pressure   47¾in (121.3cm)
   Low pressure: 74in (188.0cm)
   Piston stroke: 34in (86.4cm)

Each engine had a designed output of 3,870ihp, for a total of 11,610ihp at 135rpm.

The three phosphor-bronze propellers were three-bladed; the centre screw had a diameter of 13.2ft (4.02m) and a pitch of 17ft (5.18m); the two outer screws each had a diameter of 14ft (4.27m) and a pitch of 17.5ft (5.33m).

During her trials on 20 October/2 November 1901, *Pallada*'s engines developed 13,100ihp for an average speed of 19.17 knots at a displacement of 6,722 tons. *Diana* ran her trials a little later than her sister, and it seems there was some haste to get them completed before the winter ice set in, since they took place in weather so foul the mile-markers could not be seen and speed was measured by log; her speed of 19 knots at 12,200ihp on a displacement of 6,657 tons must therefore be considered approximate. *Avrora*'s first trials in October 1902

The overhead monorail system used for transporting ammunition along the upper deck. Seen here is a carrier (*besedka* in Russian – literally, a bosun's chair) for conveying four 6in shells and cartridge cases; it was pushed along the rail by a seaman. The 75mm guns used fixed ammunition (combined shell plus cartridge), and were supplied to the guns along the same monorail system in batches of sixteen. (Photograph by the author)

were unsuccessful, and were completed only in June 1903; at a displacement of 6,897 tons and a machinery output of 11,971.5ihp, her average speed was just 18.97 knots (19.2 knots maximum).

In all three ships the machinery exceeded the designed output, yet the speed remained about a knot short of expectations. The report on the trials of the first two ships, *Diana* and *Pallada*, noted that:

> It is difficult to explain the cause of these results, but the identical speed achieved by the two cruisers provides a basis for the belief that the same mistake was made in the hull form of both ships, identically affecting their speed …[14]

It seems plausible that the root of the problem lay in shuffling the drawings and calculations between the Baltic Works, the Experimental Basin, and the MTK's engineering section; it is noteworthy that the Experimental Basin's report in September 1895 had indicated that the hull form chosen for the 6,540-ton design would require 12,693ihp to achieve 20 knots, but the contract signed with the Franco-Russian Works that same month called for machinery with an output of only 11,610ihp. It seems remarkable that no one noticed the discrepancy.

Coal consumption proved excessive: on trials *Avrora* burned 1.44 times as much coal as had been anticipated. In service the ships consumed about 70–80 tons of coal per day at 10–12 knots, giving them a range of about 3,300nm with 972 tons of coal instead of the 4,000nm intended. However, 972 tons was a nominal figure, and coal bunkerage varied from ship to ship; instead of 800 tons in the normal condition and 972 tons at full load, *Diana* stowed 810 and 1,070 tons respectively, and is

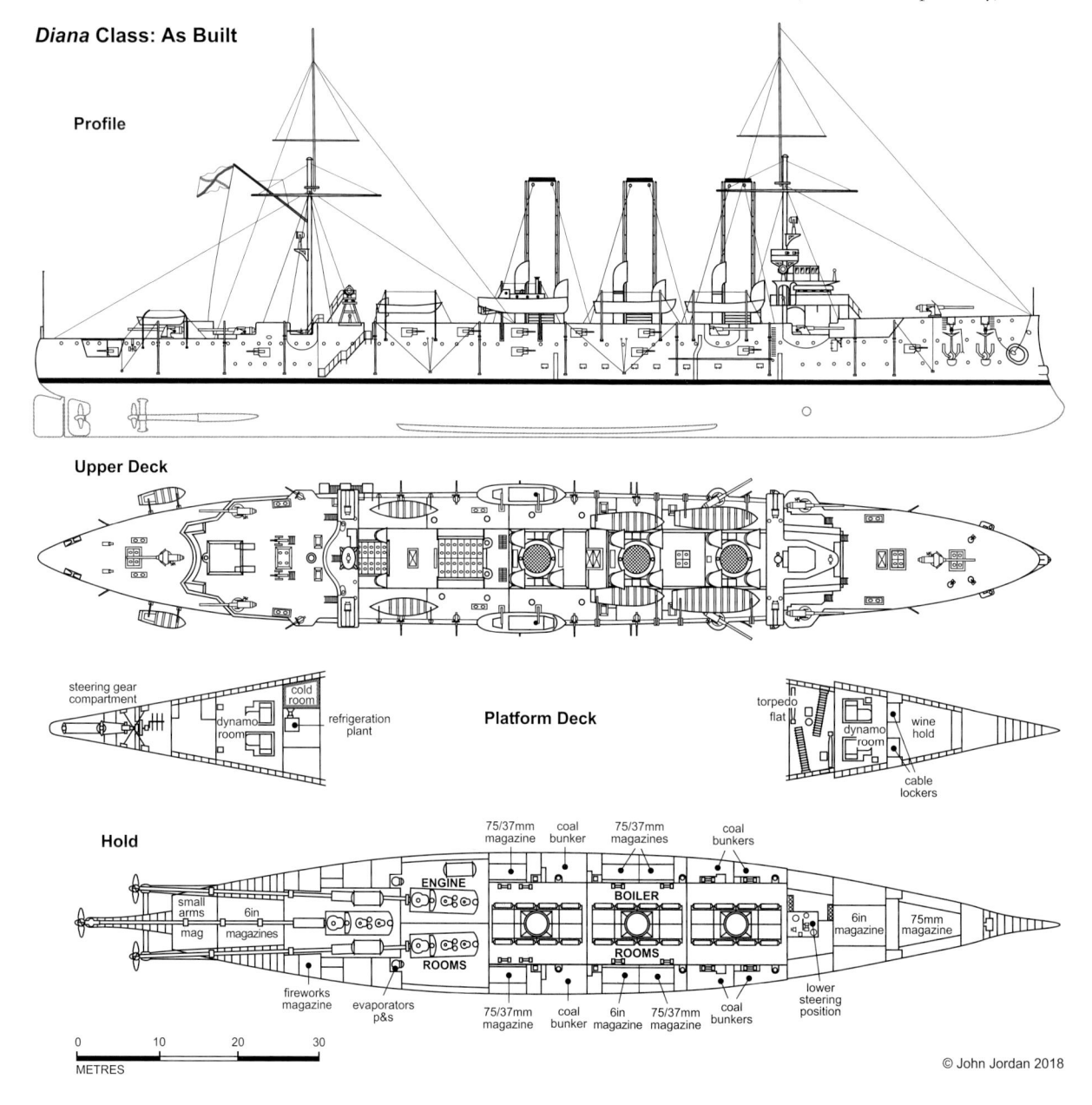

Diana Class: As Built

credited with ranges of 1,600nm at 19 knots and only 2,980nm at 11 knots. *Avrora*'s coal capacity is given as 912 tons normal, 965–970 tons full load.

Electrical power was provided by six steam dynamos, producing a total of 336kW at 105V direct current. The four 'battle' dynamos were in pairs under the armour deck on the fore and aft platform decks, output 67.2kW each, and two steam dynamos for auxiliaries, 33.6kW each, on the upper deck between the middle and after funnels. Electrically-driven systems included ventilation, steering gear, capstans, ammunition hoists, water-discharge pumps, machine tools in the ship's workshop, searchlights and shipboard lighting.

Ground Tackle and Boats

Avrora was the first Russian warship equipped with Hall anchors; it had been hoped to fit all three ships with these, but casting problems with the new type led to delays, so *Diana* and *Pallada* received the older-pattern Martin anchors. There were two anchors to starboard and one to port; the Martin anchors weighed 4.6t, the Hall anchors 4.32t. There were also four smaller anchors.

There were minor variations in the boats carried, but the standard outfit was two 30-foot steam cutters, two launches (one 18- and one 16-oared), two cutters (14- and 12-oared), two 6-oared whaleboats and a yawl.

Modifications

In April 1904, during the siege of Port Arthur, two 6in guns were removed from *Diana* and *Pallada* (the second broadside pair), as were four 75mm guns from the upper deck, all 37mm guns, and both 2.5in landing guns. *Pallada*'s guns were used to bolster the port's landward defences, while several of *Diana*'s guns were used to fill out the battleship *Retvizan*'s armament, which had been reduced while she was under repair following torpedo damage. When *Diana* returned to the Baltic after the war she still carried only 6 x 6in and 20 x 75mm.

Pallada was sunk at Port Arthur by Japanese howitzer fire on 24 November/7 December 1904 and was raised after the port surrendered. She was taken into the IJN under the name *Tsugaru*; her characteristics are given in Table 7.

In 1904, before she sailed with Rozhestvenskii's Second Pacific Squadron, *Avrora* was fitted with 1in (25mm) shields for six of her eight 6in guns. The second pair of broadside guns was not fitted with shields, because they would have hit the boat crutches above. At about the same time optical sights were installed for all 6in and 75mm guns, and a radio cabin was fitted on the upper deck between frames 87–88. She was also fitted with one or two Barr & Stroud 4.5ft (1.37m) rangefinders, and two Maxim '3-line' (.30 calibre) machine guns on her forward bridge wings.

After *Diana* and *Avrora* returned to the Baltic in April 1906 KK Ratnik, who was now the chief inspector of shipbuilding, proposed replacing the outboard engines of one of the ships with turbines in order to gain experience with this new type of propulsion plant, but nothing came of the idea. However, many other modifications were made to both ships over the following years. *Diana* underwent a two-year refit at the Baltic Works starting in 1906, during which her machinery was overhauled and her hull repaired. In addition, the fighting top on the foremast was removed, as were all the 37mm guns except two retained for arming boats. The searchlight platforms on both masts were eliminated. Four 75mm guns were removed (two from each of the upper and battery decks), and two additional 6in guns were added on sponsons abreast the mainmast. All ten 6in guns were provided with shields, although some of these had to be custom fitted due to the cramped spaces available. The over-hanging roof of the conning tower was cut away and the height of the viewing slit decreased from 12in to 3in, since experience during the war had demonstrated how easily shell splinters could be deflected through the wide aperture. The torpedo tubes were removed, as were the torpedo nets. A Barr & Stroud rangefinder, probably with a 4ft 6in base, was installed atop the wheelhouse.

Table 7: *Tsugaru* (ex-*Pallada*) in Japanese Service

Displacement:	5,830t light, 6,763t normal, 7,581t full load
Dimensions:	397ft pp, 415ft oa x 55ft x 21ft 3in (mean)
	121.01m pp, 126.49m oa x 16.76m x 6.481m
Armament:	10 x 15.2cm/45 Armstrong
	12 x 7.6cm/40 Type 41
	2 x 6.5mm MGs
	2 x 45cm submerged TT (4 torpedoes)
	4 x 75cm searchlights
Protection:	As built except for the following:
	Communications tube: 76mm (3in)
	6in gun shields: 114mm (4.5in)
Machinery:	[Original Russian machinery]
Speed:	19.27kts (trials after repairs)
Endurance:	Coal 500t normal, 989t full load; 3,378nm at 10kts
Complement:	514

Note: After her conversion to a minelayer in 1915–1918, *Tsugaru*'s characteristics were altered as follows: displacement was reduced to 6,630t, armament reduced to 5 x 15.2cm, 10 x 7.6cm, 1 x 6.5mm MG, no torpedo tubes. She could carry up to 400 Type 4 (4 *gô*) mines.

Source: Official data book (*yômokubo*), provided by Mr Ishibashi Takao via Lars Ahlberg, email to author, 12 April 2018. Note that these characteristics differ from those found in western reference books, but are almost certainly more reliable. However, there is no explanation for decrease in the thickness of the communications tube, which was originally 89mm, and the thickness of the gun shields seems unlikely.

Avrora was initially given a much more limited refit, as she was wanted as soon as possible as a training ship for midshipmen. In addition to a machinery overhaul, the 37mm guns were removed from the fighting top and two additional machine guns were installed on the after bridge. Barr & Stroud rangefinders (probably 4ft 6in) were installed atop the wheelhouse and the after bridge. Accommodations were arranged for her trainees. After conducting training cruises in 1907 and 1908 *Avrora* was again taken in hand for another year-long refit, during which she was brought up to the same standard as *Diana*. Her main battery was increased to ten 6in guns in shields, the conning tower was modified, the fighting top was removed, as were the torpedo tubes, and all the boilers were replaced. Her refit was completed in August/September 1909.

Diana underwent a machinery overhaul at the Baltic Works from the autumn of 1910 to May 1912. Although the ship was in relatively good shape, she had little combat value, and on 12/25 August 1912 the Naval Minister, Admiral IK Grigorovich, ordered that she be converted into a submarine tender. Later that same year, however, the Naval General Staff concluded that the expanding Russian Navy needed another gunnery training ship more than a submarine tender, so it was decided to convert *Diana* to this role, arming her with ten of the new 130mm/55 guns. Her boilers were also to be replaced with the new Belleville–Dolgolenko type, which featured improved water circulation. However, when war broke out in August 1914 her rearming was still incomplete, and as an emergency measure the ship was given eight 120mm guns. These were replaced in October 1914 by ten 6in guns. She also received four 75mm guns at this time.

In the winter of 1914/1915 both *Diana* and *Avrora* were fitted with mine rails on the upper decks, for laying Model 1908 mines; *Diana* could carry 126 and *Avrora* 150 mines. Both ships were also equipped with 'foretrawls', a device for clearing mines from a ship's path. A fixture below the waterline at the bow supported a long boom which, when lowered ahead of a ship, could stream two paravane-like 'buoys' from its tip, the cables of which were fitted with explosive charges. The foretrawl proved delicate in service and was subsequently removed.

In May 1915 *Diana* finally received the ten new 130mm/55 guns that had been planned for installation before the war; ammunition outfit was 150 rounds per gun. The quarterdeck gun was moved closer to the stern, while the after broadside guns were moved closer to the ship's sides to improve their arcs of fire on after bearings. Her armament also included four 75mm/50 and six .33 cal (7.62mm) machine guns. In early 1916 she was fitted with four improvised AA guns in the form of standard 75mm/50 guns on howitzer mountings, allowing a maximum elevation of 52 degrees. These were installed on the middle bridge forward of the mainmast, taking the place of two of the searchlights, which were moved to platforms on the masts. By this time the ship had two Barr & Stroud 9ft rangefinders.[15] The work was completed in August 1915, and this appears to have been *Diana*'s last substantial refit. In 1918 she was laid up at Kronshtadt, and during the Civil War her guns were removed and used to arm vessels of the Astrakhan–Caspian Military Flotilla.

Avrora, however, still had a long career ahead of her. Over the winter of 1915/1916 she was given a major refit, during which her armament was increased to fourteen 6in using four guns taken from *Diana*. Sixteen of the 75mm guns were removed (all ten guns on the battery deck and six on the upper deck), and the gunports on the battery deck were plated over. She was equipped with two 9ft Barr & Stroud rangefinders as well as a 3ft instrument.[16] She was fitted for four 75mm/50 AA guns of the same type as equipped *Diana*, but because of the shortage of guns she initially received only two of them, on the middle bridge. Two more were installed later in 1916 when the ship was sent to the Gulf of Riga. A 40mm Vickers AA gun was fitted on her quarterdeck. Her machinery was overhauled, and in May 1916 she was docked at Kronshtadt for repairs to her hull and propellers. She rejoined the fleet on 30 May/ 12 June 1916.

At the end of September 1916 *Avrora* was taken in

A somewhat indistinct photograph of *Diana* after she was rearmed with ten 130mm guns in May 1915. The boom of her foretrawl can just be made out at her bow. When lowered, the boom would stream two paravane-like 'buoys' for clearing a path through a minefield. Smaller booms on either side of the bow were rigged with cables that kept the main boom in position. Throughout the war *Diana* was assigned to the 2nd Cruiser Brigade, and the red bands halfway up her second and third funnels indicated her position as the Brigade's fourth ship. The bow of the armoured cruiser *Gromoboi*, likewise serving with the 2nd Brigade, can be seen in the left background. (Courtesy of John Roberts)

Avrora 1917

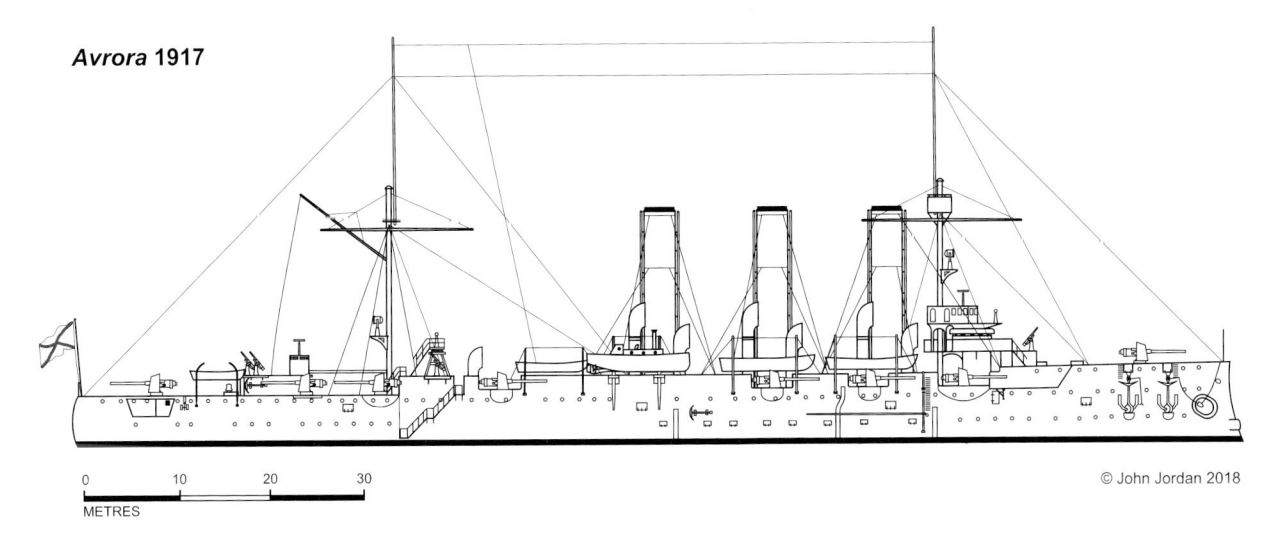

hand for a major refit, first at Kronshtadt and then at Petrograd (formerly St Petersburg). The 6in gun mountings were modified to increase their elevation to 25 degrees, and the bow and stern guns were moved closer towards the ends of the ship, increasing their arcs of fire. Guns nos 17 and 18 (at frame 109) were moved closer to the ship's sides in order to increase their arcs of fire aft and make more room inboard for working the guns. Six 3in (76.2mm) Lender AA guns were mounted: three on the after bridge, two amidships, and one on the forecastle in front of the conning tower. A special anti-aircraft central post was created in the hold. She was also fitted with Fessenden equipment for acoustic signalling to submarines. The machinery was overhauled, new boilers of the Belleville–Dolgolenko type were installed, and the teak decking was repaired. During the war *Avrora*'s crew had increased to 23 officers and 614 lower ranks.

During the Civil War *Avrora*'s 6in guns were removed for use on vessels of the Volga Flotilla. She remained unarmed at Kronshtadt until the autumn of 1922, when she was refitted as a training ship for the Red Fleet. She was rearmed with ten 130mm/55 guns, two 3in AA guns on the after bridge, and two .30 machine guns. The hoists and magazines were refitted to accommodate the new ammunition, and two Barr & Stroud 9ft (2.7m) rangefinders were installed. Her anchors, which had been lost to various causes while she was laid up at Kronshtadt, were replaced by ones salvaged from the cruiser *Oleg*, which had been sunk by a British motor torpedo boat on 18 June 1919. She began her new career as a training ship in the summer of 1923.

Over the winter of 1925/1926 four old 75mm Canet guns were installed on the upper deck (*circa* frames 55 and 86 on either beam) for training purposes. At some point during the 1930s her anchors were replaced by stockless types and hawsepipes were simplified and the hawse holes moved higher on the ship's sides. In 1933 *Avrora* was taken in hand for a refit; the hull was repaired and her machinery overhauled; the work was completed in the spring of 1935. During the winter of 1935/1936 the boilers in the centre boiler room were removed, and the ship was used as a floating school ship, being towed about as necessary. Starting in the winter of 1936/1937 she was also used as an accommodation ship for submarine crews. By 1940 *Avrora*'s metacentric height had been reduced to 2.264ft (0.69m).[17] Her ultimate fate had not been decided when the Germans invaded on 22 June 1941.

In August 1944, as the military situation improved, the decision was made to turn *Avrora* into a 'memorial-museum', and in 1946–1948 work towards that end was undertaken. Two of her engines and all the boilers were removed and scrapped. The inside of her hull was coated with a layer of ferro-concrete 50–90mm thick to keep it watertight. A modest effort was made to restore her to her 1917 appearance: ten 6in guns were installed and other superficial changes were made. But there were still many inaccurate features – the 6in guns had incorrect, coast artillery-type shields, her anchor gear was as modified in the 1930s, etc. By 1984 her underwater hull was in poor shape, and a radical reconstruction was undertaken. She was taken into drydock at the Zhdanov Works and, after the removal of equipment, funnels, and most of the bridgework, the upper hull was cut away at the level of the armour deck in four large sections and mounted atop an entirely new lower hull. The deck armour and equipment were removed from the lower hull (which was eventually sunk as part of a breakwater) and mounted in the new hull, including one engine; two replica Belleville boilers were also added. Many of the inaccurate features were removed or replaced with more appropriate fixtures, and new museum displays were created in the internal spaces. The work was completed in time for the 70th anniversary of the October Revolution in 1987.

Notes on Careers

Avróra ('Aurora', Roman goddess of the dawn): Sailed with Rear Admiral AA Virenius' squadron (*Osliabia*, *Dmitrii Donskoi*, *Almaz*) for Far East 25 September/ 8 October 1903, but when Russo-Japanese War broke out squadron returned to Baltic. *Avrora* joined Second

WARSHIP 2019

Pallada resting on the bottom of Port Arthur's inner harbour, in late 1904 or early 1905; she had been sunk by Japanese siege howitzers on 7 December 1904. She was subsequently raised and entered service with the IJN as *Tsugaru*. (Courtesy of the late Boris Lemachko)

A model of *Avrora* undergoing restoration in the workshop of the Central Museum of the Armed Forces, Moscow. Originally made for the 50th anniversary of the October Revolution of 1917, the model has been modified to show the ship's appearance at the Battle of Tsushima (14/27 May 1905). Note that small shields have been added to the 6in guns, except for the second pair of broadside guns, where space was too cramped to accommodate them. (Courtesy Sergei Vinogradov)

IN *AVRORA*'S SHADOW: THE RUSSIAN CRUISERS OF THE *DIANA* CLASS

Avrora in 1910, after her armament had been increased to ten 6in guns. The two additional guns were mounted on the sponsons formerly used for the after 75mm guns on the upper deck. Note the rangefinder atop the wheelhouse, which appears to be a Barr & Stroud 4ft 6in instrument.
(Author's collection)

Pacific Squadron, survived Battle of Tsushima 14/27 May 1905 (15 killed, 45 wounded), steamed to Manila with cruisers *Oleg* and *Zhemchug*, interned there by American authorities. Returned to Baltic April 1906. Major overhaul of hull and machinery 1906–1907, refitted August 1908 to August 1909. Carried out six training cruises between 1906 and 1914, longest to Bangkok (1911–1912).

During First World War covered minelaying operations by light forces, took part in operations Gulf of Riga 1916. Captain 1st Rank MI Nikolskii murdered 28 February/13 March 1917 during uprising on board while ship under refit Petrograd. 25 October/7 November 1917 fired single blank shot from forward 6in gun, signal for Bolsheviks to storm Winter Palace. Took part in 'Ice Voyage' from Helsingfors (Helsinki) to Kronshtadt, 22–27 December 1917 (4–9 January 1918). Remained Kronshtadt in long-term preservation. During Civil War 6in guns transferred to Volga Flotilla for use on floating batteries. Major refit November 1922 to 23 February 1923 for service as training ship. Awarded Order of the Red Banner 2 November 1927.

During Great Patriotic War guns transferred to land batteries defending Leningrad; used as floating barracks for submarine crews. Damaged by German shore-based artillery 30 September 1941, grounded Oranienbaum (now Lomonosov). Raised 20 July 1944 and towed to Leningrad, 1945; repaired Baltic Works. From 6 November 1948 moored on Great Neva River as memorial-museum. Major refit 18 August 1984 to 12 August 1987

The Red Banner cruiser *Avrora* in Kronshtadt roadstead on the USSR's Navy Day. Although the photo is dated July 1940, it may have been taken some years earlier, as Russian publications state that the boilers in her centre engine room were removed over the winter of 1935/1936, yet here that funnel is emitting smoke. Note that the hawsehole has been raised and the anchor handling gear simplified.
(Courtesy of the late Boris Lemachko)

Zhdanov Works. Towed from St Petersburg to Kronshtadt 21 September 2014 for major repair work; returned St Petersburg 16 July 2016.

Diána (Roman goddess of the hunt): Sailed for Far East October 1902 in company with sister *Pallada* and battleship *Retvizan*; joined Port Arthur Squadron 24 April/7 May 1903. During Russo-Japanese War participated in Battle of the Yellow Sea, 28 July/10 August 1904 (10 killed, 17 wounded). After battle became separated from squadron, steamed to Saigon, interned there by French authorities. Returned Russia April 1906, underwent major overhaul of hull and machinery before joining Baltic Fleet in 1908. Major refit 1912–1914. During First World War covered minelaying operations by light forces. May 1915 6in guns replaced by ten 130mm/55 guns. Operations against Swedish iron ore trade June 1916; defence of Gulf of Riga; 29 September to 6 October 1917 participated in Moonsund Operation. Took part in 'Ice Voyage' Helsingfors to Kronshtadt 22–27 December 1917 (4–9 January 1918). May 1918 put in long-term preservation at Kronshtadt; 130mm guns removed during Civil War and used to arm vessels of Astrakhan–Caspian Military Flotilla. Autumn 1922 sold to Soviet-German firm Derumetall, towed to Germany for breaking up. Stricken from Red Fleet 21 November 1925.

Palláda (Pallas Athene, Greek goddess of wisdom): Sailed for Far East October 1902 in company with sister *Diana* and battleship *Retvizan*; joined Port Arthur Squadron 24 April/7 May 1903. On night of 26–27 January/8–9 February 1904 damaged by Japanese torpedo during surprise attack (3 killed, 35 wounded); repaired by 28 April/11 May 1904. Participated in Battle of the Yellow Sea 28 July/10 August 1904 (4 killed, 10 wounded). Returned Port Arthur with main squadron. Sunk in harbour in shallow water 24 November/7 December 1904 by Japanese howitzers. Further damaged by crew 20 December 1904/2 January 1905 before surrender of Port Arthur. Raised by Japanese on 30 July/12 August 1905; renamed *Tsugaru* and rated as second-class cruiser (*nitô junyôkan*). Repaired Sasebo and Yokosuka 1906–1908. Served as machinery training ship (*kikanjutsu renshûkan*) 1911 to 1914, converted to minelayer (*fusetsukan*) July 1915 to 1918, reclassified as such 1 April 1920. Reclassified as miscellaneous harbour ship (*zatsuekisen*) 1 April 1922; stricken 1924 and sunk 27 May 1924 as target for aircraft bombing off Sarushima Island.

Acknowledgements:

I would like to express my gratitude to Sergei Vinogradov, who provided much material, including photographs of his model of *Avrora*. John Roberts allowed me to use several fine photos from his personal collection, and Aidan Dodson supplied excellent photos of *Avrora* as she appears today. John Jordan provided the superb drawings. Lars Ahlberg and Mr Ishibashi Takao provided information on *Pallada* when she served with the Imperial Japanese Navy. And once again my wife, Jan Torbet, greatly improved the article with her sharp editorial eye.

Avrora in July 2002 at her berth on the River Neva in St Petersburg. By this time the ship had been largely restored to her 1917 appearance, including the relocation of the hawsepipes. (Photograph by the author)

Avrora in September 2017 after her latest renovation. She is now painted a darker grey. (Courtesy of Aidan Dodson)

Sources:

Apal'kov, IuV, *Boevye korabli russkogo flota 8.1914g–10.1917g: Spravochnik*, INTEK (St Petersburg 1996).

Balakin, Sergei, 'Kreisera tipa "*Diana*": vneshnie razlichiia i modernizatsii', *Morskaia kampaniia no 2*, 2009, 56–63.

Burov, VN and Iukhnin, VE, *Kreiser 'Avrora': Pamiatnik istorii otechestvennogo korablestroeniia*, Lenizdat (Leningrad 1987).

Krest'ianinov, VIa. *Kreisera Rossiiskogo imperatorskogo flota, 1856–1917 gody*, part 1, Galeia Print (St Petersburg 2009).

Novikov, Viktor and Sergeev, Aleksandr, *Bogini Rossiiskogo flota: 'Avrora', 'Diana', 'Pallada'*. Iauza; Kollektsiia; Eksmo (Moscow 2009).

Ovsiannikov, SI, Bochkov, BV, and Gritsenko, VIa, 'Vosstanovitel'nyi remont kreisera "*Avrora*"', *Sudostroenie* No 10, 1987, 54–60.

Polenov, LL, *Sto let v spiskakh flota: Kreiser 'Avrora'*, Ostrov (St Petersburg 2003).

Skvortsov, AV, *Kreisera I ranga 'Avrora', 'Diana' i 'Pallada'*, Gangut (St Petersburg 2012).

Spasskii, IP (ed), *Istoriia otechestvennogo sudostroeniia*, Vol II: *Parovoe i metallicheskoe sudostroenie vo vtoroi polovine XIX v*, Sudostroenie (St Petersburg 1996).

Vinogradov, SE, 'Sovershenstvovanie artilleriiskogo kompleksa kreiserov "Avrora" i "Diana" v khode Pervoi mirovoi voiny'. In: Klimovskii SD (ed), *Kreiser "Avrora" v istorii otechestva. 1897–1916 gg: Doklady i materialy* Tsentral'nyi voenno-morskoi muzei (St Petersburg 2016), 170–179.

Endnotes:

1. Novikov and Sergeev, 5.
2. Novikov and Sergeev, 6.
3. Nicholas Papastratigakis, *Russian Imperialism and Naval Power: Military Strategy and the Build-up to the Russo-Japanese War*, IB Tauris (London 2011), 100–113.
4. Krest'ianiniv, 1:149.
5. Skvortsov, 8.
6. TNA ADM 116/878, 'Design for New Cruisers, Supplemental Programme, 1898–99', 27 September 1898, 5.
7. Krest'ianinov, 1:154.
8. Russia – War Vessels, NID Report No. 787, November 1906, 114.
9. None of the published sources mention the hoist for the forecastle gun, but it is shown and labelled in cut-away drawings, eg Polenov, 36, no 18; Skvortsov, 42–3, no 7.
10. Skvortsov, 71.
11. AM Petrov (ed), *Oruzhie Rossiiskogo flota*, Sudostroenie, (St Petersburg 1996), 65.
12. Welch, JJ, *A Text Book of Naval Architecture for the Use of Officers of the Royal Navy*, revised edition, HMSO (London 1901), 185.
13. Polenov, *Sto let v spiskakh flota*, 37, gives slightly different figures: HP cylinder: 800mm (31.5in), MP cylinder: 1,273mm (50in), LP cylinder: 1,900mm (74.8in), piston stroke: 870mm (34¼in). Since the Imperial Navy did not use the metric system at the time, the figures in the text seem more likely.
14. Krest'ianinov, 153.
15. Email, Sergei Vinogradov to author, 28 June 2018.
16. *Ibid*.
17. Wright, Christopher C, 'Soviet Cruisers', Part I, *Warship International* Vol XV No 1 (1978), 11.

PROJECT 1030: A NUCLEAR ATTACK SUBMARINE FOR THE ITALIAN NAVY

Developments in nuclear propulsion during the 1950s offered for the first time the possibility of a true submarine capable of unlimited underwater operation. **Michele Cosentino** looks at the ambitions of the Italian Navy to harness US Navy technology and create a force of torpedo-armed attack and ballistic missile submarines.

When researching another project at the Italian Navy Historical Office (INHO) recently, the author came across documents and drawings related to the design of a nuclear attack submarine for the *Marina Militare*. During the 1950s the Italian Navy embarked on several initiatives aimed at designing and building such a boat, an objective which was publicly announced in Parliament in 1959 by the then-Minister of Defence. This is the story of those initiatives and the nuclear submarine design, which involved the Constructors Department of the Italian Navy, several Italian companies, research centres and universities, and the government of the United States.

The Italian Race for Nuclear Naval Propulsion

The interest of the *Marina Militare* in adopting nuclear power for naval propulsion dates back to 1949, when the Navy began collecting technical articles and information coming from the United States. This activity was boosted by the commissioning of USS *Nautilus* on 30 September 1954 and the initiation of other submarine designs in the US.

At the time, the *Marina Militare* was debating how to create a credible submarine force, a fleet component that had been forbidden by treaty after the Second World War and was now being restored mainly by the transfer of former US submarines.[1] A key aspect that had to be taken into account was the ability of Italian shipyards and companies to build submarines and their associated equipment after the lengthy enforced lull caused by the conflict and its aftermath. The debate developed along two major lines: conventional diesel-electric and nuclear. The preliminary design of a diesel-electric submarine, to be named *Guglielmo Marconi*, was initiated within the Italian Navy's Constructors department in 1955. This was a single-hull boat displacing about 1,400 tonnes when dived, armed with four 533mm torpedo tubes and powered by two electric motors driving a single 4-bladed propeller. One year later the Italian Navy established the

The keel laying of the submarine *Guglielmo Marconi*. The event took place on 16 June 1957 at a private shipyard in Taranto. (Author's collection)

Centro per le Applicazioni Militari dell'Energia Nucleare (CAMEN), a research centre founded within the compound of the Naval Academy in Leghorn and tasked with the study of nuclear energy for military systems. The centre was initially manned by Italian Navy officers of the Naval Engineering and Weapons Corps, and its principal objective became the design and the production of an experimental nuclear reactor for naval propulsion.[2] Academics from the nearby University of Pisa and experts from the Italian government-funded Research Centre for Nuclear Energy contributed to the work performed by the Italian Navy, while an important role was played by some private companies, principally FIAT and Ansaldo, which were already working on nuclear power for civilian applications.

On 16 June 1957, an important event took place at Taranto: the first structural element, a pressure hull ring (see photo), of a submarine to be named *Guglielmo Marconi* was laid down in a private shipyard in Taranto. The lack of other information on this event, which was attended by military and civilian authorities, suggests

The building that from 1961 housed the experimental reactor Galileo Galilei at CAMEN, near Pisa. (CISAM Archive)

that the hull ring may have belonged to a future nuclear-powered submarine whose design was sufficiently advanced to justify its public disclosure. A possible hypothesis is that the Italian Navy was quite confident about the construction of this type of submarine and wished to convince the Italian government of its capabilities in the field, and that the *Marina Militare* had decided to set aside the design of a diesel-electric submarine and divert its efforts and resources towards a nuclear attack boat. The decision followed an analysis in which the *Marina Militare* took into account three key aspects: the funds already invested in nuclear research and officers' education for nuclear propulsion; the capabilities of national shipyards and suppliers of defence equipment; and the potential support provided by the United States. It should be noted that during this period the Suez crisis of October 1956 had put Britain and France at odds with the United States, a scenario that lasted for some time. The Italian government attempted to exploit this situation by proposing itself as a privileged NATO political and military partner for Washington in the sensitive Mediterranean region.

In December 1957, during a 'Subtle' meeting in Naples at CoMedCent,[3] the US Navy representatives provided unclassified information about the first operational lessons learned from *Nautilus*. This opportunity gave additional reasons for the Italian Navy to press ahead with its nuclear concepts. In June 1958 a short memorandum was sent to the Minister of Defence, explaining the importance of nuclear attack submarines in modern maritime warfare, stating the firm interest of the *Marina Militare* in this kind of naval vessel, and requesting the authorisation to develop a firm project within a short timespan.

In February 1959 Giulio Andreotti became Minister for Defence, and the Italian Navy requested, through diplomatic channels, the assistance of the United States in the construction of a nuclear submarine. Between March and July 1959, Washington announced its willingness to provide support, and requested further information in order to assess the security implications related to the transfer of technology, materiel and classified information on nuclear matters.[4]

In principle, the *Marina Militare* adopted a strategy based on three pillars that would allow the construction of a nuclear attack submarine: the Italian Navy's Contructors Department was to be responsible for the design, with the US Navy providing information and support; the construction of the nuclear reactor would be assigned to specialist Italian companies, supported and/or supplied by their US counterparts as permitted by Washington; the US government was to provide the nuclear fuel. Assured that the cooperation between Rome and Washington would pave the way for the construction of the first Italian nuclear attack boat, on 3 July 1959, during the discussion on the defence budget in Parliament, Andreotti revealed officially 'the determination of the Italian government to proceed with the construction of a nuclear-powered submarine'. The name chosen for the new boat was, again, *Guglielmo Marconi*.

Progetto 1030/SPN: Hull & Propulsion Machinery

The most important documents found at INHO are the plans of a project dubbed *Progetto 1030* relating to an 'SPN', an acronym that in Italian stands for *sommergibile*

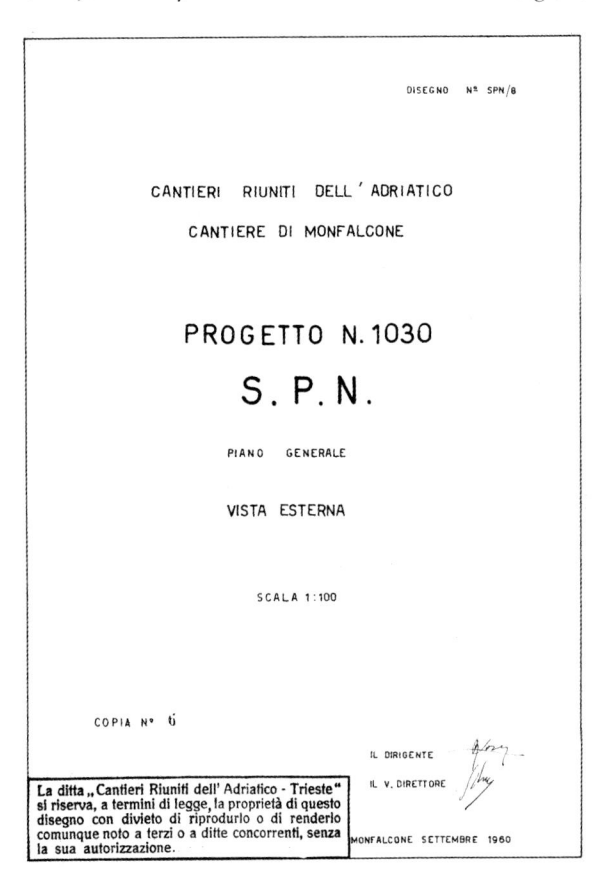

The official template used by the CRDA shipyards for each of the Project 1030 plans; the one shown here is that associated with the outboard profile (*vista esterna*) of the boat.

a propulsione nucleare (nuclear-powered submarine). These plans, comprising 14 sheets to a scale of 1:100 and an accompanying booklet, were drawn up by the Monfalcone shipyard of Cantieri Riuniti dell'Adriatico (CRDA), a Trieste-based company that had been a long-time builder of submarines for the Italian Navy.[5] The plans are dated September 1960. Comparing this date with other documents, it is safe to assume that after Andreotti's speech in Parliament (if not earlier), the Italian Navy became heavily committed to the design of a nuclear submarine, working in very close contact with the CRDA shipyard. The plans are too detailed to suggest that Project 1030/SPN was a private venture of CRDA; rather, they appear to be the result of a close collaboration between the shipyard, the Italian Navy's Constructors Department, and other organisations.

The first aspect which emerges when examining these plans is the new approach taken by the Italian Navy in designing a submarine. Project 1030 shows a boat remarkably similar in configuration to the US Navy's nuclear attack submarines of the *Skipjack* (SSN-585) class, which had been under construction since 1956.[6] At the time, the US Navy was shifting from the lessons already learned with *Nautilus* – essentially derived from the post-war *Tang*s but powered by a nuclear reactor – to experimentation with newer hull forms, including the construction and testing of the conventionally-powered experimental submarine *Albacore* (AGSS-569). *Albacore*'s pressure hull had been developed from experiments carried out at the David Taylor Model Basin, where the models tested were grouped into Series 58. The tests had revealed that a tear-drop shape with cylindrical sections allied to smoothly tapered ends would achieve the fastest possible submarine hull, and concurrent studies demonstrated that a single centreline propeller would provide the greatest propulsive efficiency.

The undisputed link between the submarines of the *Skipjack* class and Project 1030 was grounded on the close relationship and exchanges involving officers of the two navies. These were developed from around 1958, and included visits of Italian officers to US submarines both in the USA and Italy, notably in the summer of 1959, when *Skipjack* made a port call to the Italian Naval Dockyard of La Spezia.

However, Project 1030 featured a submarine slightly longer than *Skipjack*: 83.8m overall compared to 76.7m. The difference in length was due to some important modifications in the forward compartment and an unusual configuration of some spaces. As a consequence, the submerged displacement of Project 1030 would increase to 3,500 tonnes, 100 tonnes greater than *Skipjack*. External appendages, including the fin, were

The experimental submarine *Albacore*, conventionally powered but with a revolutionary 'tear-drop' hull form, slides down the launching ways at the Portsmouth US Naval Shipyard, New Hampshire, on 1 August 1953. (US Navy)

The nuclear attack submarine *Skipjack* comes alongside at the La Spezia naval base in the summer of 1959. (Italian Navy)

reduced to a minimum and carefully shaped, with the acoustic sensors housed inside streamlined domes. Project 1030 featured a mostly single-hulled submarine, in which the pressure hull had a maximum diameter of 9.6m. This allowed an internal distribution on four decks in the central compartment, because the hull shape would permit a more efficient use of its internal volume and achieve a decrease in the wetted surface area.[7]

As in *Skipjack*, Project 1030 would use a new type of steel for the 60-metre-long pressure hull, HY-80, which had been developed in the USA since the late 1940s.[8] Two main ballast tanks were located fore and aft of the pressure hull, and these were configured in order to achieve the tear-drop hull shape trialled in *Albacore* and adopted in the *Skipjack*s. Two short sections placed forward and abaft the fin were surrounded by ballast tanks. There was also a narrow boardwalk abaft the fin, providing cover for snorkel induction and exhaust ducts. The four decks, mostly located in the central compartment, were designated *1°/2°/ 3° copertino* (1st, 2nd and 3rd platform decks) and *stiva* (hold). The forward compartment had three levels, while the machinery was on two levels. Longitudinally, the pressure hull was divided into five compartments by four transverse watertight bulkheads, located at Frames 34, 52, 60, and 70, of which the centre compartment, immediately abaft the fin, housed the nuclear reactor compartment (see inboard profile).

The forward compartment housed the torpedo room, part of the crew accommodation, a space for reserve torpedoes and other equipment (see later), auxiliary equipment, and trim tanks. The central compartment

WARSHIP 2019

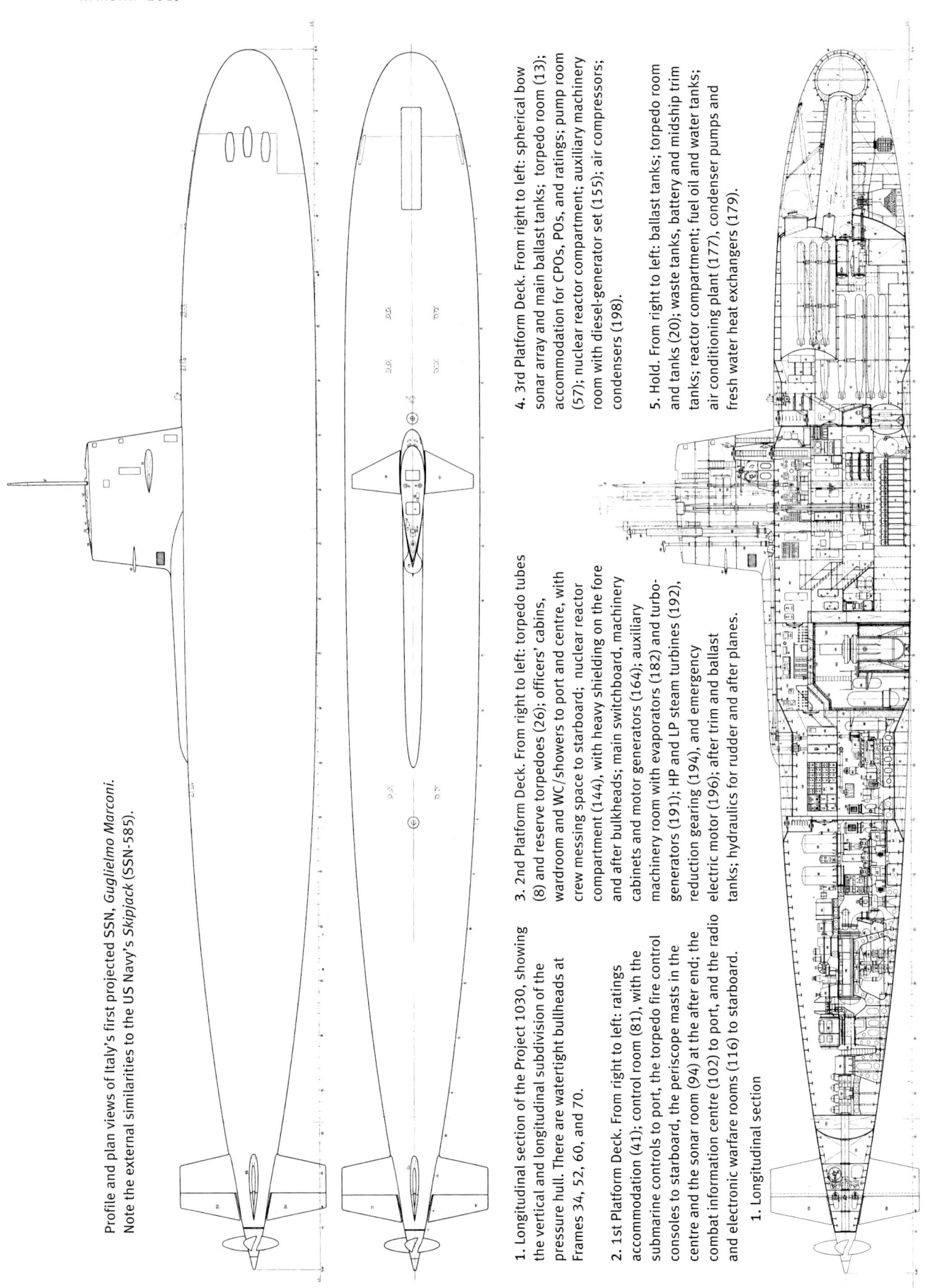

Profile and plan views of Italy's first projected SSN, *Guglielmo Marconi*. Note the external similarities to the US Navy's *Skipjack* (SSN-585).

1. Longitudinal section of the Project 1030, showing the vertical and longitudinal subdivision of the pressure hull. There are watertight bullheads at Frames 34, 52, 60, and 70.

2. 1st Platform Deck. From right to left: ratings accommodation (41); control room (81), with the submarine controls to port, the torpedo fire control consoles to starboard, the periscope masts in the centre and the sonar room (94) at the after end; the combat information centre (102) to port, and the radio and electronic warfare rooms (116) to starboard.

3. 2nd Platform Deck. From right to left: torpedo tubes (8) and reserve torpedoes (26); officers' cabins, wardroom and WC/showers to port and centre, with crew messing space to starboard; nuclear reactor compartment (144), with heavy shielding on the fore and after bulkheads; main switchboard, machinery cabinets and motor generators (164); auxiliary machinery room with evaporators (182) and turbogenerators (191); HP and LP steam turbines (192), reduction gearing (194), and emergency electric motor (196); after trim and ballast tanks; hydraulics for rudder and after planes.

4. 3rd Platform Deck. From right to left: spherical bow sonar array and main ballast tanks; torpedo room (13); accommodation for CPOs, POs, and ratings; pump room (57); nuclear reactor compartment; auxiliary machinery room with diesel-generator set (155); air compressors; condensers (198).

5. Hold. From right to left: ballast tanks; torpedo room and tanks (20); waste tanks, battery and midship trim tanks; reactor compartment; fuel oil and water tanks; air conditioning plant (177), condenser pumps and fresh water heat exchangers (179).

1. Longitudinal section

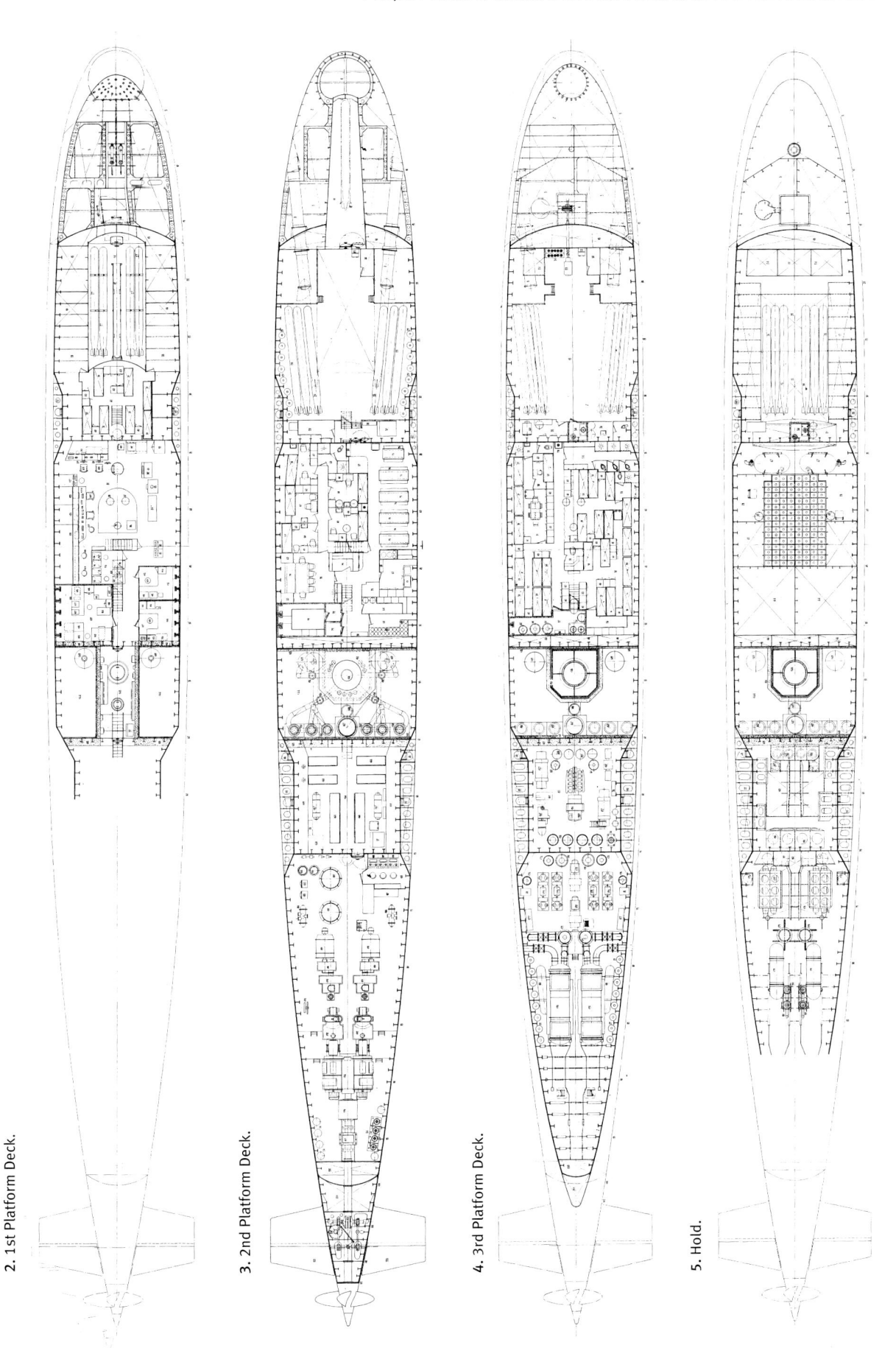

PROJECT 1030: A NUCLEAR ATTACK SUBMARINE FOR THE ITALIAN NAVY

2. 1st Platform Deck.

3. 2nd Platform Deck.

4. 3rd Platform Deck.

5. Hold.

103

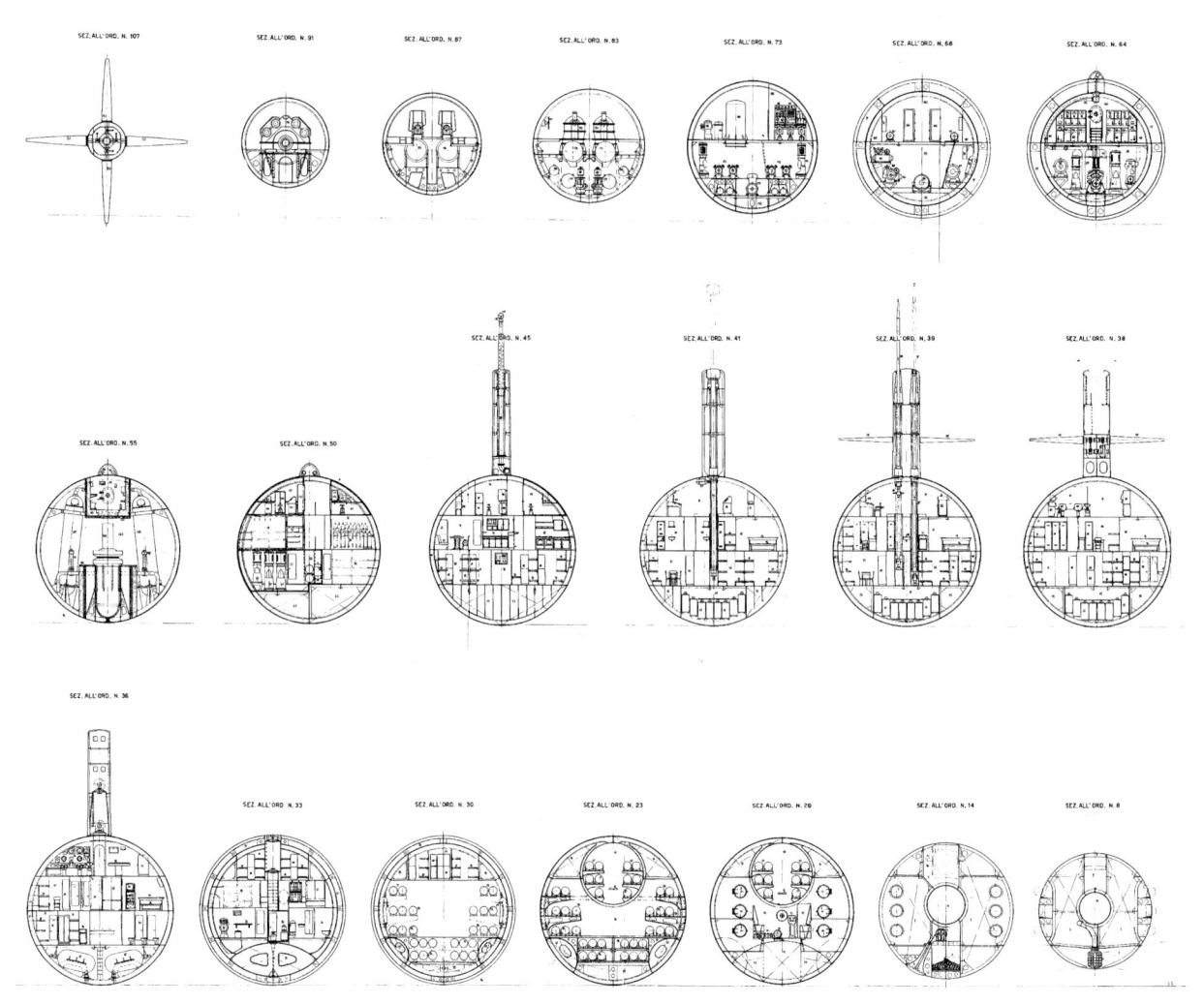

Transverse section views.

was placed directly beneath the fin: the 1st platform housed the control room and other operational spaces (with separate sonar, radar, radio, and plot rooms), while the lower decks were occupied by the accommodation for officers and CPOs/POs, auxiliary equipment, the battery room, and several tanks. The nuclear reactor compartment was provided with massive protection screens and surmounted by a tunnel for crew access. The next compartment had two levels: the upper level housed the main switchboard, cabinets for machinery equipment and the motor generators; the lower level the emergency diesel-generator set, the HP air compressors, and tanks. The aftermost compartment was the longest, was likewise on two levels and housed much of the steam propulsion machinery and the pipework from the reactor. On the upper level there were two HP and LP turbines and their reduction gearing, two turbo-alternators, the massive axial propeller shaft, two evaporators, and the machinery control consoles. The lower level housed two huge condensers, many pumps, the air conditioning plant, tanks, and other auxiliary equipment.

Unfortunately, detailed information about the power and speed of Project 1030 has not been made available. However, given that the reactor would have been designed by CAMEN but derived from the S5W Westinghouse pressurised-water model fitted in the *Skipjack*s, one can assume a power rating of no less than 50MW. Steam was to feed in parallel the two geared turbines and the two turbo-alternators, each of which was rated at 1.8MW. Available sources give a total horsepower of 15,000shp, driving a single five-bladed propeller with a diameter of 4.5m. This would provide a maximum submerged speed of about 30 knots. An emergency electrical DC motor, fed by the battery and connected to the propeller shaft between the reduction gearing and the thrust bearing, ensured a get-home capability in the event of reactor failure.

The after planes and the rudder, which were located forward of the propeller, were cruciform, and their mechanisms were cross-connected so that their movement could be controlled by a single operator. The fore planes were moved up onto the fin, as in their US contemporaries, so that they would not contribute to turbulence around the bow-mounted sonar, and so that the noise of

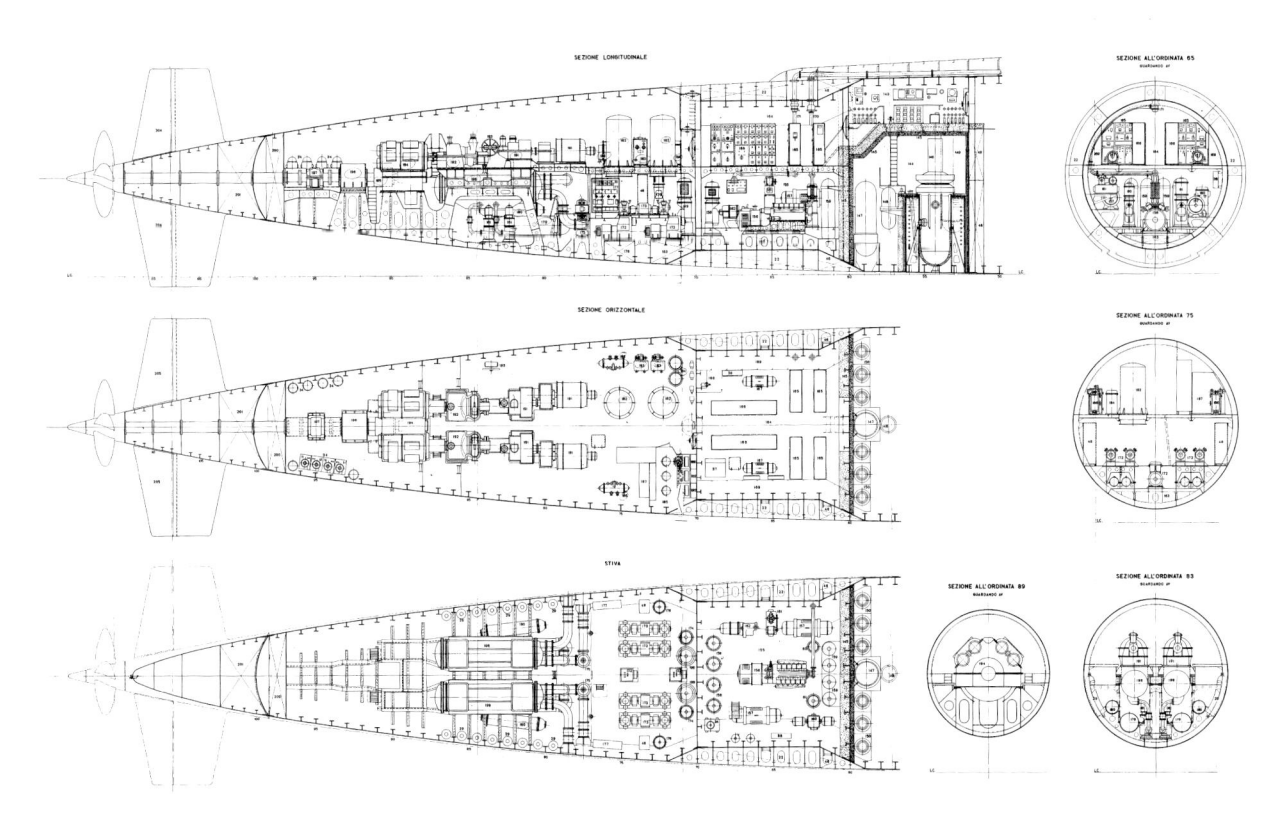

Inboard profile of the machinery spaces, with plan views of the 2nd & 3rd Platform Decks below, and transverse sections on the right. For the main items of machinery see the plans on pages 102–103.

their actuator would not mask sonar returns. Furthermore, with such a disposition the boat could change depth slowly without altering the angle of trim and possibly broaching. Depth and course would be managed by a control console placed in the control room and equipped with two aeronautical-type wheels, one for the rudder and after planes, the other for the fore planes.

As far as complement was concerned, Project 1030 provided accommodation for 11 officers, 12 CPOs, and 74 POs/ratings, a total of 97. However, it is likely that there was some redundancy as the crew of the *Skipjack* class was 93, including 9 officers. Special care was devoted to crew accommodation, including 'luxuries' such as a laundry and an ironing room. Fresh water production and air regeneration would benefit from the relative abundance of steam and electrical power available on board. The boat was to have three hatches and related escape trunks: two housed in the foremost and rearmost compartments and one connecting the control room to the bridge.

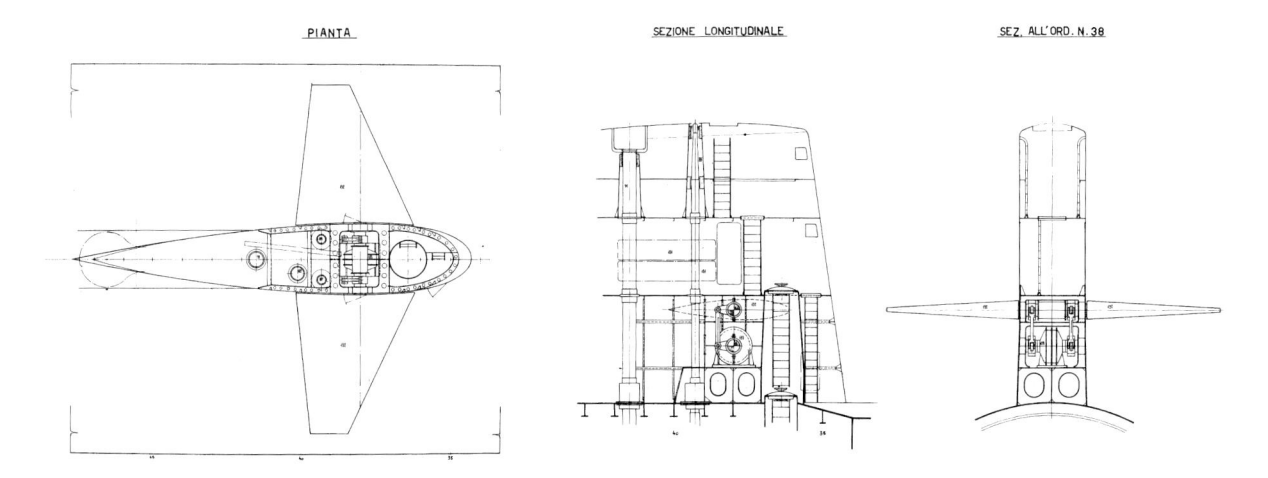

The fin, showing the arrangement of the fore planes.

Electronics & Weapons Systems

The fin was 10 metres long; its shape was consistent with the hull hydrodynamics and it was to be placed farther forward than was customary to enable the heavy reactor compartment to be located centrally. There was a comprehensive complement of masts and above-water sensors, including two periscopes (search and attack), two communications masts for VLF and UHF/MF transmission and reception, and a radar mast. An ECM antenna was mounted atop the snorkel induction mast; the snorkel exhausted via a separate mast. The snorkel induction and exhaust ducts ran from the compartment housing the emergency diesel-generator set through a fairing along the pressure hull to the top of the fin (see inboard profile). Information provided with the CRDA plans report that US-manufactured TED/AN-URR systems were to be used for UHF/VHF communications; this suggests that electronic equipment would have been procured in the US via the Foreign Military Sales channel.

Sonars were similarly located to those fitted in the *Skipjack*s, with a spherical passive transducer array placed in the big bow dome and an active transducer array in the lower part of the forward ballast tank. These systems were supplemented by a pair of passive arrays in a faired housing located on the after edge of the fin. It is likely that the *Marina Militare* gave serious consideration to the purchase of US-made systems of the AN/BQR and AN/SQS types, which were similar to those already fitted in the boats transferred from the US Navy. This was likewise probably the case for the torpedo fire control system; the ex-US diesel-electric boats were already equipped with the Torpedo Data Computer (TDC) Mk 3, so Italian submariners were familiar with this type of system. The main feature of the TDC was the ability to launch torpedoes without estimating the future position of the target. In order to generate a firing solution, it simply required own-boat speed and course (input from the log and the gyrocompass), the estimated speed/course/range of the target (achieved from sonar and/or above-water sensors), and the torpedo type. The TDC was made of two elements: the position keeper and the angle solver.

As far as weapons and their equipment, Project 1030 would have had six paired 533mm torpedo tubes in vertical rows, placed in the lower part of the torpedo room and canted outward, port and starboard. The major difference between the future *Guglielmo Marconi* and *Skipjack* lay in their complement of reserve torpedoes: the US boats were equipped with 24 torpedoes (including those in the tubes), while the Italian boat could accommodate a total of 38. This capability was made possible by what could be considered the most innovative feature of the Italian project: ten reserve torpedoes would be housed in a watertight space located in the upper part of the forward compartment and designated 'hangar' in the CRDA plans. This hangar, which was configured as a cylinder 8.5 metres long and 4.5 metres in diameter, was also equipped for the stowage of underwater vehicles for naval Special Forces operations.[9] A hinged circular watertight hatch provided access to the forward free-flooding area of the boat; a second hatch, set into the top of the free-flooding outer hull casing and operated by hydraulic actuators, would allow the passage of underwater vehicles in and out of the submarine when the latter was submerged. The concept of Special Forces operation envisaged swimmers/raiders leaving the boat via the forward escape trunk, the opening of the hatches connecting the hangar with the sea, the extraction of underwater vehicles, the execution of the mission, and return to the submarine. The existence of the hangar in Project 1030 explains two key differences between the Italian boat and the *Skipjack*s: the longer hull and the additional crew accommodation.

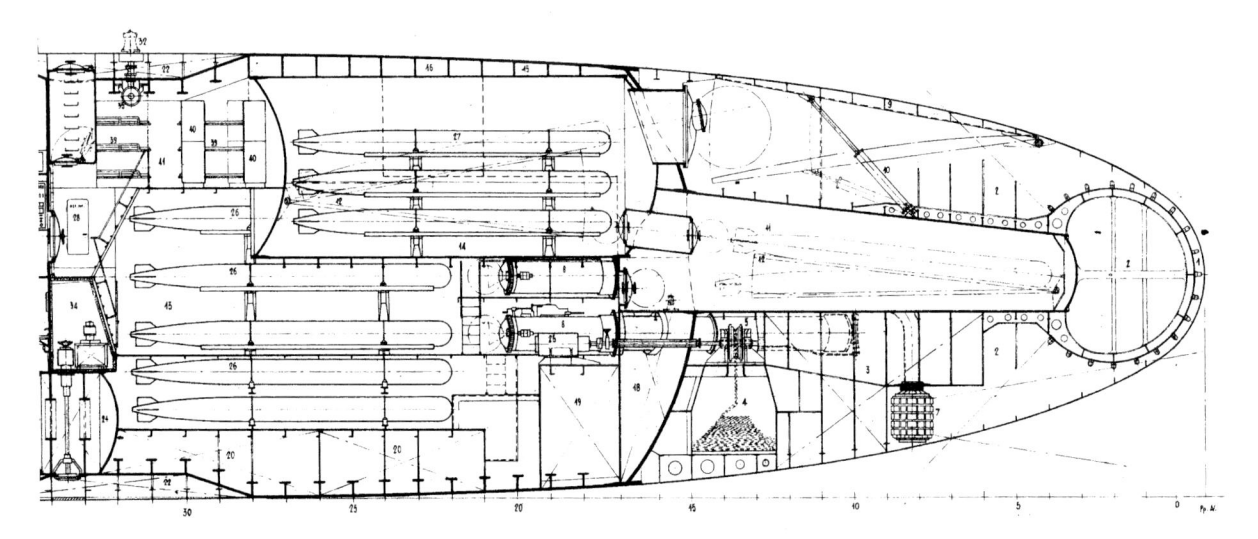

Close-up of the bow section, showing the hangar for weapons and underwater 'special vehicle' stowage, with its internal and external hatches for torpedo embarkation and the exit/entry of special vehicles. The hydraulic arms for the external hatch giving access to the free-flooding area can be seen forward of the larger of the two internal watertight hatches.

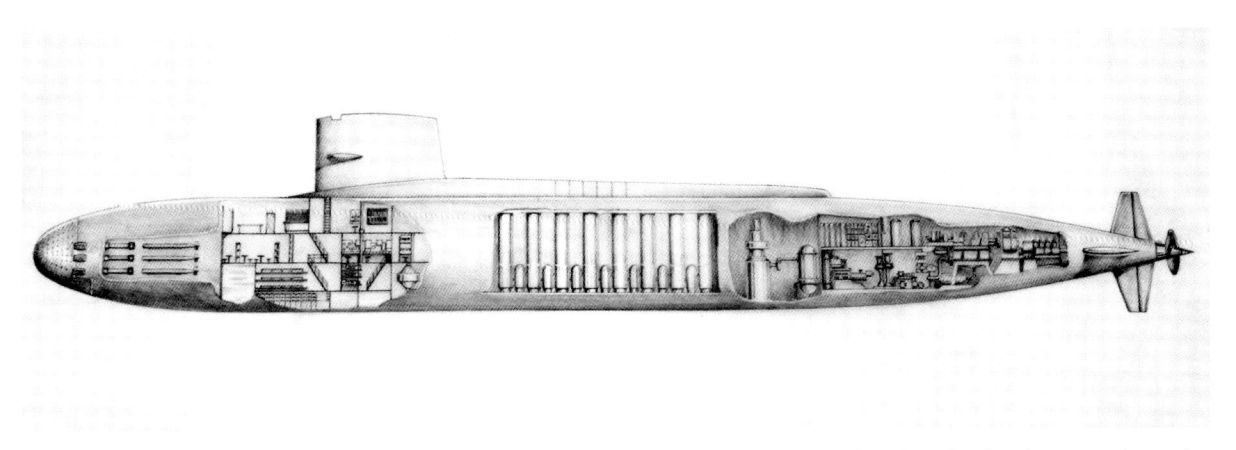

PROJECT 1030: A NUCLEAR ATTACK SUBMARINE FOR THE ITALIAN NAVY

Longitudinal section of a preliminary concept for an Italian nuclear-powered ballistic missile submarine, showing the spaces devoted to machinery and missile silos. This sketch is exhibited at the Naval Museum of Venice.

The SSBN Option

Concurrently with the design activities carried out for Project 1030, the Italian Navy elaborated concepts for another type of nuclear-powered boat, evidence of the aspiration to establish an operationally and technically advanced submarine force. Within this professional framework lay a preliminary concept study for an Italian nuclear powered ballistic missile submarine (SSBN in US/NATO jargon); the documentation for this submarine is not extensive in comparison to the SSN, and has been discovered only recently.

In May 1958, the Head of the Italian Navy sent a memo to the Ministry of Defence in which he questioned the vulnerability of the Jupiter medium-range ballistic missiles that were to be deployed in Italy in the framework of a NATO agreement. The missile pads, easily recognisable, would potentially be exposed to enemy attacks and could also lead to 'alarming internal reactions'.[10] To overcome this possibility, and to reduce the risk, the memo proposed to embark ballistic missiles on naval vessels, namely nuclear powered submarines that were able to meet the 'double requirement of security and secrecy'. The proposal made by the Italian Navy had an obvious political significance, and aimed to provide Italy with a similar underwater strategic deterrent capability to that of the United States, Britain, and France. In terms of design and technical implications, this proposal produced only a feasibility study carried out by the Constructors Department of the *Marina Militare* in accordance with clear directions from higher authorities; it was not ruled out that an Italian SSBN might be equipped with ballistic missiles of Italian manufacture, a development programme for which was currently underway. The study concept was clearly inspired by the approach followed in the US Navy, notably the insertion within the central body of a *Skipjack* class submarine of a cylindrical compartment housing 16 silos for ballistic missiles. Taking Project 1030 as a reference the concept study produced a sketch, currently exhibited the Naval Museum of Venice, for a nuclear-powered submarine with a length of 100 metres and a beam of 9.6m, fitted with 8 double rows of silos and equipped with six 533mm torpedo tubes for self-defence. Assuming that the nuclear reactor was to be the same as in *Guglielmo Marconi* and that the displacement of this SSBN would be significantly greater, it is safe to assume a speed reduction of 2 or 3 knots, which was acceptable if compared with the strategic and operational benefits provided by the embarked ballistic missiles.[11]

Enrico Fermi and the End of Nuclear Ambitions

The Italian Navy and the CRDA shipyard continued work on Project 1030 for at least two more years. In a speech given on 22 December 1962 during the ceremony for the launch of the missile cruiser *Caio Duilio*, Giulio Andreotti stated that the Italian government was 'eager to carry forward as soon as possible the design for the construction of a nuclear submarine in order to meet the requirements of the Italian Navy'. Meanwhile, there was strong opposition on the part of several officials in the United States to the transfer to foreign Nations of information, technology, and materiel related to naval nuclear propulsion. The fiercest opponent was the then-Rear Admiral Hyman Rickover, the long-time undisputed head of the US naval nuclear propulsion community. Eventually, Rickover and his followers prevailed over those in Washington who favoured military cooperation with Italy in the nuclear field. It is also possible that US opposition to the transfer of nuclear technology and information was fuelled by the political situation in Italy, where a centre-left government was about to be formed. Thus, the only achievement from the Project 1030 was the construction of two scale models, currently exhibited at the La Spezia Naval Dockyard.

When the Italian Navy was forced to abandon the design of a nuclear attack submarine, it changed its focus to a logistic support ship powered by a nuclear reactor. The design of the ship, to be named *Enrico Fermi*, was again developed by the Constructors Department of the Italian Navy. Length was to be 177m, with a beam of

A model of the nuclear attack submarine *Guglielmo Marconi*, exhibited at La Spezia. The hull configuration clearly shows her *Skipjack* origins. (M Minuto).

22.5m and a displacement of 18,000 tonnes. An 80MW nuclear reactor would supply steam for two geared turbines and turbo-alternators, developing a maximum horsepower of 22,000shp for a speed of 22 knots. The design continued for a few years, but Washington again denied its support for this programme. A last attempt to develop nuclear propulsion was made in the late 1960s, when France was asked to cooperate through its Atomic Energy organisation. When this approach failed the political conditions to go ahead with nuclear-powered naval vessels quickly vanished, and eventually the Italian Navy abandoned its nuclear ambitions.

Conclusion

The *Marina Militare* envisaged the nuclear attack submarine *Guglielmo Marconi* as an experimental and operational boat similar to HMS *Dreadnought* in Britain. Building only one such boat made little operational sense, and *Guglielmo Marconi* would probably have been followed by at least one other SSN. From an operational point of view, the Italian Navy envisaged a submarine force comprising a pair of nuclear attack boats for long-range antisubmarine warfare and special forces missions, supplemented by a larger fleet of smaller conventionally-powered submarines for the interdiction of choke-points in the central Mediterranean.

Despite the ultimate failure of the project, Project 1030 and the concept study for an Italian nuclear powered ballistic missile submarine clearly demonstrate the high-level design, technical and scientific maturity attained during this period within the Italian Navy.[12]

Sources:

Italian Navy Staff, Plans and Operations Department Collection, INHO, Rome.
Plans and Designs Collection, INHO, Rome.
Norman Friedman, *US Submarines Since 1945: An Illustrated Design History*, US Naval Institute Press (Annapolis 1984).
Michele Cosentino and Ruggero Stanglini, *La Marina Militare italiana*, EDAI (Florence 1993.

Endnotes:

[1] The transfer of US submarines to Italy began in 1954 (two boats) and continued for the ensuing twenty years, with a total of nine boats.
[2] The reactor was dubbed RTS-1 Galileo Galilei; it was later built at San Piero a Grado, Pisa, where CAMEN moved in 1961.
[3] 'Subtle' was a working group established within AFMED, the Allied Forces Mediterranean Command, and was composed of submariners from the NATO Mediterranean Navies, the Royal Navy, the US Navy, and the RAF. CoMedCent was the then-NATO Command, Central Mediterranean Area, based in Naples and headed by an Italian admiral.
[4] The US Department of State designated the US Embassy in Rome as the entry point for the requests advanced by the Italian Navy.
[5] The shipyard had been heavily bombed in 1943–44 by Allied

Cutaway model of the Italian nuclear attack submarine, showing details of the reactor, machinery spaces, and accommodation. (M Minuto)

air forces and was totally destroyed at the end of the Second World War. However, its workers had contributed to the recovery of many wrecks sunk in the ports of Trieste and Monfalcone.

6 Between May 1956 and October 1961, US naval and private shipyards built five *Skipjack* class submarines.

7 Reducing the wetted surface area of the hull contributed to limiting the submerged displacement of the submarine; together with the greater thrust provided by the axial single propeller, this ensured a higher submerged speed.

8 HY-80 steel has a yield strength of 80,000psi, a figure which, associated with a suitable thickness and safety coefficient of the pressure hull, made possible an operational depth of no less than 600 feet (183m).

9 Underwater operations with Special Forces are a long-standing tradition in the Italian Navy.

10 The reference was to the then-strong Italian Communist Party, which was opposed to the missile's deployment. Thirty ballistic missiles, operated by the Italian Air Force under a US-Italian double key arrangement, were eventually based in Puglia and Basilicata, in Southern Italy. They, and those deployed in Turkey and Britain, would be removed in the spring 1963 in the aftermath of the Cuban missile crisis.

11 The Italian Navy developed a 'surface' variant for a deterrent naval force, fitting the refurbished missile cruiser *Giuseppe Garibaldi* with four silos for ballistic missiles. The work was actually carried out at the La Spezia Naval Dockyard and the missiles would initially be transferred from the US Navy. Trials with dummy and actual weapons (without warheads) took place successfully in the early 1960s, but Washington subsequently vetoed the transfer of Polaris and this programme had to be abandoned.

12 The name *Guglielmo Marconi* was finally assigned to a conventionally-powered submarine of the *Sauro* class, which served in the *Marina Militare* from September 1982 to October 2003.

THE 340mm COAST DEFENCE BATTERY AT CAPE CÉPET

The battery at Cape Cépet, employing some of the largest French naval guns, was a key element in the defences of the naval base of Toulon during the 1930s, and under the German occupation was integrated into the coastal defence network known as the *Südwall*. **John Jordan** looks at the development and subsequent history of this impressive complex.

Batteries of heavy guns had traditionally been a key part of France's coastal defences. In combination with *défenses mobiles*, comprising coast defence battleships and massed torpedo craft, they were intended to deter and defeat an attempt at close blockade or an assault on French ports and harbours from the sea. They often utilised naval guns from decommissioned battleships, which were mounted in specially-commissioned, purpose-built turrets, and the large-calibre guns generally had naval crews.[1] Such was the importance of coast defence for French security that it was funded from a separate budget, *Défense des côtes*.

The year 1920 saw the most ambitious French coast defence programme to date. The centre-piece was a proposal to deploy five turrets each mounting two 340mm (13.4in) Mle 1912 guns – the same model carried by the three 'super-dreadnoughts' of the *Bretagne* class – in the area around Toulon,[2] with additional guns of the same calibre for Bizerte. The guns were to come from the uncompleted battleships of the *Normandie* class, each of which was to have carried twelve 340mm Mle 1912 guns in three quadruple turrets.

In the event the 1920 programme was severely curtailed due to budgetary constraints, and only two twin turrets were ordered for metropolitan France, with a further two for the El-Metline battery at Bizerte. The former were to be located on the heights of the south-eastern corner of the Saint-Mandrier peninsula, and were intended to protect the approaches to Toulon. Mounting the guns in naval-type turrets meant that they provided 360-degree coverage, and were therefore equally capable of being trained on the anchorage and inland beyond Toulon itself.

The two turrets at Cap Cépet were the visible element of a large and extensive network of underground concrete fortifications to house the gun crews, the ammunition, and a two-part Battery Command Post (*Poste de batterie*), all of which required major and costly excavation work. There was sufficient separation between the turrets – in excess of 300 metres – for them not to constitute a single target for aerial or sea attack. The western turret complex was designated *Ouvrage C*, and was intended to cover Cavalas Bay to the south of the peninsula; the eastern complex, *Ouvrage F*, covered the southeast of the peninsula. Some 800 metres to the south of them, on the tip of Cape Cépet east of the existing fort, were a fire control post (*Ouvrage D*) and a command post (*Ouvrage E*) that served both turrets. The two turret blocks were connected by an underground gallery with an above-ground entrance half-way between them. Communications with the Battery Command Post were by radio or telephone only.

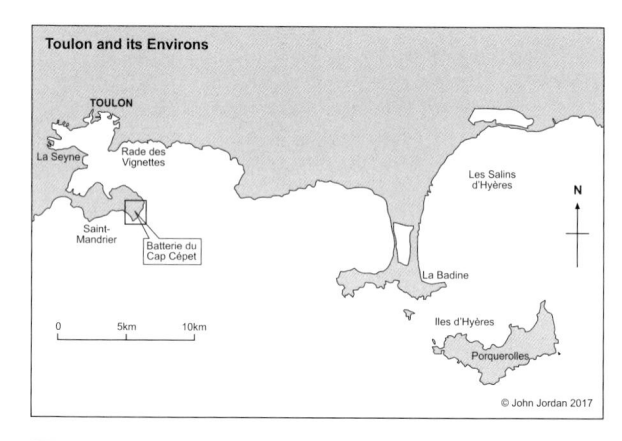

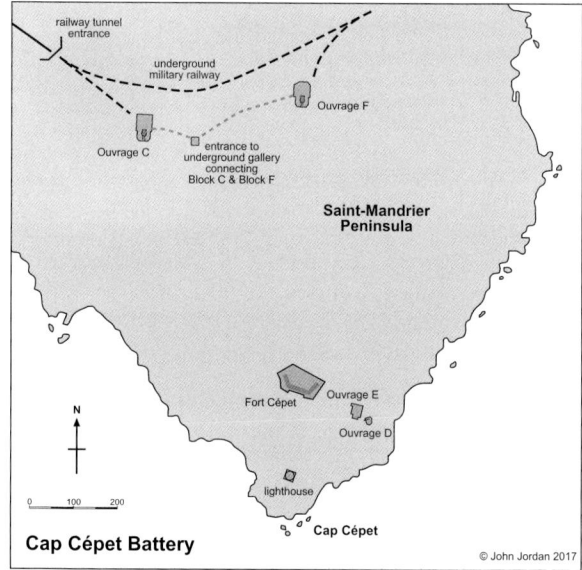

The contract for construction was finally awarded in 1928, and mounting of the turrets began two years later:

34cm Turret: Cap Cépet

Profile View

View from Front

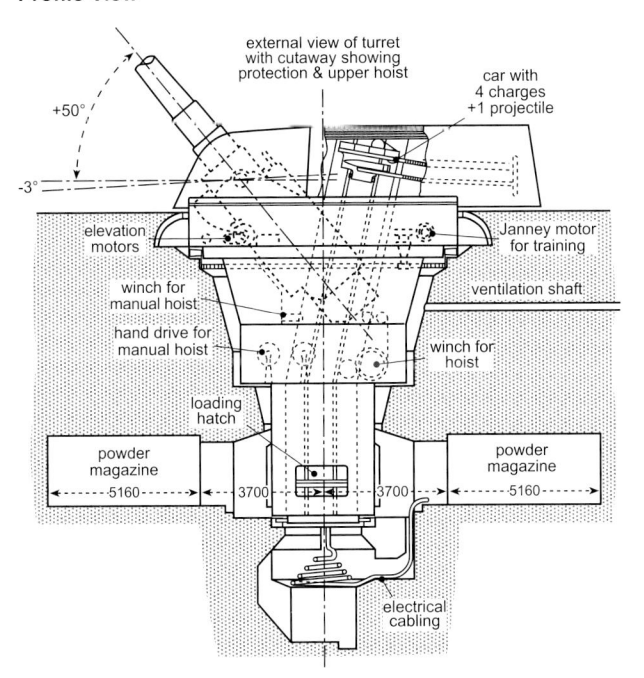

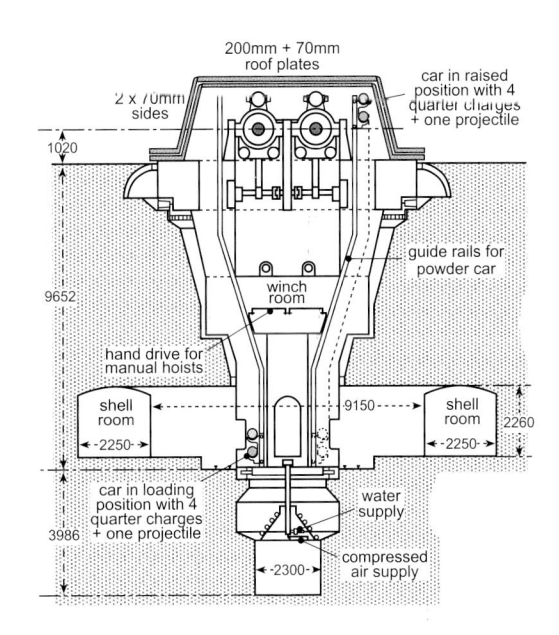

Note: Adapted from official plans.
All measurements are in millimetres.

© John Jordan 2016

the first turret was installed in February 1931, the second in the autumn of the same year. The first test firing of turret C took place on 15 December 1931. Changes and improvements were subsequently made. Work was completed in September 1933, and the battery was declared operational in March 1934.

The Guns and the Turrets

The 340mm Mle 1912M guns intended for the battleships of the *Normandie* class were built by Ruelle, the Navy's own artillery works, and by the private company of Saint-Chamond. It is not clear how many were ordered before the programme was suspended, but we know that twelve guns were commandeered by the Army with a view to mounting them on trains (ALVF: *artillerie lourde sur voie ferrée*) – six were delivered during 1918 – and that the remaining guns were retained in storage; they were subsequently used to rearm the battleships of the *Bretagne* class during refits that took place between 1931 and 1935.

Individual guns were designated using the initial(s) of the manufacturer, the year of manufacture, and a gun number sequence relating to the year. The guns mounted in turret F were SCD 1913 no 1 and 1914 no 2 built by Saint-Chamond, and were from the original *Normandie* order. However, the guns in turret C were R.1922 nos 1 & 3, and were therefore clearly part of a supplementary order placed with Ruelle post-war. All four guns were on a newly-developed mounting designated TC Mle 1924.

The twin turrets were designed and manufactured by Schneider, and received the designation 340 *tourelle* C Mle 1924. Like their naval counterparts, the turrets were powered electrically in train; Janney motors were used for elevation. Protection was optimised for direct attack from the sea or from the air. The turrets had two layers of armour plating, with the thickest protection reserved for the face and roof of the gunhouse. The armour on the sides and the rear of the gunhouse comprised two plates 70mm thick (total 140mm), and this was extended outward over the turret roller path. For the face and the roof of the turret the upper layer was replaced by 200mm plates, giving a thickness of 270mm. The total weight of armour was 163 tonnes.

The rotating part of the turret comprised the armoured gunhouse and the trunk for the ammunition hoist, which was secured to the gunhouse platform and turned with it. Total weight of the revolving structure was 602 tonnes. Access to the gunhouse was via a rectangular opening in the rear wall of the turret that was closed by a sliding door.

The twin hoists emerged outside the guns, and the upper part of the hoist was angled at 12 degrees so that the APC Mle 1924 projectile, which weighed more than half a tonne, could be slid backwards onto the waiting tray. There was a loading carriage (*chariot de chargement*) on the axis of each gun with separate compartments for the projectile, the four-part bagged charge and the rammer.

Finally, each of the turrets had a shielded 100mm Mle 1897 gun mounted atop the gunhouse for sub-calibre firing.

Ammunition and Handling

The armour-piercing shell fired by the 340mm Mle 1912 guns of the battleships of the *Bretagne* class when they first entered service was a standard short-bodied steel projectile with a hardened cap and a 21.7kg bursting charge of picric acid and dinitronaphthalene (*mélinite*). The guns in these ships could elevate to only 12 degrees, limiting the maximum range of engagement to 14,500m. During the postwar period the elevation of the guns was increased, first to 18 degrees, then to 23 degrees, and the Mle 1912 shell was given a ballistic cap in the shape of a truncated cone (see accompanying drawing), which increased the length in calibres from 3.17 to 3.7, becoming the Mle 1912–1921. Maximum range was now 23,000 metres, but even this was well below the maximum range desired for the new 340mm defence batteries, which being fixed could not close their target. The new Mle 1924 mounting allowed a maximum elevation of 50 degrees, and the modified projectile (also designated Mle 1924) was even more aerodynamic, with a long ogival cap that increased the length of the shell from 1,260mm to 1,496mm (4.4 calibres). This gave the Cap Cépet battery a maximum theoretical range of 35,400 metres using the same 155kg powder charge as the battleships.

The replenishment and loading mechanisms were specially developed to be able to handle this exceptionally long projectile; the battleships of the *Bretagne* class, which had two-stage hoists with upper cage hoists that ran on curved rails directly behind the breech, were unable to accommodate it, and retained the Mle 1912–1921 throughout the remainder of their service lives.

In the Cap Cépet installation, the Mle 1924 projectiles were stowed on tiered horizontal racks in two shell rooms, each of which held 90 projectiles. The shell rooms were on the upper level of the complex and were located on either side of a central handing room; the 576kg projectiles were lifted from the racks and moved to one of the eight waiting trays in the handing room using a winch and overhead rails. The powder magazines, each with a capacity of 400 bagged quarter charges, were on the south and north sides of the turret, so that the passageways connecting the shell rooms and the magazines with the central handing room were in the form of a cross (see plan view of the installation). The powder charges were stowed in metal cases, and each of the magazines had two sets of chutes with rollers on which the charges were transferred to fixed waiting trays in the handing room. As a fire precaution, the chutes passed

One of the 340mm turrets of the Cap Cépet battery during the mid-/late-1930s. Note the shielded 100mm Mle 1897 mounting on the turret roof used for sub-calibre firing. (Courtesy of Philippe Caresse)

THE 340mm COAST DEFENCE BATTERY AT CAPE CÉPET

340mm OPf Mle 1912–1921
with detachable ballistic cap

340mm OPf Mle 1924
with extended ballistic cap

- ballistic cap [detachable]
- fixing ring
- armour piercing cap
- shell body
- bursting charge: 21.69kg mélinite
- driving bands
- base fuse

1260mm | 1496mm

shell weight: 554kg | shell weight: 576kg

© John Jordan 2017

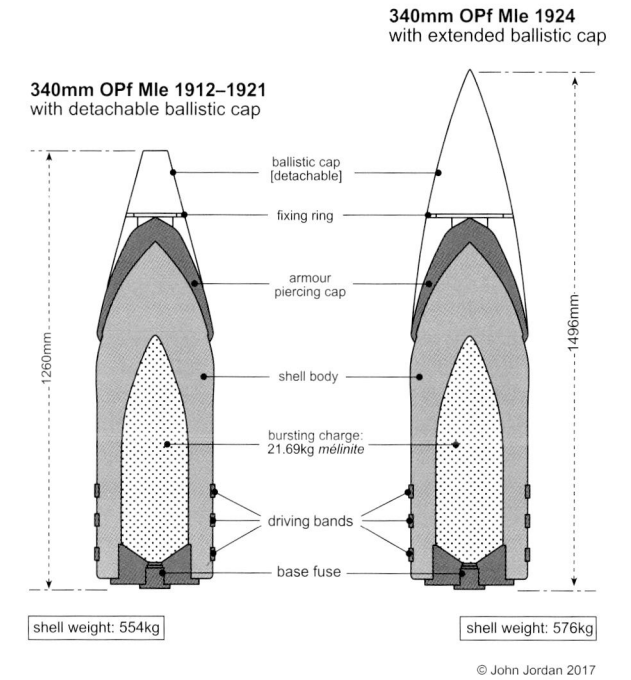

through double flash doors set into the walls of the magazine that operated in such a way that only one was open at a time. The magazines were also equipped with a sprinkler system. In the handing room, two mobile replenishment carriages were used to transfer the projectiles and charges to the hoist cages.

Turret Blocks

The two turret blocks were constructed using the open excavation method. Concrete for the blocks was poured into the excavation and covered by the excavated earth, so that each block formed a small hill topped by the twin turret. The turret block surrounds comprised inner and outer concrete walls 1.5 metres thick with a 2-metre layer of sand between; the wall around the shell rooms and magazines was 2 metres thick, and the roof over the section of the block around the turret was of the same thickness, again with alternate layers of sand and concrete beneath (see drawing). The concrete walls and roof served as a burster, and the purpose of the sand was to absorb any fragments that penetrated the outer layer. Each of the two-storey blocks had a depth of 17 metres; both were approximately 50 metres long, but the overall width varied from 30 metres in *Ouvrage C* to 35 metres in *Ouvrage F*.

The forward end of each block was occupied by the shell rooms and magazines for the turret. Directly behind these, there was an upper level housing the turret commander's quarters and rest and ready rooms for the turret crew, who were accommodated in shelters around the site; all the working and living spaces were ventilated and equipped with gas filters. Stairs at the rear of the block provided access to the lower level, which housed the generators for electrical power and lighting, the switchboard, the fresh water tank, the fuel tanks, the air compressor and the pumps – in French fortification parlance the whole was termed the *usine* or 'factory'.

To the north of the turret blocks a military railway line ran from west to east; spurs connected it to each of the blocks (see map). The section north of the turrets was in a tunnel to provide protection for the ammunition and supplies that it carried. Each block had a long gallery connecting it to the unloading platform for the railway; once off-loaded, the shells and powder charges were moved into the block on wagons that ran on twin Decauville tracks similar to those used for minelaying. In Block C the gallery emerged at the upper level, where the shell rooms and magazines were located. In Block F, however, it emerged on the lower level, and the ammunition was then raised to the shell rooms and magazines via lifts (see drawing). On the opposite side to the gallery with the tracks was a second gallery with flights of stairs to the main ground-level entrance, and a third which provided access to the 370-metre underground gallery that connected the two blocks.

Command Post

The *Poste de batterie*, located close to the tip of the Saint-Mandrier peninsula, to the east of Fort Cépet, comprised the *Poste Central* (PC – *Ouvrage D*) and the *Poste de Direction de Tir* (PDT – *Ouvrage E*). The PDT was above ground, and had a similar function to the conning tower on board a battleship; it comprised an observation chamber constructed of thick concrete and protected with armour plates from the recently-stricken pre-dreadnought battleships *République* and *Patrie*, topped at its rear end by a long-base rangefinder in a rotating turret. The observation room had three CT-style slits, and was equipped with periscopic sights for target bearing, for the Control Officer and for an inclinometer. The rangefinder cupola was accessed by a short vertical ladder to the rear of the observation chamber. The rangefinder was an SOM coincidence model with a 10-metre base. The rangefinder turret was protected by 30mm plating, and steel plates were welded to the lateral arms of the rangefinder and painted to simulate a thick plant.

If the PDT was the conning tower, the *Poste Central* was the transmitting station, its function being to calculate a fire control solution for the guns and to transmit gun-range and angle of train to the turrets using a follow-the-pointer (FtP) system. The PC was housed in an underground concrete block 28.5 metres by 20.3 metres with a height of 4.4 metres (see drawing). Besides the fire control computer in the *poste à calcul*, there was a generator group to provide power and lighting, ventilators and filters, a damage control room, the officers' quarters, a rest room and POs' mess for the PC personnel, and two W/T rooms. The PDT could receive inputs from auxiliary *postes de direction* (PDTA) located around Toulon; target range could then be calculated using triangulation.

The main entrance to the *Poste Central* from ground

Coast Defence Battery at Cap Cépet

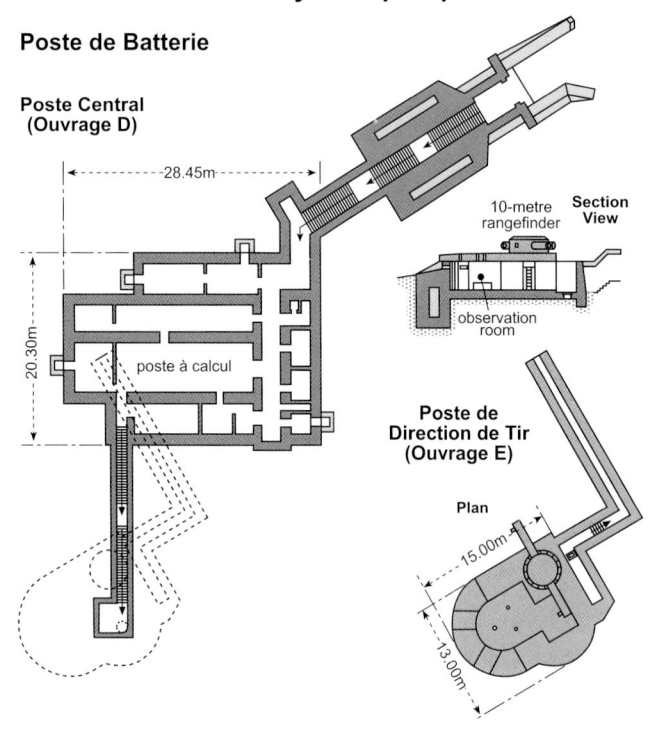

Poste de Batterie

Poste Central (Ouvrage D)

Ouvrage E: Poste de Direction de Tir (PDT)
– observation room with three slits; sights for target bearing, Control Officer and inclinometer
– cupola with 10-metre SOM coincidence rangefinder
– PDT protected by armour plates from battleships *Patrie* & *République;* cupola had 30mm plating and plates secured to outer arms of RF to simulate thick plant

Ouvrage D: Poste Central (PC)
– *poste à calcul* in central bay
south side
– generator group with ventilators/filters
– rest room for personnel
– POs' mess
– two W/T offices
north side
– damage control room
– officers' quarters
– two W/T offices

Ouvrage F: Tourelle de 340mm
upper level
– handing room, shell rooms (each 90 projectiles) & powder magazines
– turret commander's quarters
– POs' mess
– seamen's mess
– sanitation
– fresh water tank
– ventilation machinery room
– emergency post
lower level
– main generator group
– auxiliary machinery room with converters, electrically-driven pump, air compressor, 45hp generator for lighting, battery for emergency lighting & electrical switchboard
– fresh water tanks + water valves
– fuel tanks
– compressed air bottles
– two lifts

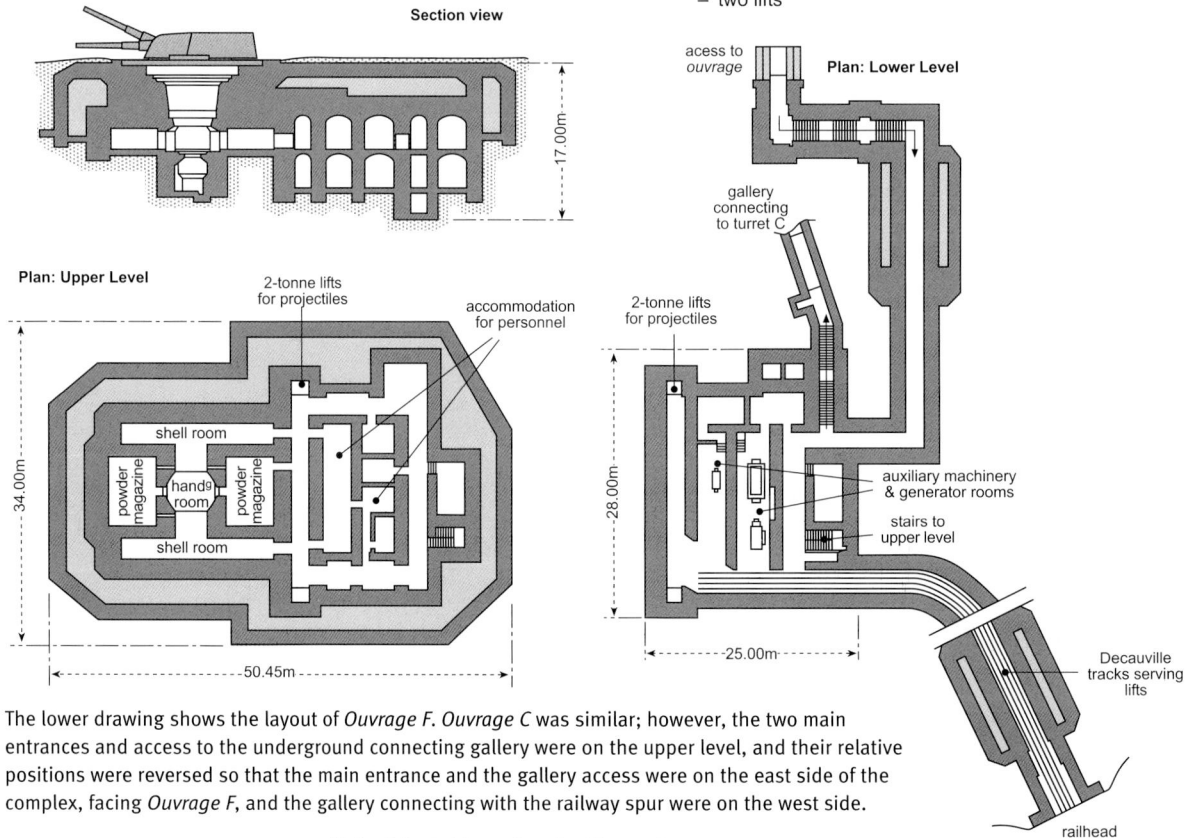

Ouvrage F

The lower drawing shows the layout of *Ouvrage F*. *Ouvrage C* was similar; however, the two main entrances and access to the underground connecting gallery were on the upper level, and their relative positions were reversed so that the main entrance and the gallery access were on the east side of the complex, facing *Ouvrage F*, and the gallery connecting with the railway spur were on the west side.

© John Jordan 2016

Note: Adapted from plans by J Laurent, published in Chazette & Gimenez, *Südwall*.

THE 340mm COAST DEFENCE BATTERY AT CAPE CÉPET

level was in the northeast corner of the block. Access was via an armoured door, followed by a gas-tight door; a staircase 13.5 metres high then led down to the north corridor of the PC. The PDT, which was offset to the southeast, was accessed by a vertical ladder at the end of a gallery staircase two metres high and 1.1m wide.

In Service

Although the Cap Cépet battery was operational from 1934, the guns were never fired in anger during the early phase of the Second World War. Under the terms of the Armistice of June 1940 the battery was to be demilitarised. However, the British unprovoked assault on Mers el-Kebir prompted the Germans to reconsider plans for the defence of Toulon, and following German pressure on the Italians, the battery remained in service under the Vichy Government, albeit with reduced manning.

When the German Panzers seized Toulon on 27 November 1942, the French commander ordered the battery to be sabotaged. Explosive charges placed in the guns of turret F damaged the rifling of both guns, but the guns in turret C suffered only minor damage, and trial firings of the latter turret were carried out under German supervision from 8 December 1942. The guns in turret F were so badly damaged that it was decided to replace them with two guns from the sabotaged battleship *Provence*, suitably modified by Ruelle.

On 15 December 1942 the Italians took charge of coast defence for the Toulon area, but showed little inclination to bring the Cépet battery up to full strength. Turret F remained out of action, and this became a source of concern for the Germans, who saw the battery as key to the protection of the base of 29.U-Flotilla. It was only at the beginning of August 1943 that the *Direction des industries navales* (DIN)[3] Toulon received an order to replace these guns. A few days later, following the start of Operation 'Achse', the Cépet battery was taken over by the German Captain-Lieutenant Ludwig Rubenbauer.

The takeover was hindered by the complexity of the 340mm mountings and their associated equipment, and by a shortage of suitable personnel. Despite their efforts, the Germans had difficulty finding an interpreter familiar with the technical aspects of the battery; moreover, there was only a single copy of the instruction booklet, which therefore had to remain with the battery. However, work to bring turret F back into service began again on 22 September 1943. The first of the guns salvaged from *Provence* left Toulon for Ruelle on 9 November; it returned around mid-January 1944 to be mounted immediately in turret F (now re-christened *Friedrich*[4]), from which the sabotaged guns had been removed in

A photo of *Turm Cäsar* taken after the surrender of the battery. The turret training mechanism had been jammed following a direct hit by a 2000lb bomb on 16 August. Part of the 200mm plate at the corner of the turret is missing, and the 70mm side plates have been stove in and dislodged. The following day the Germans sabotaged the undamaged right-hand gun by placing 50–60kg of explosive in the breech. The US officer standing by the side of the gun gives an indication of the size of the 340mm turrets. (Courtesy of Philippe Caresse)

A more distant view of *Turm Cäsar*, showing the earth surrounding the complex pock-marked by shells and bomb craters. (Courtesy of Philippe Caresse)

December 1943. The second *Provence* gun returned from Ruelle in mid-February and was in place by the end of the month. The first firing trials of turret *Friedrich* duly took place on 13 April 1944, but they were not entirely satisfactory, and there was another firing on 4 May to make the necessary adjustments.

Each of the two turret complexes was provided with land defences in the form of heavy machine guns and Pak 112 anti-tank guns, and with air defences – a mix of French 25mm Hotchkiss Mle 1938 and German 37mm FlaK guns. There were also two 60cm searchlights to illuminate surface and aerial targets at night. Each of the turret complexes was manned by around 140 officers and men. However, the latter were of variable quality: there was a lack of homogeneity in the ages, skills, and service of the personnel. There is no record of firing exercises, and it is reported that the Germans found the unfamiliarity and complexity of the gun mountings daunting.

Attempts at Sabotage

As the Allied invasion of southern France drew near, the French liaison team from DIN Toulon led by Engineer Henri Lafaurie attempted to sabotage the guns of turret 'F', on which they had continued to work under German supervision. The run-out cylinder of the left-hand gun was found to be defective due to a leaky joint and, as there was no spares, the French workshops were directed to produce a joint that would deteriorate rapidly on firing. The Germans spotted that this joint was not functioning properly during trials on 14 July and summoned Lafaurie for an explanation. However, acts of attempted sabotage continued until 12 August, three days prior to the invasion. The French chief of works deliberately misdirected an operation by the German gun crew that made it impossible to repair the run-out cylinder on site. At the same time one of the elevation motors burst into flames, provoking considerable displeasure on the part of the Germans.

It proved impossible to get the damaged run-out cylinder through the turret door because the guns from

A close-up of *Turm Friedrich* taken after the surrender. The left-hand gun had long been immobilised due to a defective recuperator. However, the right-hand gun succeeded in getting off more than 200 rounds against the Allied land and naval forces. (Courtesy of Philippe Caresse)

Provence had a different configuration to the originals. The Germans had to cut a hole in the armour at the rear of the turret. The run-out cylinder eventually arrived at a workshop, but the specialist had injured himself and was replaced by an unqualified colleague. The run-out cylinder was finally put back into place on 11 August but there was a further leak, and the French informed the Germans that a new run-out cylinder would have to be found.[5] From 15–20 August the Germans were working feverishly on this gun under the instructions of an engineer, but when the Allied fire support ships arrived off Saint-Mandrier the left-hand gun of turret F was immobilised, the work having effectively been halted by firings with the right-hand gun and by the constant aerial and surface bombardments.

Operation 'Dragoon'

The first firings of the battery were against Allied troops advancing on Toulon on 16 August. However, at 1704 on that day turret *Cäsar* was struck on the front left-hand side by a 2000lb bomb and immobilised, with the guns still pointing north. The bomb struck the extension to the side armour covering the roller path close to the face of the turret. The armour extension was deformed by 30cm, and part of the 200mm face armour was torn off and projected several metres away. The roller path itself was deformed and the shield for the left-hand gun ripped away. There were cracks in the trunnion support of the latter gun, and the barrel rifling was damaged; the explosion also resulted in five casualties among the gun crew. Turret C was unable to take any further part in the campaign, and on 17 August the Germans sabotaged the right-hand gun using 50–60kg of explosive placed in the breech. The resulting explosion tore off the rear door of the turret and lifted the roof plating.

During the period 14–26 August no fewer than 800 Allied bombs (550 tonnes) and 1,400 shells (350 tonnes) were directed at the two turrets alone. The *Ouvrage F* complex was struck by 14 bombs, but there were no direct hits on the turret except for a hit on the roof from a small-calibre shell that caused only minor damage. The right-hand gun continued to engage the ships of the Allied naval bombardment force throughout the campaign, firing an estimated 218 rounds. The fire from this single gun was largely ineffectual, although several ships reported being 'bracketed'.[6] However, it compelled the ships to keep their distance and to manoeuvre behind smoke screens, which had a negative impact on the accuracy of their own fire. The gun crew of turret *Friedrich* finally surrendered on 26 August, having broken the German record for coast defence firings.

Post-War

There were various proposals to refit turret F after the war, but budgetary constraints precluded this. Turret C was removed in 1951, and it was proposed to refit *Ouvrage C* for the sector defence of Toulon and the *Poste à calcul* as a fire control and information centre (*Centre de Renseignement et de Direction de Tir* or CRDT). The project to refit turret F was abandoned in 1954 and the turret was removed. *Ouvrage C* was subsequently used for the stowage of ammunition between 1957 and 1965. In 1966 there was a new proposal to refurbish *Ouvrages C & F*, but in 1969 *Ouvrage C* was filled with earth, the connecting gallery was sealed off and the original entrances to F were blocked with rubble. The main entrance was unblocked in 1991, and *Ouvrage F* is still used by the Commando Hubert; the *Poste de batterie* has also survived.

Sources:

Chazette, Alain & Gimenez, Pierre, *Südwall: Les batteries côtières de Marine françaises et allemandes*, Editions Histoire & Fortifications (Vertou 2010).

Kaufmann JE & Kaufmann HW, *Fortress France: The Maginot Line and French Defenses in WWII*, Stackpole Military History Series (2007).

Zaloga, Steven J, *The Atlantic Wall (3): The Südwall*, Osprey (London 2015).

Endnotes:

[1] One of the technical specialisms in the Navy was *canonnier sédentaire*, charged with the maintenance and operation of coastal batteries; applicants followed a four-month course at the *Artillerie de côte* battalion attached to the port in which they were recruited.

[2] The installations in the Toulon-Hyères area were to be at Cap Cépet, Sicié, Giens, Porquerolles and Bénat.

[3] The DIN was formed during the war by merging the *Direction de l'artillerie navale* (DAN) *and the Direction des constructions navales* (DCN).

[4] Turret C was renamed *Cäsar*.

[5] Chazette & Gimenez (*op cit*) recount a further series of efforts on the part of the French workforce to sabotage the elevation mechanisms and electrics of the gun; however, these were all discovered and frustrated by the Germans.

[6] The battleship *Lorraine* reported one shell landing only 35 metres away, with another missing by 150 metres; the *contre-torpilleur Le Malin* reported a shell missing by 300 metres.

The French cruiser *Montcalm* under fire from the German-manned batteries at Saint-Mandrier on 20 August 1944. The shell splashes indicate a smaller-calibre weapon than the 340mm gun. (Author's collection)

POWDER MAGAZINE EXPLOSIONS ON JAPANESE WARSHIPS

The Imperial Japanese Navy experienced more than its fair share of magazine explosions. **Kathrin Milanovich** looks in detail at the cases involving major vessels, and recounts the efforts of the various Investigation Committees set up to determine the causes.

Incidents involving an explosion in a magazine were a common occurrence in many of the major navies before the chemical stability of the new propellant powders developed during the late 19th century was improved, and control and handling methods refined. Naturally there were strict control procedures in every navy,[1] and all concerned paid the utmost attention to avoiding a catastrophic explosion. However, such explosions were a regular occurrence in Japanese warships, and the IJN arguably suffered the most serious losses among the principal naval powers.

The first major explosion took place on the flagship of the Combined Fleet, *Mikasa*, at the end of the Russo-Japanese War (RJW), in September 1905. The battleship sank to the bottom of Sasebo naval port. Soon afterwards the former flagship during the Sino-Japanese War (SJW), *Matsushima*, now a cadet training ship, sank in the port of Makô (Pescadores Islands) in April 1908. In October 1912 the battleship *Mikasa* suffered from a second magazine explosion off Kobe but did not sink. This was followed in November by the explosion of a powder magazine on board the armoured cruiser *Nisshin* in Shimizu port; as with *Mikasa* she did not sink. In January 1917 the battle cruiser (formerly armoured cruiser) *Tsukuba* sank in Yokosuka naval port, and in July 1918 the battleship *Kawachi* shared her fate off Tokuyama, both as the result of a powder magazine explosion.

Each case was investigated and countermeasures were taken, particularly after the later sinkings of *Tsukuba* and *Kawachi*, and it appeared that the issues which had been responsible for this series of accidents had been finally

Table: Magazine Explosion Accidents of IJN Warships

Date/time	Name (displacement)	Location	Cause	Dead[1]
11 September 1905 0130	*Mikasa* (15,360 tons)	Sasebo naval port	Fire started 0020, explosion of the after magazine 0130; sunk, salvaged (8 Aug 1906) and repaired.	251
30 April 1908 0408	*Matsushima* (4,210 tons)	Makô (Pescadores)	Explosion of the black powder magazine for 32cm Canet gun; sunk, total loss.	207
3 October 1912	*Mikasa* (15,360 tons)	Kobe	Fire in forward magazine; extinguished at 1843 by flooding; local damage, repaired.	1 (9 injured)
18 November 1912 1850	*Nisshin* (7,750 tons)	Shimizu	Fire in after magazine; local damage, repaired.	1 (18 injured)
14 January 1917 1515	*Tsukuba* (13,750 tons)	Yokosuka naval port	Explosion of the forward magazine; sunk, total loss.	181[2]
7 July 1918 1551	*Kawachi* (20,800 tons)	Off Tokuyama	Explosion of the forward magazine; sunk, total loss.	621
8 June 1943 *ca* 1210	*Mutsu* (39,130 tons)	Hashirajima	Explosion of No 3 main gun magazine; sunk, total loss.	1,680

Notes:
1 The number of dead is the total. It includes victims of rescue, fire fighting & other parties sent from other ships or land.
2 Sources differ; also 164 dead, 35 injured.

resolved. However, 25 years later, in June 1943, the battleship *Mutsu* sank in Hiroshima Bay following the explosion of a powder magazine.

In these seven cases five warships sank with great loss of life, and among them only one was salvaged and recommissioned (see Table).

Explosion on *Mikasa*

Mikasa, Japan's most modern battleship and the fleet flagship during the RJW, entered Sasebo naval port on 9 September 1905 following the completion of repairs. On the following day C-in-C Tôgô Heihachirô left for Tôkyô with his staff. At about 0020 on 11 September an explosion took place near the mainmast. The fire raged uncontrolled and eventually reached the after powder magazine, which exploded at 0137. One hour later the battleship rested on the seabed with only the mainmast and two funnels above water.

The Imperial General Headquarters (IGHQ) announced the accident on the next day, stating that irrespective of the dispatch of fire fighting brigades from the ships anchored in the port and units stationed on land, all efforts to control or extinguish the fire had been in vain because the source of the fire could not be located. At 0137 the fire reached the after powder magazine and the explosion blew a hole in the hull plating to port below the waterline. Large quantities of water flooded in and the ship sank to the bottom of the sea at 0230. The communiqué continued that 'an investigation committee will be established and begin its work immediately …'. The number of casualties was stated to be 599, among them 232 dead or missing, of whom 206 were from *Mikasa*.[2] The crew members who had been near the source of the fire were all dead; the middle and lower decks were impassable. The forward powder magazine had been flooded but no water had entered the after magazine until after the explosion.

Two investigation committees were established. The first was chaired by the CO of the First Fleet, Vice Admiral Hidaka Sônojô, the second by the CO of Yokosuka Navy Yard, Vice Admiral Inoue Yoshika. The second committee began its work after the salvage of the battleship, which was completed on 8 August 1906, and identified the cause of the fire as 'the corded propellant [British cordite Mk I] stowed in the port-side magazine of the secondary [6in] guns. Natural decomposition had resulted in a change in the properties of this powder, and this resulted in an increase in temperature that caused the fire and subsequently the great explosion …'.

This conclusion was in conflict with the account of the Executive Officer of *Mikasa*, Commander Matsumura Tatsuo, who stated: '… around 0020 I was woken by an explosion. It is reported that a fire had broken out some ten minutes earlier. There is a widely-held view that this was a result of arson by a member of the crew …'.

This seems more likely than the official conclusion. It was reported that some members of the crew had decided to hold a party in the powder magazine. In order to

Salvage work is carried out on the battleship *Mikasa* (Takagi Hiroshi & Ushioshobô Kôjinsha)

remove the poisonous part of some methyl alcohol (used for flares) they had stolen, they set light to it in the corridor to the powder magazine and there was a spillage that led to the explosion. The story was told to a medical orderly by a heavily injured member of the crew immediately before he died. The medical orderly initially regarded this confession as confidential and did not immediately report it to his superior. The account was therefore not considered by the investigation committee but emerged later.

Two years and seven months after her sinking, *Mikasa* again became the flagship of the First Fleet on 30 April 1908. It was the day the Navy suffered another catastrophe.

Explosion and Loss of the Cruiser *Matsushima*

The three ships of the *Matsushima* class (*Matsushima*, *Itsukushima* and *Hashidate*) formed the training squadron for FY 1908 and departed Yokosuka for the Indian Ocean on 25 January 1908 for a long-distance cruise, with 176 cadets of the 35th class on board. After calling at Hong Kong, Saigon, Singapore, Colombo, Trincomalee, Batavia and Manila the ships anchored off Makô (Pescadores Islands) on 27 April. The squadron was scheduled to return to Japan via Port Arthur and Dalny and enter Yokosuka on 31 July, thus completing the six-month training cruise.

The day before departure from the Pescadores, 29 April, the promotion of petty officers was unofficially broadcast and there were several celebrations under the supervision of the duty officer. 'This evening the lower deck will not be quiet!' a petty officer is quoted as declaring. At dawn on the 30th a medical officer who was sleeping in a hammock on the lower deck saw a thin wisp of smoke passing over him. At almost the same time he heard a voice in the direction of the entrance to the 32cm Canet main gun magazine shouting 'It's in the hull bottom!' When he came down from his hammock he saw the duty officer ordering members of the crew to take water in buckets.

Thinking nothing of it, he visited the heads and went to the wardroom; the clock in the wardroom showed a few minutes past 0400. On the way back to his cabin to continue his sleep he heard the words 'It's the powder magazine!' near the entrance to the said magazine. He now recognised the danger and made for the sick bay on the upper deck near the bow. Before he entered the compartment a terrible explosion happened in the after part of the ship. It was now 0408.

The explosion was in the powder magazine of the main 32cm gun, which was located aft. It opened up large holes and cracks in the outer plating on both sides of the hull and the ship's bottom, and huge quantities of water flooded in. The ship listed 5 degrees to starboard almost immediately and the stern settled on the bottom. The bow followed; the ship was now resting on the bottom of the anchorage with only the upper part of the mast and the funnel above sea level.

Of the complement of 382 men and 57 officer cadets 207 died,[3] including Captain Yashiro Yoshinori, *Matsushima*'s Commanding Officer since 28 September 1907. Losses among the cadets – 33 out of 57 (60 per cent) – were particularly high; it constituted the biggest tragedy in the history of the long-distance training cruises of the IJN.

The Investigation Committee (*Samon Iinkai*) concluded: 'It is supposed that something combustible was in the vicinity of the magazine and caused the explosion of the powder nearby'. Evidence cited to justify this conclusion was as follows:

Matsushima was one of three coast defence cruisers (*Kaibôjunyôkan*) designed by Emile Bertin for Japan in the late 1880s (see the article by Jirô Itani, Hans Lengerer and Tomoko Rehm-Takahara in *Warship 1990*). Intended as a counter to the armoured ships of the Chinese fleet, they were armed with a single powerful 32cm (12.6in) Canet gun but had only light deck protection over their machinery and magazines. Unlike the other two ships of the class *Matsushima*, which was built at La Seyne in France, had the Canet gun aft and her QF battery of twelve 12cm (4.7in) guns forward. *Matsushima* became a training ship in 1906. (Naval History and Heritage Command, NH 58805)

The upperworks of *Matsushima*, photographed after she settled on the bottom following the explosion. (*Ships of the World*, June 1914)

- all 53 cases containing the powder bags in the magazine of the 32cm main gun were damaged
- the brown hexagonal powder was considered to be very stable
- the keys for the shell room and the powder magazine were kept in a safe in the Captain's cabin, so it was unlikely that anyone could have gained unauthorised access to the magazine; also, before the explosion smoke had been seen in the handing room (*Kyûyo-shitsu*) – see drawing.

It appears that blankets and 24 cloth cases for the 57 cadets were stowed directly outside the powder magazine; however, the cause of the fire in the handing room was not discovered.

According to the investigation report the stowage of blankets in the handing room was requested by an officer who had taken part in an earlier long-range training cruise, and had been permitted by the chief gunnery officer (*Hôjutsu-chô*). However, neither numbers nor content could be determined – evidence that stowage was not closely monitored. In addition, the container for the oily cloths used for cleaning the shell hoist was also stored here. Therefore many dangerous, flammable items were stored in the space next to the powder magazine. According to the evidence of a cadet, candles, matches and open beer bottles were hidden in this room, which was apparently used for parties and celebrations, and there may have been a party on the evening in question. However, it is unlikely that the smoke seen rising from the hoist opening of the handing room and thought to be from two or three cigarettes was directly related to the explosion of the powder in the magazine, which was separated from the handing room by a bulkhead with a wooden lining 76mm (3in) thick.

The Investigation Committee could find no other cause than the supposition outlined above, and the explosion in the powder magazine of *Matsushima* remains unexplained.

The Powder Investigation Committee

When the IJN used mechanically mixed powder[4] no accidents took place; however, the situation changed following the adoption of chemical powders with a volatile or non-volatile solvent. The primary advantages

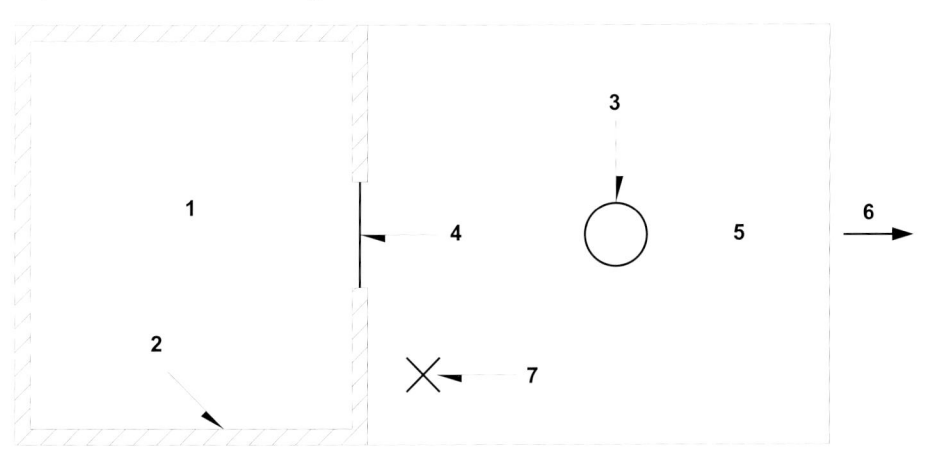

Rough drawing of *Matsushima*'s powder magazine and handing room for the 32cm Canet main gun. (Drawn by Waldemar Trojca)

Key:
1 32cm powder magazine
2 Interior trim (76mm wood)
3 Shell hoist
4 Access door
5 Handing room
6 Bow
7 Stowage of woollen goods, etc

of the new chemical powders were the lower generation of smoke and nearly complete residue-free combustion in the firing chamber, which contributed to increased firing cycles and an improvement in the hitting rate. However, while the older powders did not decompose naturally, a downside of the new smokeless chemical powders was a tendency to natural decomposition after long periods of storage. This feature was dependent on a number of factors, of which the most influential were (i) the chemical composition of the powder, and in particular the addition of a stabiliser, (ii) the temperature in the powder magazine, and (iii) stowage and handling procedures.

Following the *Mikasa* and *Matsushima* accidents and similar incidents in other navies, the Navy Ministry established, in May 1908, the Powder Investigation Committee with the aim of preventing further disasters. The committee comprised 31 members with the Head of the Navy Technical Department, Vice Admiral Kataoka Shichirô,[5] as chairman, Rear (later Vice) Admiral Kawashima Reijirô as the chief secretary and Captain Takamatsu as the special secretary. The members of the committee were specialists from all branches and their brief was to study ways of improving the stability of the powder, to investigate inspection procedures for the stowage of propellant and the layout of the magazines, and to propose improvements. As a result, the basic testing method formerly used, the heat resistance method associated with Abel, was superseded by the silver bottle testing method,[6] refrigeration of the magazines was improved,[7] the procedures for raising the powder bags from the magazine to the loading tray of the guns were modified, and improvements to the stability of the propellant proposed, but the latter proposals were somewhat vague.

Following these recommendations the IJN established the Powder Testing Laboratory (*Kayaku Shikenshô*) in Kure Navy Yard on 28 April 1909. Its brief was to investigate methods of powder stowage and to conduct relevant trials, to study methods of chemical analysis, and to oversee the heating and other physical tests for all naval powders. On 2 December 1912 Chief Tanaka was transferred to the Shimose Powder Factory and Lt-Cdr Hatano Sadao took over the post. The activities of the Powder Testing Laboratory were then expanded so that there were similar investigation and testing facilities in all of the Japanese naval dockyards.

The intensified research and testing quickly brought about the appearance of a new stabiliser called *jara jara* (beta naphthol methyl ether) that was introduced by Kususe Kumaji in about 1912/13. After extensive testing Two-Year-Type Powder (*Ni nen shiki kayaku*) was formally adopted on 12 September 1917. This powder (abbreviated C_2/T_2),[8] composed of 30% nitro-glycerine, 65% gun cotton, 3% mineral jelly and 2% *jara jara*, constituted the first Japanese-type powder[9] to be manufactured purely as a result of independent Japanese research and development.[10]

Explosion and Loss of *Tsukuba*

Before the adoption of the new powder, on 14 January 1917, the battle cruiser *Tsukuba*, while anchoring in the naval port of Yokosuka, experienced an explosion in the forward powder magazine for the secondary guns at 1515, closely followed by the explosion of the powder magazine for the main guns. *Tsukuba* heeled over to port immediately and settled on the bottom, bow first; 151 members of her crew died.

The investigation committee excluded natural decomposition of the powder and malfunction of the electrical system as causes; the investigation found that temperature and humidity in the magazines had been normal. After further investigation attention was focused upon a particular member of the crew. His body had not been found inside the ship, and it was surmised that his body had probably been pulverised by the explosion. In the morning of that day he had been reproached severely by his superior for theft. Judging from his behaviour, the probability was that this was a planned suicide. After investigating his character, behaviour, and his relation-

Tsukuba was the first of four powerful armoured cruisers built in Japan and completed between 1907 and 1911 (see the article by Jirô Itani, Hans Lengerer and Tomoko Rehm-Takahara in *Warship 1992*). They had a main armament of four 12in guns in twin turrets fore and aft; *Tsukuba* and her sister *Ikoma* had a secondary casemate battery of twelve 6in QF guns. All were reclassified as 'battle cruisers' in 1912. (Naval History and Heritage Command, NH 58800)

The explosion on board *Tsukuba*. (Kure Maritime Museum)

Tsukuba following the explosion. (Kure Maritime Museum)

ships with his family, the investigation committee concluded that the magazine explosion had been caused by intentional suicide.

The committee directed its attention to the ease with which members of the crew could enter even a locked powder magazine, and the regular occurrence of irregularities such as drinking inside the powder magazine. However, it was recognised that under normal circumstances concern for their own lives meant that crew members would know better than to put the ship at risk. This was the first time that these irregularities had been acknowledged, and they cast doubt on the claims in other reports that access to magazines was strictly regulated.

Explosion and Loss of Battleship *Kawachi*

The dreadnought-type battleship *Kawachi*, registered as a ship of the Yokosuka Naval Station and assigned to the First Fleet, departed Yokosuka in July 1918 together with the flagship *Yamashiro*, *Fusô*, *Ise* and *Kawachi*'s sister *Settsu* to conduct training in the Kyûshû area.

The scene of the rescue operation of *Kawachi* directed by Captain (later Rear Admiral) Kutsumi Masao. He was the commander of the Kure naval port (*Kure-Kômu-buchô*) at that time. In the background the salvage vessel (*Zatsuekisen* = multi-purpose vessel) *Itabashi Maru* can be seen. She was a war prize from the Russo–Japanese War of 1904–05 and the former German sullage lighter *Industrie*. (*Ships of the World*, May 2009)

Kawachi's capsized hull seen from the port side, with the bow on the right. The battleship *Fusô* is on the left of the picture. The position in which *Kawachi* sank is documented as 60 degrees NE of Hebijima at Tokuyama Point and 42 degrees NW of Kurokamijima Point. *Ships of the World*, May 2009)

Kawachi entered Tokuyama Bay (Yamaguchi prefecture) in the evening of 11 July. In the early hours of the following morning the crew prepared for the towing of a target for torpedo firings (*Gyorai-hyôteki*). However, the firings were abandoned due to an unfavourable weather forecast, and the crew was engaged in routine tasks after finishing preparations against the incoming storm.

At 1551, when the crew was putting everything in order, a small explosion was heard in the forward main gun turret to starboard, followed by a bigger explosion two minutes later. Yellow smoke rose between the first and second funnels, soon followed by brown smoke and fire from the said turret and from all three funnels, and hot gases emanated from all openings in the upper deck.

The explosion caused a large fissure in the hull with the position of the powder magazine of the turret at its centre, as shown in the drawing. *Kawachi* began to heel over to starboard and capsized quickly into the mud of the sea bottom with only the port-side bilge keel, propeller and one half of the rudder above the sea level.

Only four minutes passed between the explosion and the sinking, so there was no time for countermeasures by the crew. The sea was quite rough; however, 433 men were rescued by the efforts of other ships in the bay.

Among the survivors were men who escaped through openings in the hull up to four hours after the ship capsized. There were also crew members trapped inside the hull who answered the knocking from outside; however, these could not be rescued and died. A total of 600 men became victims of the explosion, of whom 109 were missing presumed dead because their bodies were never found.

The C-in-C of the First Fleet set up an Investigation Committee on 13 July. Because the flag staff of the First Fleet was fully engaged in the training exercise, the C-in-C Kure Naval Station, Vice Admiral Katô Teikichi, was asked to take the chair on the advice of the Bureau of Naval Affairs. The members of the committee considered all the possible causes and investigated them. Ignition by natural decomposition of the powder was thought to be the most likely cause, so this was the primary focus of the preliminary investigation.

The MDC powder used as propellant for the shells of the main guns was known to be prone to natural decomposition when stored over a long period of time. The powder was mixed with a mineral jelly stabiliser; however, this stabiliser was by no means perfect, so the normal procedure was to perform a stability test every six months, with further checks at shorter intervals.

Kawachi's capsized hull seen from the opposite side with the stern on the left and the bow facing 62 degrees southwest. As in the previous photos several vessels can be seen participating in the rescue work. Apart the salvage ship, various tugs (*Eisen*), small transports (*Unkasen*), station vessels (*Denmasen*) and other miscellaneous craft took part. The battleship on the left of the photo is either *Fusô* or *Ise*. (*Ships of the World*, May 2009)

POWDER MAGAZINE EXPLOSIONS ON JAPANESE WARSHIPS

This photo was taken from a position near the stern of the capsized hull, looking towards the bow. *Kawachi* capsized shortly after the explosion, and many members of her crew were trapped within the hull. In the foreground there is an oxygen flask. Holes were bored in the hull at several points in order to rescue trapped crew members and oxygen was blown into the hull. However, most of these attempts were unsuccessful. (*Ships of the World*, May 2009)

It was found that one part of the powder that had been stowed in the magazine was between five and seven years old; however, it had shown no sign of any change when its stability had been tested in January/February of that year. Temperature and humidity inside the magazine were also found to be normal.

The most widely-published image of the capsized hull of *Kawachi*, showing the rescue teams on her bottom and various types of craft assisting. Note the tent and the trestle in the foreground. The purpose of the latter is uncertain; it may have been used to measure the list, which gradually increased. To the extreme right of the photo a destroyer of the *Nara* class is partially visible. (Kure Maritime Museum)

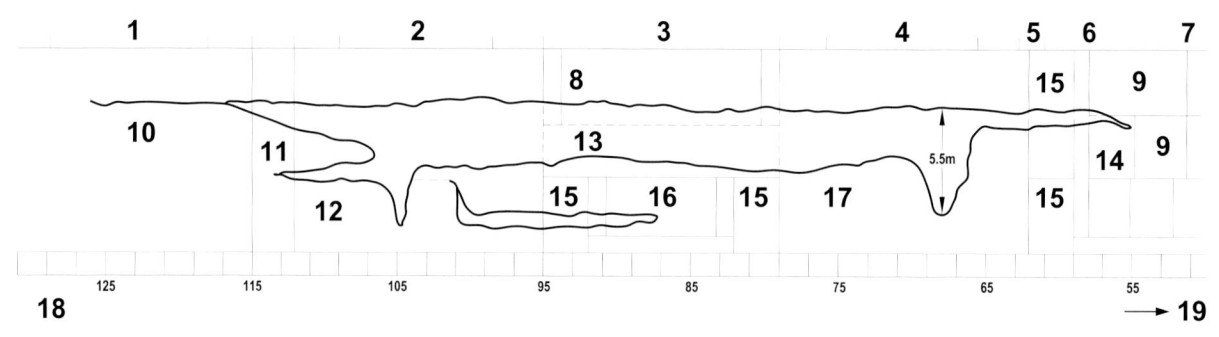

The state of damage to the hull after the explosion. (Drawn by Waldemar Trojca)

Key:
1 No 3 funnel
2 No 2 funnel
3 Upper forward part of std gun turret
4 No 1 funnel
5 Foremast
6 Upper bridge
7 Lower bridge deck
8 Forward part of std 12in shell room
9 Forward part of 4.7in shell room
10 After boiler room
11 Pumping station
12 Centre boiler room
13 Forward part of std 12in magazine
14 Forward part of 4.7in magazine
15 Coal bunker
16 Forward part of std auxiliary magazine
17 Forward boiler room
18 Frame number
19 Bow

On the day the accident took place the ammunition handlers in the shell and powder magazines of the main gun turret in question had undertaken a replenishment practice, but at the time of the explosion the training session had been completed and the magazine doors had been closed and locked. Defective procedures in propellant handling during training could therefore be ruled out. Any attempt at arson would have involved a member of the crew gaining unauthorised access to the magazine.

As decomposition of the propellant and a handling error were ruled out, the focus of the *Kawachi* Explosion Investigation Committee (*Gunkan Kawachi Bakuchin Jiken Sateisho*) now shifted to possible arson. The character and behaviour of every member of the crew, including the survivors and those who had died, was investigated, and the civilian and military police (*Kenpeitai*) were called in. More than 350 offices in Tôkyô and other cities in 25 prefectures were instructed to investigate the background of the crew. Some 1,400 items of mail originating from survivors of the explosion were examined, and members of the powder handling crew of No 1 turret, who were prime suspects, were repeatedly screened for security and shadowed for weeks. However, no possible suspects were uncovered, and the committee returned to the powder.

The powder that had been stowed in the magazine was found to have experienced no malfunction in the annual firing exercise (*nendo shageki*), and had passed all the standard tests. However, because of its age (more than five years) the committee cast suspicion on the powder and concluded that it was the most probable cause of the accident.

Following the publication of the committee's findings, magazine procedures were revised and supervision of the keys became more rigorous. Only ammunition handlers were permitted access to the magazines.

Following the sinking of *Kawachi* a Rescue Unit (*Kyûnan-tai*) was organised and headed for the spot on board the salvage ship (*Kyûnansen*) *Itabashi Maru*. The Navy Ministry ordered an investigation into the possibility of salvage on 18 July. The report was submitted on 6 August, and confirmed that salvage would be a very difficult and dangerous undertaking. Other negative factors were (i) the postponement of the construction of a new battleship for more than a year because of the repair effort, and (ii) the shortening of the life of the ship by three to four years compared to her sister *Settsu*, which had been completed around the same time. It was therefore decided to dismantle the wreck, and *Kawachi* was removed from the Navy List on 21 September.

The Second Powder Investigation Committee

The successive explosions on board the battle cruiser *Tsukuba* and the battleship *Kawachi* were a greater shock than the sinking of the battleship *Mikasa* and the loss of the coast defence ship *Matsushima* some years previously, as it was believed that, given the changes in the composition of the propellant and stricter stowage and handling procedures, this kind of catastrophic explosion was no longer possible. The Navy Ministry ordered the establishment of a second Powder Investigation Committee immediately after the explosion of *Kawachi*, to be chaired by Admiral Murakami Kakuichi. Detailed investigations were undertaken into the production, storage, and handling of propellant, and as a result the specifications of the materials, powder production, storage facilities and handling procedures were again modified. Research into improved stabilisers, and the development of means to prevent premature explosion such as ignition safety were practised, and new regulations promulgated concerning quality control, test production, stability and trajectory tests, inspections and monitoring of propellant storage on board ship. A systematic control system[11] was devised to prevent further accidents; naturally, it took time to implement all this, but they appear to have been effective.

Explosion and Loss of the Battleship *Mutsu*

The loss of the battleship *Mutsu* during the Second World War was the IJN's biggest accident with regard to the displacement of the ship and number of dead.[12]

On 8 June 1943 *Mutsu* was moored to the buoy for the flagship near Hashirajima. At about 1210 she was shaken by a major explosion in the after part of the ship. The hull was severed abaft the No 3 main gun turret by the force of the explosion. The fore part capsized to starboard and sank. The stern section rose out of the water and sank more slowly. Four hours after the explosion *Mutsu* had disappeared beneath the waves.

When *Mutsu* exploded and sank the following four ships were in the vicinity: the battleship *Fusô*, the light cruiser *Tatsuta*, and the destroyers *Wakatsuki* and *Tamanami*. At first it was thought that the ship had been torpedoed by an enemy submarine. Consequently, the destroyers were ordered to drop depth charges, while at the same time the survivors were rescued. In addition to her own complement of 1,321, a further 153 men of the 10th Harbour Defence Unit were on board for training and of the total of 1,474 officers and men, 1,121 became victims of this catastrophe and only 353 survived.

An investigation committee with Admiral Shiozawa Kôichi in the chair was established and began its enquiry. Basing on the testimony of survivors, the committee concluded that if a fire had broken out in the neighbourhood of the No 3 and No 4 main gun turrets and caused such a catastrophic explosion, it must have begun in the powder magazine of No 3 main gun turret. The most likely cause was self-ignition of the Type 3 shell (3 *Shiki-Dan*) stowed in that magazine.

The Type 3 shell was a shrapnel (*Sankai-Dan*) shell recently developed for defence against air attacks. Inside the body of the 40cm shell there were 735 hollow steel tubes 20mm in diameter with a length of 90mm stacked in layers and filled with white phosphorus. The incendiary bodies were expelled forwards from the detonation point and formed a cone-shaped danger zone with a diameter of up to 240 metres at the end of their trajectory. However, despite the spectacular visual effects, the danger to an attacking aircraft was low.

Large numbers of Type 3 shells were stowed in the magazines of battleships and heavy cruisers but, following the destruction of *Mutsu*, all had to be landed by order. The investigation committee conducted repeated experiments at the Kamegakubi Experimental Range of Kure NY in order to confirm the self-ignition of this shell as the most probable cause of the explosion. Models of the original size were produced and numerous experiments carried out. In parallel a colour-burning experiment was executed in the presence of several dozens of survivors. The colour of the smoke generated by the burning of the powder of the Type 3 shell was white, while that emitted by the propellant for the standard projectiles (common and AP) was brown. *Mutsu's* survivors confirmed that the smoke emitted when the magazine exploded was brown. The tests also failed to generate self-ignition of the Type 3 shell, and it was absolved as the cause of the loss of *Mutsu*; as a result, the Type 3 shell was again embarked on major IJN warships.

A later investigation by divers found that the third

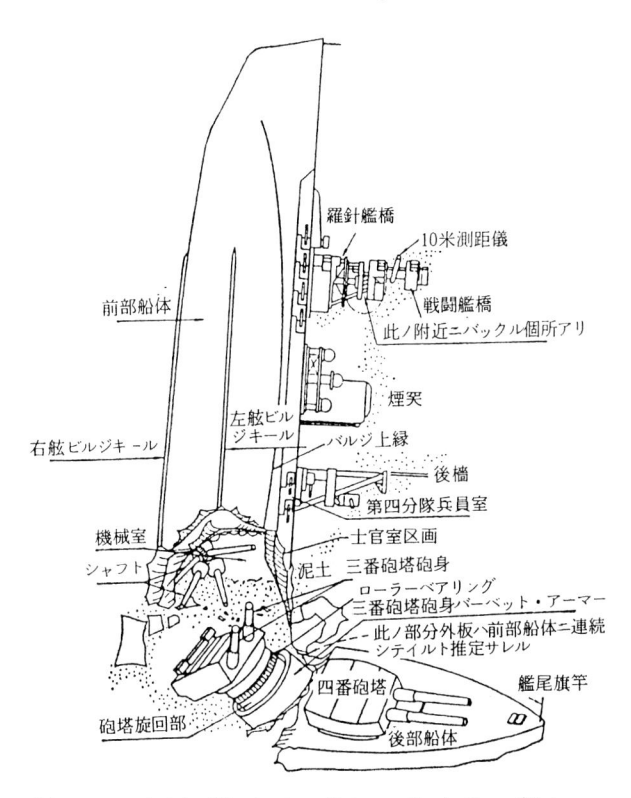

A Japanese sketch of the broken *Mutsu* on the bottom. (Mutsu Memorial Museum)

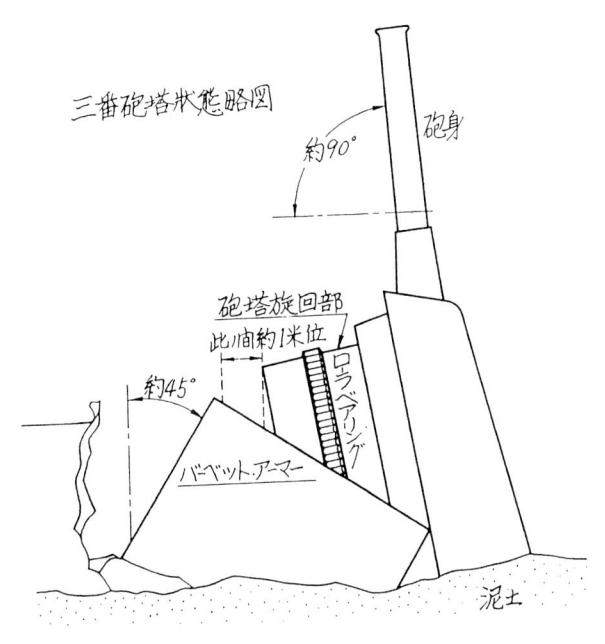

A Japanese sketch of *Mutsu's* No 3 main gun turret as found. Gun, turret and barbette were later salvaged and are currently exhibited. (Mutsu Memorial Museum)

main gun turret and its barbette had been separated from the hull and was damaged. This discovery served to confirm the assumption that the explosion had taken place in the powder magazine below No 3 main gun turret. However, the true cause of the explosion could not be established and is still unknown; again, arson or decomposition of the propellant were suspected.

The IJN investigated all items relating to naval powder from the administration to the production including handling and stowage inside the ship, and there were further attempts to improve safety. A plan to increase powder production by the simplification of the processes was also postponed, meaning that the magazine explosion on board *Mutsu* did have an impact on powder production planning in the IJN.

Epilogue

Determining the cause of a magazine explosion is extremely difficult, and conclusive evidence is difficult to come by if the ship sinks in a very short time after the explosion with heavy loss of life. The conclusions of the various investigation committees were therefore provisional, and the official verdict was generally either natural decomposition of the powder or 'Unknown'.

However, the explosions of the powder magazines of the battleship *Mikasa* and the armoured cruiser *Nisshin* in 1912 were probably the result of sabotage by members of the crew. Six years before the explosion on *Nisshin*, in November 1906, it had been determined that a fire on the armoured cruiser *Aso* (ex *Bajan*) had been started by a crew member who planned to commit suicide; in this case an explosion of the powder magazine was narrowly prevented.

Arson by crew members was suspected in the cases of *Mikasa* (1905), *Matsushima* (1908) and *Tsukuba* (1917). With the latter ship the crew member in question appeared to have gained unauthorised access to the magazine. Following the loss of *Kawachi* to an explosion the following year strict regulations came into force and only selected personal were allowed access to the magazines. These measures are generally considered responsible for the pause in this type of accident during the interwar period. When the *Mutsu* incident took place in the middle of the Pacific War, the investigation committee again concluded that a crew member might have committed arson.

In the end only in two of the seven cases – those in which the ships did not sink – could the cause of fires be ascertained. In the other five cases, which resulted in the total constructive loss of four ships, suspicion of arson was supposed as most probable, but the true cause was never established.

Endnotes:

[1] The findings of the *Tsukuba* Investigation Committee suggest that the word 'strict' is an over-statement as far as the IJN was concerned.

[2] The others were members of the fire fighting brigades dispatched from other ships: the battleships *Shikishima*, *Asahi*, *Fuji*, etc.

[3] The losses were 24 officers, 150 ratings and 33 out of her 57 cadets (midshipmen). Lacroix & Wells II (*Japanese Cruisers of the Pacific War*, 656) give only 23 midshipmen, but this is apparently a printing error. The total of 207 is confirmed by several Japanese sources and by a contemporary ONI report that states that one of the officers killed was the eldest son of Vice Admiral Uriû and one of the midshipmen the son of Prince Oyama.

[4] The black powder used as propellant in naval guns was a mix of potassium nitrate (75%), charcoal (15–12.5%) and sulphur (10–12.5%).

[5] Kataoka directed the Navy Technical Department (NTD) from 22 November 1906 until 28 August 1908.

[6] Abel's method continued to be used, but for reference purposes only.

[7] Several possible solutions were proposed, but the final decision was to institute magazine refrigeration and the maintenance of a constant temperature for the propellant.

[8] The 'C' designation was used for Cordite, 'T' for Tubite, to show the shape of the powder.

[9] In the British MDC powder (the Japanese warships used the imported Mark I during the RJW), mineral jelly (vaseline) was used as a stabiliser, but stability in storage was not good and natural decomposition was an issue. Like other navies, the IJN considered improvements in the storage stability of the smokeless powder a crucial measure.

[10] In 1906 the Japan Explosive Co (*Nihon Bakuhatsubutsu*) was established as a 'joint venture' of the IJN and the British Armstrong Co, Nobel Co, and Vickers Co, with its main office in London and a factory at Hiratsuka to produce smokeless powder using the techniques of the Nobel Co. While the setting up of this company provided the foundation for these developments, the contract included the clause that the company was to be suspended after 10 years and purchased by the IJN. The main purposes of this agreement were (i) to obtain the knowledge necessary to build a factory, (ii) to introduce advanced production techniques, (iii) to apply the results of testing, and (iv) to educate and train the required engineers and factory workers. Construction of the factory began in September 1905; the buildings were completed in December 1906 and, following the installation of the necessary equipment and the training of the workforce, production began in December 1908. This factory produced British-type smokeless powder for the IJN until 31 March 1919 without incident.

[11] Out of each lot of the all-important raw materials one part had to be preserved as a sample by each producer or manufacturer, and every stage of production was strictly recorded so that in the event of a problem the specialists could check back through the entire production process.

[12] See Mike Williams, '*Mutsu*: An Exploration of the Circumstances Surrounding Her Loss' (*Warship* 2009, 125–142). In a more recent article to be published in *Contributions to the History of Imperial Japanese Warships*, compiled and written by Hans Lengerer and by Lars Ahlberg, Williams concludes that a shell may have been inadvertently dropped on its nose, and that this started a chain reaction leading to the explosion in turret No 3. The first part was published in issue 17 (September 2018). Orders for the *Contributions* may be placed with lars.ake.ahlber@telia.com or hans.lengerer@ gmx.de.

BEYOND THE KAISER:

THE IMPERIAL GERMAN NAVY'S DESTROYERS AND TORPEDO BOATS AFTER 1918

Following on from his earlier study of the light cruisers of the Imperial German Navy in the wake of the First World War, **Aidan Dodson** reviews the careers and ultimate fates of the destroyers and torpedo boats in service or building in 1918, which included the unlikely transformation of two unfinished destroyers into fully-rigged sailing ships.

At the time of the Armistice on 11 November 1918, the German High Seas Fleet deployed eight flotillas of what it classified as 'large torpedo boats', but were equivalent to the destroyers of the Royal Navy.[1] They ranged from the 1,350-tonne *B97* type built in 1914–15, through the ongoing 1,000-tonne 1916Mob/1917Mob types, to the 500–960t vessels of prewar design.[2] Away from the main fleet, there were various other ships in local defence roles, modified as fast minesweepers or serving as training vessels. Specifically built for coastal service were the 100–330t A-I, -II and -III types. In addition, undergoing trials were the first ships of the large 2,000-tonne, 15cm-armed *S113* type. The last 20 ships of the 1,020/1,061-tonne Type 1917Mob had been laid down or were fitting out, and a number of the new standard 1,268-tonne Type 1918Mob were on the stocks.

Under paragraph 23 of the Armistice, 'fifty destroyers of the most modern types' were to be included among the 'German surface warships … designated' to be 'interned in neutral ports or in default of them in allied ports to be designated by the Allies and the United States'. Thus, along with the battleships, battlecruisers and light cruisers that arrived in the Firth of Forth on 21 November were vessels from the I., II., VI., VII. and IX. Flotillas (less *V30*, which was mined and sunk *en route*; she was replaced by *V129*, which arrived with the battleship *König* and the cruiser *Dresden* on 6 December). Of the ships left in Germany, a number (led by the big *V116*, and including at various times *V26*, *V28*, *V79*, *S63*, *S133–35*, *S139* and *H146*), were formed into the *Eisernen Flottille* ('Iron Flotilla'), a 'voluntary' formation under the auspices of the new Provisional *Reichsmarine*, which played a security role during the political and social upheavals of the spring of 1919. Like the cruisers remaining in German hands, all had their torpedo tubes removed.

Along with the rest of the fleet, the Scapa destroyers were scuttled on 21 June 1919, but many were beached by the British before they sank, some simply drifting ashore when their cables were parted. Two were sold by the British Admiralty[3] in 1920: *G89* to A Young for £500, who broke her up locally at Stromness,[4] and *V83*

SMS *V156* (1908), a typical German destroyer of the era (officially designated 'large torpedo boat'), armed with two 8.8cm guns and three single 45cm torpedo tubes. The well forward of the bridge would be a characteristic feature of German destroyers until the latter years of the First World War. *V156* was one of the last German destroyers to be powered with reciprocating engines, but remained in service (latterly under the designation *T156*) until 1945. (US Naval History and Heritage Command, NH 65793)

The Eisernen Flotille hoists the old Imperial flag on its establishment on 24 February 1919. Second from the left is V79, the remaining ships being S133, S134, S135 and S135; note the absence of torpedo tubes. (Author's collection)

to the East Coast Wrecking Company for £120, which then re-sold her to Peter Kerr of Aberdeen. *V83* remained intact, however, and the option was transferred by the Admiralty to Cox & Danks on 25 January 1924, Kerr receiving a refund. Cox & Danks, which would salvage a large proportion of the former German wrecks, at the same time purchased the sunken *S53*, *S55*, *V70* and *G91* for £200 each (along with the battlecruiser *Hindenburg* for £3,000), and acquired options on *S32*, *S36*, *G38–40*, *S52*, *S56*, *S65*, *V78*, *G86*, *G101*, *G103–104*, *B109–112*, *V129*, *S136*, *S138* and *H145*; *S54* was purchased on 30 November 1927. The latter was blown up *in situ* in 1931, but the remainder were all raised between 1924 and 1926, and either broken up locally at Lyness (*G38*, *G39*, *S53*, *S56*, *V70* and *S136*), or sold on for scrapping at Scottish yards. The remaining four destroyer wrecks, *V45*, *S49*, *S50* and *S131*, had been sold on 26 April 1923 to the Scapa Flow Salvage Company, and subsequently refloated and scrapped (*S49*, *S50* at Scapa; *V45* at Troon; *S131* at Granton).

Under the Treaty of Versailles, signed in June 1919, Article 181 restricted the new German Navy to twelve destroyers and twelve torpedo boats, and Article 190 fixed the displacement of replacements at 800 long tons (810 metric tons/tonnes) and 200 tons respectively – half the size of the ships in these same categories that were building at the end of the war. Article 184 required the surrender of the Scapa vessels, and Article 185 that of 'forty-two modern destroyers and fifty modern torpedo boats, as chosen by the Governments of the Principal Allied and Associated Powers'. Following the scuttling at Scapa, 41,800 tons of small floating docks and dockyard cranes were demanded to make up for the lost destroyers (five additional cruisers and heavy dockyard equipment were required to make up for the remainder of the losses). In practice, all surviving destroyers of the *V25* and later classes were to be handed over, as were all type A-II (*A26*) and A-III (*A56*) coastal torpedo boats. In addition, a number of older vessels were also to be surrendered, particularly those that had hitherto been used for minesweeping duties.[5]

Following long-drawn-out negotiations,[6] a division of surrendered German and Austro-Hungarian vessels was agreed between the Allies and the USA, under which the majority of ships would be disposed of, but each principal nation was allocated a battleship, a cruiser and three destroyers for short-term 'propaganda' or experimental purposes before being sunk or broken up. Except for Italy, which received three Austro-Hungarian vessels, the 'propaganda' destroyers were all ships beached at Scapa: the UK took *V44*, *S82* and *V125*, France *V46*, *V100* and *V126*, Japan *S60*, *V80* and *V127*, and the USA *V43*, *G102* and *S132*.

The ships allocated to Japan were not taken over by their new owners and were immediately sold for scrapping at Dordrecht in the Netherlands. However the others were earmarked for weapon trials. The American *G102* and *S132* were sunk by bombs on 13 and 15 July

BEYOND THE KAISER: THE IMPERIAL GERMAN NAVY'S DESTROYERS AND TORPEDO BOATS AFTER 1918

From left to right: *S51*, *S137*, *G89*, *V80*, *S65*, *V82* and *S54*, beached off the island of Fara, following the scuttling of the High Seas Fleet in June 1919. All had been launched during 1915/16 (except for *S137*, launched 1918), and were typical of the High Seas Fleet's destroyer flotillas. Following refloating by the Royal Navy, *S51* and *S137* were sold for scrap at Grangemouth in February/March 1921; *G89* was sold at Scapa in 1920 and broken up there; *V80* was handed over to Japan as a 'propaganda' ship, but sold for scrap in June 1920; *V82* was used as a target off Portsmouth in October 1920, the hulk later being beached in the harbour; and *S54* was sold to Cox & Danks in November 1927, by then ashore at Flotta, the remains ultimately being blown up in 1931. *S65* was refloated by Cox & Danks in May 1925 and broken up at Granton. (US Naval History and Heritage Command, NH 426)

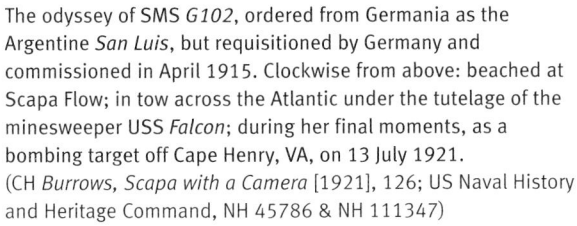

The odyssey of SMS *G102*, ordered from Germania as the Argentine *San Luis*, but requisitioned by Germany and commissioned in April 1915. Clockwise from above: beached at Scapa Flow; in tow across the Atlantic under the tutelage of the minesweeper USS *Falcon*; during her final moments, as a bombing target off Cape Henry, VA, on 13 July 1921.
(CH Burrows, *Scapa with a Camera* [1921], 126; US Naval History and Heritage Command, NH 45786 & NH 111347)

S132, a Type 1916Mob vessel commissioned in 1917, was another ship beached at Scapa. She is shown under tow in the Atlantic, *en route* to the USA, where she was sunk as a gunfire target by USS *Delaware* (BB-28) and *Herbert* (DD-160) on 15 July 1921. (US Naval History and Heritage Command, NH 111341)

The three US 'propaganda' destroyers following their arrival at New York. From the left: *V43*, *G102* and *S132*, with one of their escorting minesweepers outboard, showing the exceptional size of the ex-Argentine ships as compared with the standard German destroyer. (Library of Congress)

V82, a Type Mob vessel commissioned in 1916, is fired on by HMS *Terror* in October 1920. (Adapted from RN Gunnery Manual 1921, Fig 34)

1921, and *V43* sunk by gunfire from the battleship *Florida* on the 15th. Two of the UK's ships were used in gunnery trials, with the monitor *Terror* as firing ship (*V82* on 13 and 15 October 1920, *V44* on 8 December); both were subsequently beached at Portsmouth, and although sold to TW Ward on 30 March 1921, they remained in place and were re-sold to Pounds in 1927; although stripped *in situ*, parts of their hulks are still visible at low tide.[7] *V125* was not employed in the trials and was sold to John Cashmore on 20 July 1921 for £1,010, being towed away from Portsmouth, bound for Newport, on 2 September. Of the French ships, *V100* was scrapped in 1921, with the other two lasting as trials vessels until 1924 (*V46*) and 1925 (*V125*).[8]

A Reprieve for Some

A small number of ships were allocated to particular nations for future operational use. France had a desperate need for modern destroyers, having not laid down any such vessels in her own yards since 1913. The only ships to join the fleet during the war had been two of the 1913 ships, four ships that had been laid down for Argentina, and a dozen ordered in Japan as an emergency measure. Italy had maintained a healthy destroyer programme during the war, but for political reasons demanded equal treatment with France in the share-out of ex-enemy vessels. Thus, it was agreed that each nation should receive ten destroyers, taken from the vessels surrendered by Germany under Article 185 of the Versailles Treaty and/or those given up by Austria-Hungary under Article 136 of the Treaty of Saint-Germain. In addition, coastal torpedo boats were to be allowed to Brazil (six ex-German, disarmed for police duties), Greece (seven ex-Austro-Hungarian: one for any purpose, six disarmed for police duties), Poland (six ex-German or ex-Austro-Hungarian, disarmed for police duties), Portugal (six ex-Austro-Hungarian torpedo-boats, disarmed),[9] Romania (seven ex-Austro-Hungarian, six disarmed) and the Serbo-Croat-Slovene State (later Yugoslavia: twelve ex-Austro-Hungarian).

The 42 Article 185 destroyers included *B98*, which had been used to carry post between Germany and the interned fleet and had been seized by the British when she arrived at Scapa the day after the scuttling. However, while in tow to Rosyth she broke her cable and grounded on the beach at Lopness, Sanday (Orkney) on 17 February 1920. She was sold to the East Coast Wrecking Company on 25 June, although parts of the wreck,

All the surviving units of the later types of coastal torpedo boats (A-II and A-III) were, with the exception of a number taken over directly by Belgium, surrendered at Rosyth during August/September 1920. This is *A68*, which became the Polish *Kujawiak*. (Author's collection)

133

including turbine casings, remain visible to this day.[10] Of the remaining 41, these would be pooled with the eight surrendered Austro-Hungarian destroyers; ten ships from that pool were to be taken for commissioning by each of Italy and France; the rest would go to the UK for scrap. Twenty of what were judged to be the best German ships (mainly ex-*Eisernen Flotille*) were accordingly delivered to Cherbourg during May/July 1920 to allow Italy and France to make a selection. France eventually took nine German and one Austro-Hungarian vessels, with Italy obtaining three ex-German and seven ex-Austro-Hungarian. All other ships due to be surrendered by Germany were delivered to Rosyth during August/September for further distribution or disposal.

Poland

As noted above, Brazil and Poland were each allocated six torpedo-boats, to be disarmed for police duties, and while Poland had an option to take Austro-Hungarian vessels, both nations' shares were in fact taken from the 50 such vessels delivered to Rosyth, the remainder of which were listed for scrapping in the UK. Brazil did not take over her ships (*A65*, *A74*, *A78*, *A93*, *V105* and *V106*), instructing the British Admiralty to sell them for scrap, which they did at Rosyth on 13 June 1921 (*A93* to W Duguid of Bo'ness, the rest to James W White of Queensferry, for £170 each – except for *V105*, which went for £550 and *V106* for £610). However, in the event, *V105* was swapped for *A69*, which had been allocated to Poland but proved to be in very poor condition and went for scrap in her stead. *V105* thus commissioned into the Polish Navy in 1921 as *Mazur*, alongside her sister *V108* (*Kaszub*), and the smaller *A59* (*Ślązak*), *A64* (*Krakowiak*), *A68* (*Kujawiak*) and *A80* (*Góral*, renamed *Podhalanin* in 1922). *Kaszub*, *Krakowiak* and *Kujawiak* were refitted at Rosyth Dockyard between December 1920 and August 1921. It was envisaged that they proceed under their own power to Poland; however, breakdowns *en route* meant that all eventually arrived at Danzig under tow, *Kaszub* on 3 October 1921; *Mazur*, *Ślązak* and *Góral* had already been delivered to Poland under tow in September 1921.

The ships received a proper armament only in 1925, when they were fitted with two 75mm guns, two 450mm torpedo tubes and mine rails. On 20 July 1925 *Kaszub* was broken in half by the explosion of her forward boiler in the Neufahrwasser at Danzig; the forepart sank but the after part of the ship remained afloat. The bow was refloated a week later and the whole ship docked; she was subsequently broken up. The other ships survived into the 1930s, *Mazur* being extensively rebuilt during 1935–37 and still in service as a gunnery training ship at the time of the German invasion; she was, however, bombed and sunk at Oksywie on 1 September 1939. *Krakowiak* had been stricken in October 1936 and broken up, while *Ślązak* had become a target ship for aircraft in 1937; captured by the Germans in 1939, she subsequently sank while under tow. *Kujawiak* and *Podhalanin* had been stricken and converted to oil hulks

The two completed units of the 2,060-tonne *S113* type, *S113* and *V116*, went to France and Italy respectively; *Amiral Sénès* (ex-*S113*) is shown here. (*Author's collection*)

in 1939; they were sunk respectively by bombs at Oksywie on 3 September, and in tow between Jastarnia and Hel on 24 September.

France and Italy

The French and Italian ships from the Cherbourg hand over were each headed by one of the two *S113*-type large destroyers that had commissioned for trials before the Armistice. *S113* herself became the French *Amiral Sénès*, and *V116* the Italian *Premuda*. Italy received two further German ships, *B97* and *S63*, plus the seven ex-Austro-Hungarian vessels, taken over at Adriatic ports. France took just one ex-Austro-Hungarian ship but eight further ex-German vessels: *V79* (*Pierre Durand*), *V130* (*Buino*), *S133* (*Chastang*), *S134* (*Vesco*), *S135* (*Mazaré*), *S139* (*Deligny*), *H146* (*Rageot de la Touche*) and *H147* (*Delage*). The ships were little altered by their new owners, the main exception being the ex-*B97*, which became a trials ship in 1932; her after boiler was replaced with a gyro-stabiliser and the armament modified. All served into the 1930s, and were disposed of only when new tonnage was delivered to replace them.

Leftovers

As already noted, the UK was to scrap all unallocated surrendered ships, and thus sold *T159*, *T160–161*,

France also received smaller destroyers. This is *Buino* (ex-*V130*), one of five Type 1916Mob vessels taken over. (*Author's collection*)

T163–166, *T169*, *T173–174*, *T176*, *T178*, *T179*, *T180*, *T182–184* and *T186* (for £600–£1,030 each) at Rosyth during February/March 1921. The small *A27–29*, *A31*, *A33–39*, *A41*, *A44–46*, *A48–49*, *A52–55*, *A61–63*, *A66*, *A70*, *A75–76*, *A81*, *A86*, and *A87–95* were similarly disposed of, all but one of them going to John Jackson & Co for £150 (February) or £120 (March) each, for scrapping at Bo'ness. The exception was *A81*, lying at Limekilns, slightly further up the Forth estuary, which went to Thomas Round of Sunderland on 30 March

Delage (ex-*H147*) was one of two Type 1917Mob ships that joined the French fleet. They were among the last German destroyers to commission, and their design formed the basis of the post-war *Möwe* class. (*Author's collection*)

For political reasons, France also insisted on one of the eight surrendered Austro-Hungarian *Tátra*-class destroyers, the other seven going to Italy. Accordingly, *Uzsok* (880 tonnes, 1917) became *Matelot Leblanc*, which served until placed in reserve on 1 December 1935; she was stricken on 30 May 1936 and sold for scrap. *(Author's collection)*

1921 for £150. The UK also sold *S24*, *V26*, *V28*, *T189*, *T192–193*, *T195* and *T197* at Cherbourg on 22 October, the ships being 'rejects' from the Franco-Italian allocations.[11]

In addition to the ships handed over to victorious nations through the process described above, Belgium had seized three Type A-I small torpedo boats (*A4*, *A12* and *A14*) that had been found at Antwerp following the German evacuation, and also took over another six A-Is (*A5*, *A8*, *A9*, *A11*, *A16* and *A20*) and five Type A-IIs (*A30*, *A40*, *A42*, *A43* and *A47*) on 25 June 1919, when these ships, in accordance with Article 184 of the Versailles Treaty, were handed over by the Netherlands from internment at Hellevoetsluis, having taken refuge there after the Armistice on 16 November. They remained in Belgian operational service only until 1927, when the Belgian navy was abolished. Most were then scrapped, but some lasted longer as training vessels – one (the former *A20*) survived to be captured by Germany in 1940, re-enter service with her former owners, and be broken up post-war.

One other A-boat served during the Second World War. This was the former *A32*, which had been wrecked off the Baltic island of Saaremaa in October 1917, salvaged by Estonia in October 1923, and commissioned as *Sulev* in August 1924. She became the Soviet *Ametist* in 1940, following the annexation of the Baltic States, and survived as a patrol vessel and finally a tender until scrapped in the 1950s.

Destroyers to Merchantmen[12]

Article 186 of the Treaty of Versailles required 'the breaking up of all the German surface warships now under construction', but on 13 February 1920 the German Foreign Office wrote to the Naval Inter-Allied Control Commission (NIACC) asking for a ruling on the definition of 'breaking up'. The Germans proposed that it be understood as 'so stripping such vessels of their characteristics as war vessels that re-construction of war vessels would be impossible', the intention being that such vessels could be used as the basis for merchant ships. This was referred to the Allied Conference of Ambassadors with a recommendation for acceptance, as it achieved the objectives of the clause while also meeting a requirement under Article 189 that material arising from the breaking up of warships must be used for industrial or commercial purposes.

Although capital ships, cruisers and destroyers were all put forward for potential mercantile completion,[13] it was only in the case of destroyers that actual adaptations were put in hand (although only completed in four cases). Two basic schemes were drawn up and approved by the NIACC: one for a conventional coaster with a diesel engine aft, and one for a four-masted schooner with auxiliary diesel propulsion. Both these schemes involved cutting off the bow at the forward boiler room and the stern at the after engine room bulkhead, and adding new ends to the midship section. The latter comprised principally the machinery spaces, which

BEYOND THE KAISER: THE IMPERIAL GERMAN NAVY'S DESTROYERS AND TORPEDO BOATS AFTER 1918

S113 class

V117 & V118: schemes for conversion to coasters

Type 1917Mob

H166–69: scheme for conversion to schooners

Type 1918Mob

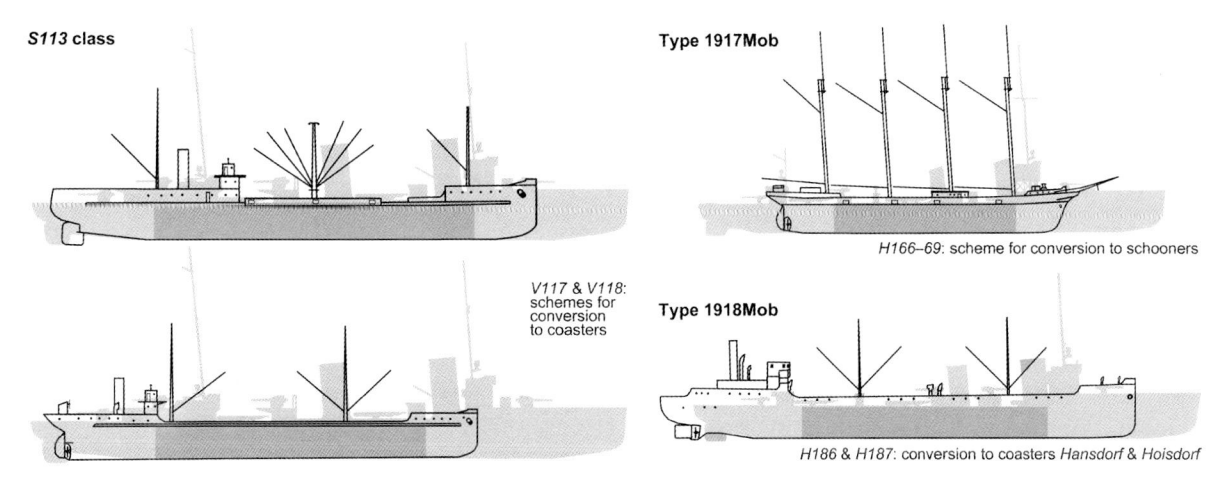

H186 & H187: conversion to coasters Hansdorf & Hoisdorf

The three types of German destroyer under construction at the end of the war, showing the two variants of the coaster scheme drawn up for the big S113 type, the schooner scheme as applied to the conventional Type 1917Mob, and the actual coaster conversion of two of the new Type 1918Mob, none of which had been completed as destroyers. The dark shading indicates the section of destroyer hull retained in the mercantile conversion. (Author's drawings)

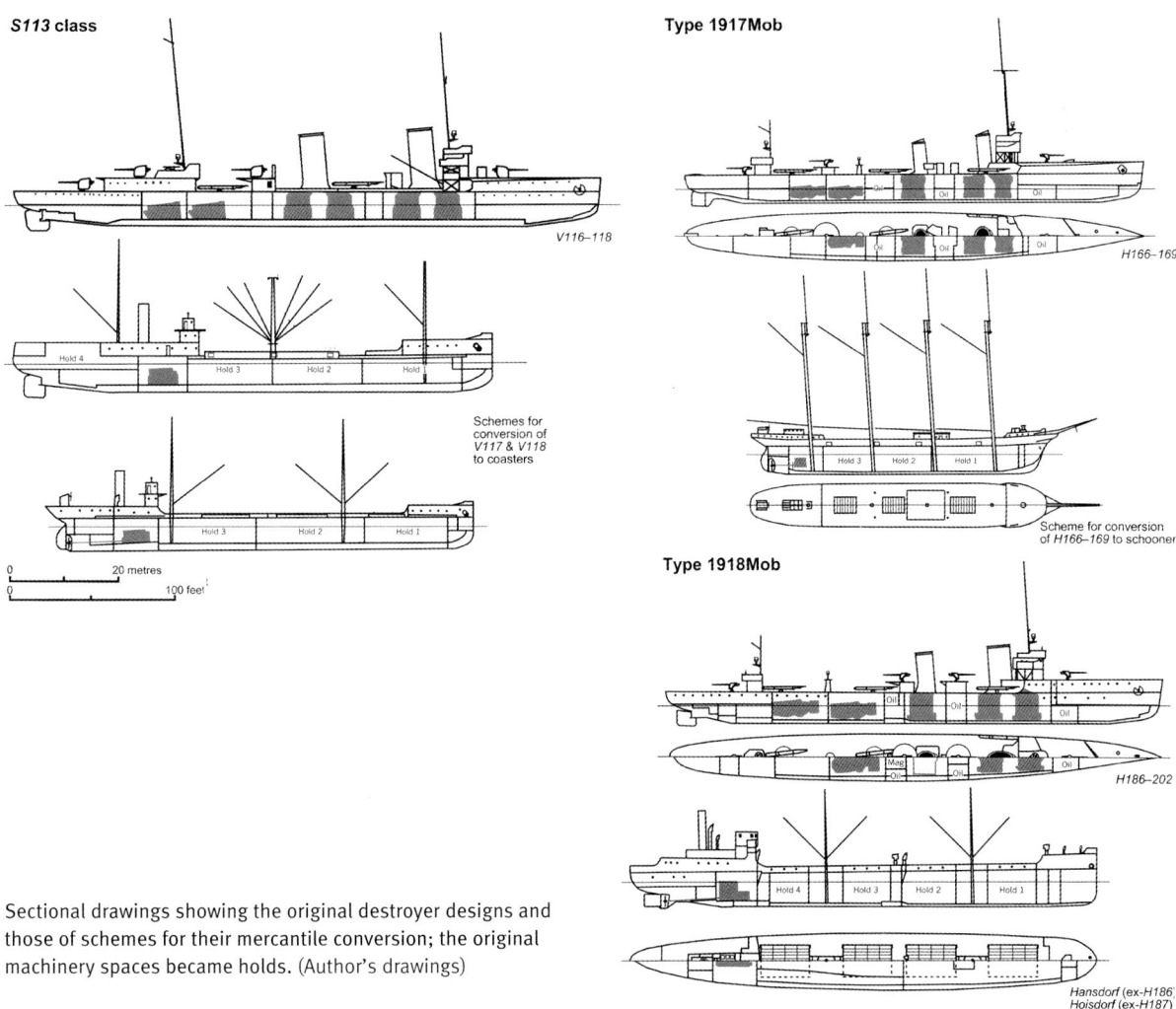

S113 class

V116–118

Schemes for conversion of V117 & V118 to coasters

Type 1917Mob

H166–169

Scheme for conversion of H166–169 to schooner

Type 1918Mob

H186–202

Hansdorf (ex-H186) Hoisdorf (ex-H187)

Sectional drawings showing the original destroyer designs and those of schemes for their mercantile conversion; the original machinery spaces became holds. (Author's drawings)

would be emptied of boilers and turbines and would now accommodate the holds of the new merchantman, whose propulsion machinery would be housed in the brand-new stern section.

The removal of the original bow and stern was necessary to meet a key Allied requirement that the hull-form of any converted vessel be no longer suitable for high speeds: '[i]n the case of Torpedo Craft a complete recon-

137

struction of the Bow and Stern would do away with the character of a fast warship'.[14] It was on this basis that a design for the mercantile completion of the large *S113*-type *G119–121* was rejected on 7 September 1920 as retaining whole underwater form. However, a modified scheme, apparently including the requisite truncation, was approved at the beginning of October for *G119–121*, together with the very similar *S114–115*, *V117–118* and *B122–124* (of which *B122* was then being employed as an fuel hulk for the generators at the Blohm und Voss shipyard at Hamburg). Plans for converting the four A-III type torpedo boats fitting out at the Armistice (*A67* and *A83–85*) were also approved, but none of these even got as far as finding a potential mercantile converter.

The concern to remove any chance of the ships ever being completed or rebuilt as warships led to demands by the NIACC that bows and sterns be cut from all unfinished destroyer hulls that were not being scrapped immediately, whether or not a contract yet existed for a mercantile conversion to be carried out. Eventually, a compromise was reached in July 1921 whereby the bow and stern of still-extant unfinished destroyers could be cut off, but only as far as the waterline, to allow the hulks to be stored afloat until they could be converted; shaft brackets had also to be cut away and all original machinery cleared from the hull to fully meet NIACC requirements.

By the summer of 1921, however, while various unfinished destroyers had had their bow and/or stern cut away and their machinery removed, substantive conversions had been restricted to four ships, all of the 1918Mob type. Two (*S178* and *S179*) had been building at Schichau, Elbing; following the removal of their bows, they had been moved successively to Danzig, then to Bremen, where they were completed to the schooner scheme, and entered service with the Bremerhaven firm of F Kimme in 1921. The former *S178* became *Franziska Kimme*; in 1926 she was renamed *Kapitän J Frobeen*, finally becoming the Brazilian *Captain Alfredo Kling II*. Later still she became *Ajuricaba* and finally *Gonza* in 1933; she disappears from lists in 1936. The ex-*S179* entered service as *Georg Kimme*, becoming *Anneliese Rathjen* in 1927 and the French *Zazpiakbat* in 1928; she was scuttled at Martigues, near Marseilles, on 21 August 1944.

The remaining pair were the Howaldtswerke (Kiel) *H186* and *H187*, which were completed as coasters in early 1921, under the names *Hansdorf* and *Hoisdorf* respectively. In 1924 they became *Dietrich Bohmekamp* and *Hermann Bohmekamp*, being sold to a Brazilian owner as *Peryneas II* and *Peryneas* in 1930. The former *H186* was sold there for scrap in 1935, but her sister passed under the Newfoundland (1931) and British Honduran (1933) flags before being broken up in 1933.

The same basic coaster drawings were approved for the conversion of both the large *S113* type and the smaller 1917Mob-type *S152–157*, *V158–163* and *H166–169*. Schooner drawings were also provided for *H166–69*, but neither coaster nor schooner conversions were taken forward in these cases. A proposal of 5 March 1921 (approved 12 March 21) was that *H166–169* be converted to oil lighters without propulsion engines but with auxiliary boilers to drive oil pumps and to warm oil. Although bows and sterns had been removed by April, work was then suspended following the receipt of additional requirements from Germanischer Lloyd (the national maritime classification society). This proved to be the end of the vessels, the hulks being scrapped at Kiel before the end of the year. Scrap also proved to be the ultimate destiny of all the remaining would-be-mercantile conversions from the summer of 1921 onwards, since although a further handful had been sold for conversion, a shipping slump meant that the reconstructions were simply no longer economically viable.

The *Reichsmarine*

Under the Versailles Treaty, the new German Navy, the *Reichsmarine*, was allowed to have twelve destroyers and twelve torpedo boats in commission at any one time. An additional allowance of a further four of each type in reserve, without stores or ammunition but with guns on board, was added under an agreement with the Allies dated March 1920 – this also allowed two reserve battleships and two reserve cruisers. Initially, the operational torpedo boats were to be the 310-tonne *T99*, *T101*, *T102*, *T103*, *T104*, *T105*, *T106*, *T107*, *T108*, *T109*, *T110* and *T113* (1900–02), with the 142/147t *T88* and *T89* (Kiel) and *T86* and *T85* (Wilhelmshaven) – built 1897–98 – as reserve vessels. However, it was then agreed in early 1921 that, given the limited military value of these ancient vessels, twelve plus four ships of the 412–665t *T132–168* series (1906–11) would become the *Reichsmarine*'s 'torpedo boat' allocation. The vessels selected were *T139*, *T141*, *T143*, *T146*, *T149* and *T168* (Baltic operational), *T144* and *T155* (Kiel reserve), *T151*, *T153*, *T154*, *T156*, *T157* and *T158* (North Sea) and *T148* and *T152* (Wilhelmshaven reserve).

The dozen operational 'destroyers' were originally to be *T185*, *T190*, *T196* (650/660t, 1911), and *V2*, *V3*, *V5*, *V6*, *G8*, *G10*, *G11*, *S18* and *S19* (570t, 1911–13). However, the list was adjusted to take into account the state of the various units, with ships swapped with others from a pool of 'alternatives' that initially comprised *S23*, *T151–156*, *T158*, *T167*, *T168*, *T170* and *T175*. Thus, in March 1920, *G7* was substituted for *V6* and *S23* for *S19*, the latter joining the group of four ships that by the summer were listed as the four allowed long-term reserves (*S19* and *T175* at Kiel; *V6* and *T170* at Wilhelmshaven). Further changes then occurred: *V1* was withdrawn from the scrap pool to replace *T185* on the 'active' list, the latter going into the reserve pool in place of *T170* at Wilhelmshaven, which was now joined there by *T175*, *V6* going to the 'active' list and replaced by *T190*, which went to the Kiel reserve alongside *S19*. *T170* was stricken on 22 March 1921, along with many of the remaining old torpedo craft in excess of the Versailles allocations (others had already been stricken

BEYOND THE KAISER: THE IMPERIAL GERMAN NAVY'S DESTROYERS AND TORPEDO BOATS AFTER 1918

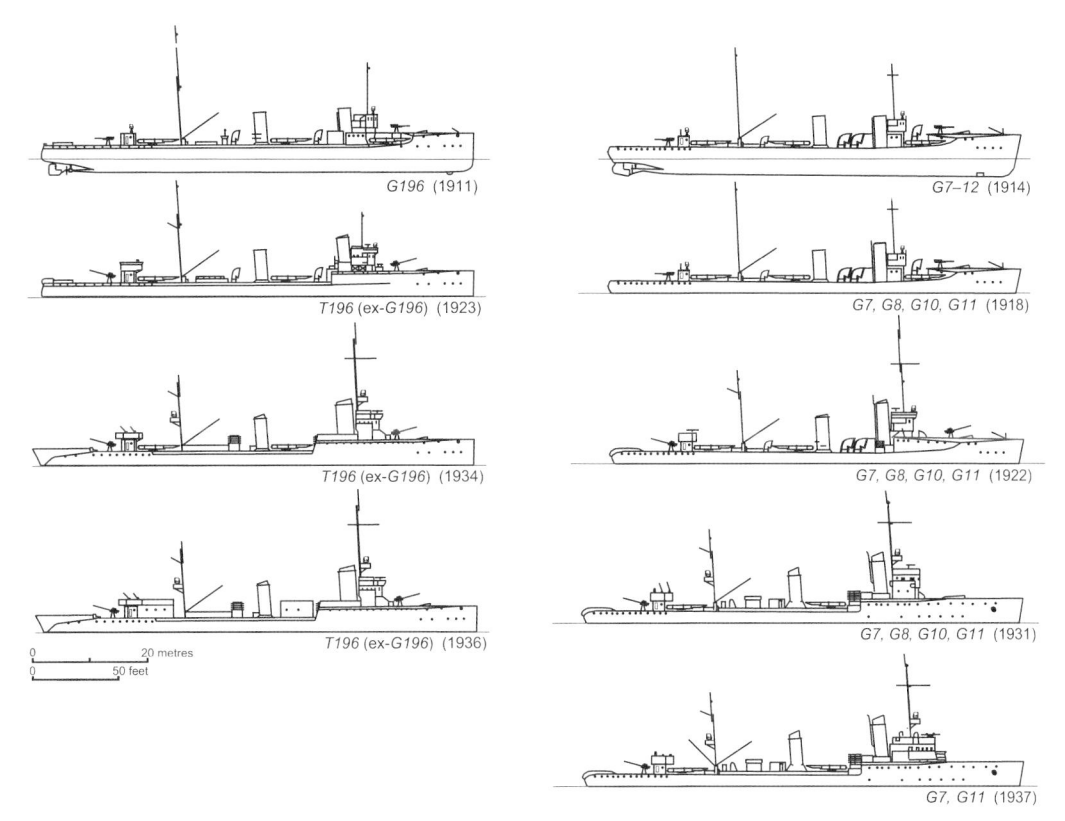

The successive changes in appearance of *T196* (ex-*G196*), one of the largest destroyers retained by Germany after the First World War, and the surviving *G7* series, the newest boats left to the new German navy. Both underwent major modernisations, which followed a basic pattern common to most of the old destroyers and torpedo boats retained beyond 1927–28, with a unified profile, main guns of an increased calibre, oil-fired boilers and the elimination of the well-deck forward of the bridge that had been characteristic of German destroyers up to the mid-war period. The *G7*s were also lengthened during their second rebuilding. (Author's drawings)

Part of the *Reichsmarine*'s destroyer force in the late 1920s, with *S19* and *T185* in the foreground and three of the *T151–158* series beyond them. *S19* was the second newest destroyer left to the German Navy after the First World War, commissioned in 1913. Despite this, *S19* was stricken in 1931 and broken up in 1935, while the older (but larger) *T185* survived as a control vessel for the radio-controlled targets (ex-battleships) *Zähringen* and *Hessen* under the name *Blitz* (ii) until 1945, when she became the Soviet *Vystrel*. (Author's collection)

139

The first six Versailles replacement 'destroyers' (regarded by the *Reichsmarine* from the outset as torpedo boats) commissioned during 1926–27. Four of the first series, *Greif*, *Falke*, *Möwe* and *Kondor* are shown here in 1937, moored alongside the brand-new *Z1* (*Leberecht Maass*), flying the pennant of the Leader of Torpedo-boats, and the first of the *Kriegsmarine*'s full-size (2,600-tonne) destroyers. (*Author's collection*)

during 1920). The surviving modern Type A-I boats were stricken in May 1922, and the few remaining older vessels were gone by the end of that year.

Although divided between 'destroyers' and nominally smaller 'torpedo boats' by the Versailles Treaty (which also limited replacement 'destroyers' to 800 tons and 'torpedo boats' to 200 tons), most of the ships listed as 'destroyers' were actually smaller than some of the 'torpedo boats'. This had at its root a decision that torpedo boats built under the 1911 programme should be smaller than those of the 1910 programme, which were seen by some as too large for effective service with the battle fleet. The 14% reduction of displacement (from 660 tonnes to around 570 tonnes) resulted, however, in an unacceptable loss of seaworthiness, and the ships of the 1913 programme (*V25*–*S36*) were enlarged to *ca* 800 tonnes, with succeeding classes showing further growth. As it was the 1913 and later ships that were surrendered under the Versailles Treaty, it was the small vessels of the 1911 and 1912 programmes that represented the most modern torpedo vessels left to Germany, and thus most appropriate to the higher-ranked 'destroyer' category.

In allocating ships to Treaty headings, a key criterion also seems to have been propulsion, all the 'destroyers' being turbine-powered, while of the 'torpedo boats' only *T168* had turbines, the others all having reciprocating engines. Nevertheless, in practice, the 'destroyers' and 'torpedo boats' were managed as a single pool, all units of which continued to be designated *Torpedoboot* by the *Reichsmarine*: it was not until the 1930s that the formal classification of *Zerstörer* was introduced into the German Navy (see below).[15]

The first 'destroyers' to recommission were *G7*, *G8* and *G11* on 22 March 1921, joining the battleship *Hannover*, commissioned as the first ship of the *Reichsmarine* on 10 February. They were followed on 25 May by *S18* and *S23*, and later in the year by *V5* and *G10*. By the beginning of 1923 *G7*, *G8*, *G10*, *G11*, *S18* and *S23* were allocated as the Baltic 'destroyers', with *T139*, *T141*, *T143*, *T144*, *T146* and *T149* as the 'torpedo boats', supporting *Hannover* and the cruisers *Medusa*, *Berlin* and *Thetis*. Attached to the North Sea station were *V1*, *V2*, *V3*, *V5*, *V6* and *T196* as 'destroyers' and *T151*, *T153*, *T154*, *T156*, *T157* and *T158* as 'torpedo boats', alongside the cruisers *Hamburg* and *Arcona*, and the battleship *Elsaß*. Reserve vessels remained as previously listed.

All recommissioned vessels had been refitted at Wilhelmshaven, a common pattern of initial modifications being adopted across the 'destroyer' fleet, with 10.5cm guns substituted for their original 8.8cm weapons; in compensation their torpedo complement was halved from four 50cm tubes to two. This latter modification also allowed the ships' forecastles to be extended aft, thereby filling the characteristic 'well' in front of the bridge. The 'torpedo boats', however, initially retained their original armament but received enlarged bridges and funnel caps. During 1922–24,

T151, *T153*, *T155–158*, *T185*, *T190* and *T196* were converted to oil firing; the last three were also reboilered and had their forecastle further extended. *G7*, *G8*, *G10* and *G11* were then rebuilt during 1928–31, being lengthened amidships by 4.7 metres. In this form they carried two 10.5cm/45 guns, plus two single 50cm TT; *T185*, *T190* and *T196* received the same guns, but carried two twin 50cm TT mountings. These seven vessels thus represented the most effective of the old destroyers available to the *Reichsmarine* at the beginning of the 1930s and, with the exception of *T185*, would remain in front-line service until the latter part of the decade.

The End of an Era

The days of the aged warriors were, nevertheless, drawing towards their close. In 1932, the I. Torpedo Boat Flotilla still comprised *G8*, *G7*, *G10* and *G11* (1. Half-flotilla) and *T151*, *T156*, *T158* and *T153* (2. Half-flotilla), but the II. Flotilla was now made up of eight modern vessels of the *Möwe* (Type 1923) and *Wolf* (Type 1924) classes. Under the Versailles Treaty, destroyers and torpedo boats could be replaced fifteen years after launch, and as the oldest ships had been launched back in 1907, new ships could be begun in 1922. However these should have been 200-ton 'torpedo boats', so in the event the first ships to be 'replaced' were the smallest 'destroyers', launched in 1911, and thus replaceable in 1926.

The design of the new Type 1924/1925 'destroyer' was based on that of the *H145–147* series, with various modifications, including a distinctive profile that was also adopted in the modernisation of older ships – and would continue into the Second World War. Twelve were built and commissioned between October 1926 and August 1929. They were nominal replacements for the dozen 'destroyers', although of the ships actually stricken in compensation, only six (*T175*, *V1*, *V2*, *V3*, *V5* and *V6*) were from the declared 'destroyers' list, the other six being nominally 'torpedo boats'. Evidently it was a case of clearing out the least capable vessels rather than strict adherence to the letter of the Versailles Treaty, particularly since further replacements would be subject to the 200-ton 'torpedo boat' replacement limit, a figure less than a third of the displacement of the ships that would need to be taken out of service in exchange. Doubtless it was for this reason that the building of the *Möwe/Wolf* classes was accompanied by the rebuilding of *G7*, *G8*, *G10*, *G11*, *T185*, *T190* and *T196*, to allow an adequate fleet of torpedo craft of useful size to be maintained until such time as the restrictions of Versailles could be eased or disregarded.

On the other hand, the 200-ton limit pushed the *Reichsmarine* towards considering less conventional ways of providing a second echelon of torpedo craft. This resulted in the development of the big motor torpedo boat *S1* (ex-*W1*, ex-*UZ16*), commissioned in August 1931, and the precursor of the extremely effective vessels deployed by the German Navy during the Second World War.[16]

With the commissioning of the new ships during the late 1920s, some old vessels began to reduce to secondary duties in preference to disposal. Already in 1927, *S139* and *S141* had been disarmed and converted (under the names *Pfeil* and *Blitz*) to act as control vessels for the former battleship *Zähringen*, which had just completed reconstruction as a radio-controlled target. *Blitz* was replaced in 1932 by *T185*, which became *Blitz* (ii); *S23* (renamed *T23* in 1932 and *T123* in 1939, when a new *T23* was projected) also became a control vessel (*Komet*)

As new ships came into service, the old destroyers and torpedo boats passed into secondary roles. One of the oldest of all, *T151* (first commissioned as *S141* in 1907), became in 1927 the control-ship for the newly-converted radio-controlled target battleship *Zähringen*, with the name *Blitz*. Shown here in 1928, she was sold for scrap in April 1933, her name and role being taken by the modernised *T185*. (US Naval History and Heritage Command, NH 88048)

GERMAN DESTROYERS AFTER 1919: CHARACTERISTICS & FATES

The table provides a summary of the names, general characteristics and fates of the German destroyers and torpedo boats that remained operational after the implementation of the Treaty of Versailles, either with the *Reichsmarine* or the navies to which they had been allocated. Not included are the small A-series torpedo boats taken directly by Belgium.

Name in 1919	Later name(s)	Launched	Displ^{mt}	Armament in 1919 Main guns	TT	Engines/ power	Speed
T139 (ex-*S139*)	*Pfiel* (Aug 27)	12 Nov 06	533t	2 x 8.8cm/35	3 x 45cm	R/11,000	30kt
T141 (ex-*S141*)	*Blitz* (i) (Aug 27)	7 Feb 07					
T143 (ex-*S143*)		6 Apr 07					
T144 (ex-*S144*)		27 Apr 07					
T146 (ex-*S146*)		27 Jun 07					
T149 (ex-*S149*)		19 Oct 07					
T151 (ex-*V151*)		19 Sep 07	558t	2 x 8.8cm/45		R/10,900	
T152 (ex-*V152*)		11 Oct 07					
T153 (ex-*V153*)	*Eduard Jungmann* (29 Aug 38)	13 Nov 07					
T154 (ex-*V154*)		19 Dec 07					
T155 (ex-*V155*)		28 Jan 08					
T156 (ex-*V156*)	*Bremse* (1944)	29 Feb 08					
T157 (ex-*V157*)		29 May 08					
T158 (ex-*V158*)	*Prozorlivyi* (1946), *Araks* (1950)	26 Jun 08					
T168 (ex-*S168*)		16 Mar 11	665t			Tu/17,500	32kt
T175 (ex-*G175*)		24 Feb 10	700t		4 x 50cm	Tu/15,000	
T185 (ex-*V185*)	*Blitz* (ii) (Oct 32); *Vystrel* (1946)	9 Apr 10	650t			Tu/18,000	
T190 (ex-*V190*)	*Claus von Bevern* (29 Aug 38)	12 Apr 11	666t				
T196 (ex-*G196*)	*Pronzitelnyi* (1946)	25 Apr 11	660t			Tu/18,200	
V1		11 Sep 11	569t			Tu/17,000	
V2		14 Oct 11					
V3		15 Nov 11					
V5		25 Apr 13					
V6		28 Feb 13					
G7	*T107* (23 Apr 39) *Porazaiuskii* (1946)	7 Nov 11	573t			Tu/16,000	
G8	*T108* (23 Apr 39)	21 Dec 11					
G10	*T110* (23 Apr 39)	15 Mar 12					
G11	*T111* (23 Apr 39)	23 Apr 12					
S18		10 Aug 12	568t	2 x 10.5cm/45		Tu/15,700	32.5kt
S19		17 Oct 12		2 x 8.8cm/45			
S23	*T123*, *Komet* (23 Apr 39)	29 Mar 13					
S63	*Ardimentoso* (1920)	25 May 16	919t	3 x 10.5cm/45	6 x 50cm	Tu/24,000	34kt
V79	*Pierre Durand* (1920)	18 Apr 16	924t			Tu/23,500	
B97	*Cesare Rossarol* (1920)	15 Dec 14	1374t	4 x 10.5cm/45		Tu/40,000	36.5kt
V105	*Mazur* (1920)	26 Aug 15	340t	2 x 8.8cm/45	2 x 45cm	Tu/5,500	28kt
V108	*Kaszub* (1920)	12 Dec 14					
S113	*Amiral Senès* (1920)	31 Jan 18	2,060t	4 x 15cm/45	4 x 60cm	Tu/45,000	34.5kt
V116	*Premuda* (1920)	2 Mar 18					
V130	*Buino* (1920)	20 Nov 17	924t	3 x 10.5cm/45	6 x 50cm	Tu/23.500	34kt
S133	*Chastang* (1920)	1 Sep 17	919t			Tu/24,000	32kt
S134	*Vesco* (1920)	25 Aug 17					
S135	*Mazaré* (1920)	27 Oct 17					
S139	*Deligny* (1920)	24 Nov 17					
H146	*Rageot de la Touche* (1920)	23 Jan 18	990t			Tu/24,500	
H147	*Delage* (1920)	13 Mar 18					
A59	*Ślązak* (1920)	13 Apr 17	330t	2 x 8.8cm/30	1 x 45cm	Tu/6,000	28kt
A64	*Krakowiak* (1920)	30 Mar 18					
A68	*Kujawiak* (1920)	11 Apr 17	335t				
A80	*Góral* (1920); *Podhalanin* (1922)	24 Oct 17	330t	3 x 8.8cm/30	Nil		

Fate

Not traced after 1944
Sold 28 Apr 33
Str 10 May 27; sold 25 Mar 30
Str 8 Oct 28; sold 10 Apr 29
Str 8 Oct 28; sold 10 Apr 29
Str 16 May 27
To USA 4 Jan 46; sold 1948
Str 31 Mar 31; BU 1935
To USA 22 Dec 45; BU 1949
Str 8 Oct 28; BU 1935
Scuttled Swinemünde 22 Apr 45
Foundered in tow North Sea 10 Jun 46
Mined Danzig Neufahrwasser 22 Oct 43
To Soviet Union 13 Feb 46; str 31 May 61
Str 8 Jan 27
Str 23 Sep 26
To Soviet Union 1946
To USA 1945; scuttled Skagerrak 16 Mar 46
To Soviet Union 13 Feb 46; str 30 Apr 49
Str 27 Mar 29
Str 18 Nov 29; sold 25 Mar 30
Str 18 Nov 29; sold 25 Mar 30
Str 18 Nov 29; sold 25 Mar 30
Str 27 Mar 29
To Soviet Union 13 Feb 46; str 12 Mar 57
To UK 6 Jan 46; BU
Scuttled Travemünde 5 May 45
Bombed Kiel 3 Apr 45; wreck blown up by UK 14 Dec 45
Str 31 May 31; BU 1935
Str 31 May 31; sold 4 Feb 35
To Soviet Union 1945
To Italy 1920; str 4 Feb 39
To France 1920; Str 15 Feb 33; sold 1934
To Italy 1920; str 17 Jan 39
To Poland 1920; bombed Oksywie 1 Sep 39
To Poland 1920; internal explosion Danzig Neufahrwasser 20 Jul 25; salved 29 Jul 25 & BU
To France 1920; sunk as target 19 Jul 38
To Italy 1920; str 1 Jan 39
To Italy 1920; str 15 Feb 33
To France; str 17 Aug 33
To France; str 24 Jul 35; BU 1936
To France; str 24 Jul 35; BU 1936
To France 1920; str 17 Aug 1933, BU 1934
To France 1920; str 15 Feb 1933, BU 1934
To France 1920; str 15 Feb 1933, BU 1934
Target ship 1937
Str Oct 36; BU
Bombed Oksywie 3 Sep 39
Bombed between Jastarnia and Hel 24 Sep 39

in 1939, following the addition of another ex-battleship, *Hessen*, to the radio-control fleet, as a replacement for *Pfeil*, which became a torpedo-recovery vessel (TRV); *Komet* followed suit in 1943.

Initially, only the dozen operational 'destroyers' were replaced by the Type 1924/1925 vessels. The 1920 agreement had been unclear with regard to replacement of the four reserve vessels when they became over-age. This led to a disagreement between France and the UK when Germany projected replacements for the two permitted reserve battleships in its 1931 Naval Programme.[17] The German position was that not to allow such a replacement of reserves with new construction would mean that the reserve ships would be so much older than the operational ships that they could in no way substitute for them when required, a view with which the British expressed sympathy. Duly encouraged, the *Reichsmarine* programmed four further 'destroyers' of significantly enhanced displacement, in the expectation that further relaxations of the Versailles regime could be achieved. The characteristics of the new Type 1934 would thus be kept secret until Adolf Hitler's denunciation of the military clauses of the Treaty on 16 March 1935. The ships finally emerged as 2,500-tonne giants with an ancestry going back to the *S113* type of 1918; they would be the first German vessels to be designated 'Zerstörer'.

The commissioning of these ships and their immediate Type 1934A successors during 1937–38 meant that there was no longer a need to keep any of the older destroyers as potential operational vessels. However, none was disposed of; rather, they began to be adapted for a range of support duties. *G7*, *G8*, *G10* and *G11* became training ships from 1936 onwards with their forward guns removed and bridgework extended for instructional purposes; in April 1939 they were renamed *T107*, *T108*, *T110* and *T111* in order to free their names for the new (but ultimately abortive) *Geleitboot* (escort vessel) programme; *G10* had the distinction of carrying out the German Navy's first seaborne radar trials in 1938. *T155–158* became submarine tenders and TRVs from 1936, *T151* served as a fast tug and TRV from 1937, and *T153* became a rangefinding training vessel (under the name *Eduard Jungmann*). *T190* became an experimental vessel (*Claus von Bevern*), and *T196* a minesweeper command ship from 1938.

Restricted by their duties to home waters, the old ships survived essentially unscathed until 1945, with the exception of *T157*, mined in the Danzig Neufahrwasser on 22 October 1943. However, as the war drew to a close, *T111* was bombed at Kiel on 3 April 1945, *T155* was scuttled on the evacuation of Swinemünde on 22 April, and *T110* at Travemünde on 5 May. Although the fate of *Pfeil* after 1944 is unknown, the remaining ten old warriors nevertheless survived to be surrendered to the Allies. *T108* and *T151* were broken up respectively by the UK and USA; *Eduard Jungmann* went to the USA, but served as a German Minesweeping Administration buoy tender until scrapped in the Netherlands in 1949;

and *Claus von Bevern* was handed over to the USA and scuttled in the Skagerrak. *T156*, found derelict near Bremen, was also scheduled to be scuttled there, but foundered in tow *en route*, her wreck being unexpectedly found during the search for wrecks from the Battle of Jutland.[18] The other five ships went to the Soviet Union, being delivered in January 1946 and commissioned the following month. *Blitz* (ii), now renamed *Vystrel*, still acted as control ship for *Hessen* (now the Soviet *Tsel*). *T158* was renamed *Prozorlivyi*, *T196 Pronzitelnyi* and *T107 Porazaiuskii*; *Komet*'s new name (if any) is unknown. The former *T196* was stricken in 1949, but the former *T158* and *T107* were further renamed *Araks* and *Kazanka*, respectively, in 1950, on going into reserve until 1957. The ex-*T107* was then stricken, but the ex-*T158* then became the trials vessel *UTS-67*, until finally stricken in 1961. The final fates of the former *Komet* and *Blitz* remain obscure, but if the latter remained in service as long as their associate *Tsel* (ex-*Hessen*), it would have been only in 1961 that the last two of the former Imperial German Navy's destroyers finally went to the scrapheap, after five decades of service.

Endnotes:

[1] For destroyers from 1914–39, including an in-depth treatment of German vessels, see H Fock, *Z-vor! Internationale Entwicklung und Kriegseinsätze von Zerstören und Torpedobooten 1914 bis 1939*, Koehlers Verlagsgesellschaft (Hamburg, 2001).

[2] In 1918, German large torpedo craft were numbered in two series. Ships built since 1911 had a number allocated in a single sequence, prefixed by a letter denoting their builder (B = Blohm & Voss, Hamburg; G = Germania, Kiel; H = Howaldtswerke, Kiel; S = Schichau, Elbing; V = Vulcan, Stettin; Ww = Wilhelmshaven Dockyard); while the basic characteristics within a type were similar across all builders, each used its own design, giving a range of profile differences. Older vessels had previously been so numbered, but had progressively been renamed, with a 'T' replacing their builder prefix as new ships were ordered with their old names; thus *G197* was renamed *T197* in February 1918 when the new *H197* was programmed.

[3] To which all German ships sunk at Scapa had fallen by Allied agreement.

[4] According to SC George, *Jutland to Junkyard*, Patrick Stevens Ltd (Cambridge, 1973), 54, her 'boiler tubes were polished and cut up and sold in thousands for curtain rods'. George covers the destroyers at Scapa on pages 54–79 and 171–72 of his book, with some corrections required on the basis of the Admiralty Sales Register for the period (held by Naval Historical Branch, Portsmouth).

[5] *T159–160*, *T165*, *166*, *169*, *T174*, *T181–184*, *T192*, *T195*, *T197* and *V106*.

[6] See A Dodson, 'After the Kaiser: The Imperial German Navy's Light Cruisers after 1918', *Warship 2017*, 142–43. A full account and documentation of these discussions, subsequent allocations and their implementation is in preparation: A Dodson and S Cant, *Spoils of War: the Fates of the ex-Enemy Fleets After the Two World Wars*, Seaforth Publishing (Barnsley, due to be published 2020).

[7] See S Fisher and J Whitewright, 'Hidden Heritage: The German Torpedo Boats in Portsmouth Harbour', *Warship 2017*, 166–70.

[8] The boilers from *V100* and *V126* were used to reboiler the French destroyers *Aventurier* and *Intrépide* respectively during 1924/27.

[9] Portugal was also awarded an ex-Turkish gunboat, which was never delivered following the failure of the Treaty of Sèvres.

[10] http://canmore.org.uk/site/102230/sms-b98-west-langamay-bay-of-lopness-sanday-orkney-north-sea.

[11] *S24* and *T189* both went ashore off Torquay *en route* to breakers at Teignmouth on 12 December 1920, *T189* breaking her back on rocks near Roundham Head, where her remains can still be seen; *S24* was, however, towed off and broken up.

[12] Most of this section is based on UK National Archives files ADM116/1994, ADM116/1992 and ADM116/2113.

[13] For proposed capital ship conversions, see A Dodson, *The Kaiser's Battlefleet: German capital ships 1871–1918*, Seaforth Publishing (Barnsley 2016), 149–52; for cruisers see the author's article in *Warship 2017*, 143–45.

[14] ADM 116/1994, 7 February 1920.

[15] Some of the larger First World War vessels had nevertheless been colloquially referred to as such.

[16] For the early history of the S-boats, see P Schmalenbach, 'The Genealogy of the Schnellboot', *Warship International* VI/1 (1969), 10–23.

[17] TNA ADM 116/2945.

[18] I McCartney, 'Scuttled in the Morning: The Discoveries and Surveys of HMS *Warrior* and HMS *Sparrowhawk*, the Battle of Jutland's Last Missing Shipwrecks', *International Journal of Nautical Archaeology* 2018, 10–12, with further discussion in Dodson and Cant, *Spoils of War*. Contrary to what has frequently been stated in print, there is no evidence that any of the ex-German warship scuttlings in the Skagerrak in 1946 were part of the chemical weapons disposal programme under which merchant ships were sunk carrying such loads. Rather, the ships were scuttled to ensure their destruction prior to an inter-Allied deadline for the disposal of all ships not capable of repair.

EARLY BRITISH IRON ARMOUR

In a follow-up to his article in *Warship* 2017, **David Boursnell** looks at the manufacture and testing of early British iron armour plate.

At the end of the Crimean War the alliance between France and Britain came to an end. Both countries now looked to influence the development of the emerging national movements in Germany and Italy, and also to extend their colonial empires in Africa and the Far East. This mutual flexing of muscles led the French Emperor, Napoleon III, to develop the armoured batteries of the Crimean War into seagoing ships. The *Gloire*, launched in November 1859, was a 5,500-ton wooden-hulled ship covered in iron plates. Although technologically she was not a huge advance over the earlier batteries, her 120mm (4.7in) armoured protection and thirty-six 164mm (6.46in) guns made her powerful enough to threaten the British fleet and its home ports. The French also planned a programme of larger iron-hulled armoured frigates, the first of which, *Couronne*, was laid down in early 1859.

Lord Palmerston, the British Prime Minister, took the view that Napoleon was building up the French Navy 'for the purpose of keeping us in check and overawing us upon some occasion', and so the government strengthened the coastal defences and decided to build two large iron-hulled armoured frigates to 'enable us to have an opinion on matters which may seriously affect our interests'.[1] In February 1859 Robert Napier & Sons were invited to submit a design and tender for a shot-proof frigate of 36 guns, cased with 4½in armour from the upper deck to 5ft below the waterline and capable of 13½ knots. The contract for the first frigate, *Warrior*, was awarded to Thames Ironworks & Shipbuilding Co, but Napier's won the order for the second ship, *Black Prince*.[2] *Warrior* and *Black Prince* were laid down in August and October 1859 respectively.

Although both shipbuilders had submitted 4in test plates for trials in 1856, they submitted further trial plates to the Committee on Iron Plates and Guns, which had been set up in July 1859. A sample plate from Thames Ironworks for *Warrior* was tested in May 1860 and was judged 'not satisfactory', but by August the company had produced a plate judged 'equal to anything tested at Portsmouth'. In June Napier's sent a plate for testing: it was judged to be tough, but not as good as plates from the Thames Ironworks.

The Committee's final report in March 1860[3] concluded:

– Thin plates could break up cast iron projectiles, but fragments from the plate and projectile were very destructive.
– Rolled iron plates 4½in thick were 'for all practical purposes' invulnerable against any current projectile.
– The plates should be strongly backed and fixed firmly.

Both of the companies building the new frigates made their armour by hammering, and of the two the Thames Ironworks plates were more successful, as only five of their finished plates were rejected; a higher proportion of the plates for *Black Prince*, made by Rigby and Co, who had taken over the running of Napier's Parkhead ironworks, were rejected. After testing plates from the two ships in 1863, Captain Hewlett of HMS *Excellent*, the Royal Navy's Gunnery School in Portsmouth, concluded that 'the *Warrior* is by far the better protected of the two'.[4] Because of these difficulties Samuel Beale & Co of Park Gate Iron Works in Rotherham, who had been rolling armour plates since 1855, manufactured the plates fitted between the gun-ports on *Black Prince*.[5]

The technology of making wrought iron armour plates was, however, still poorly understood, and as a result the War Office and the Admiralty set up a new 'Special Committee on Iron', which was given a much more scientific brief than the previous committee. It was asked to:

– ascertain the thickness, size and weight of iron plates of specified composition and manufacture that will resist shot of given shape, weight and material, at given velocities
– test the resisting powers of different qualities of iron plates, following the history of the manufacturing process in each plate
– measure the chemical composition, specific gravity, fibrousness and density of different plates, and
– record every particular of each experiment, including the velocity of shots fired at the plates.

Captain (later Rear Admiral) Dalrymple Hay chaired the committee, which also included three members of the Royal Society: John Percy, William Fairbairn and William Pole. Fairbairn, an engineer, industrialist and shipbuilder, who had been pushing for the use of iron in naval shipbuilding for some years, commented in his memoirs:

> On finishing the *Warrior*, the Government bethought themselves that it would be desirable to do what they ought to have done five years before, namely to ascertain something about the principles which should guide the design of iron armour.[6]

The committee commissioned plates from a number of ironworks for testing. In their four reports they described experiments carried out on armour plates by 23 named British companies, two French companies and the Royal Dockyards at Chatham and Portsmouth. The companies were spread across the heartlands of Victorian industrial Britain as follows:

Scotland	2
North East	2
West Yorkshire	3
South Yorkshire	4
Merseyside	1
East Midlands	1
West Midlands	3
South Wales	4
East London	3

In total 290 iron or steel plates by named companies were submitted for firing tests:

– 119 (46%) were thin plates of under 4in thickness.
– 139 (49%) were either 4½in or 5½in thick.
– Eight (2.8%) were between 6in and 7½in thick.
– One plate of 11in was tested.

More than half of these armour plates were supplied by five companies (see Table 1).

Making Iron Armour Plates

There were two main methods of manufacturing iron armour plates: rolling or hammering. In truth most companies used both techniques at some stage of the process and the term adopted generally refers to the method used at the final stage of the process. In January 1861 the Committee interviewed several of the manufacturers about the processes they used to make their plates.[7]

William Hardy of Thames Ironworks described to the Committee how they manufactured the plates for HMS *Warrior*:

– Scrap iron was rolled into bars and piled with bars of puddled Yorkshire iron.
– Each layer of bars was laid at right-angles to the previous layer so that the fibres of the metal ran in both directions.
– Scrap iron bars were placed on the outside edge of each layer as they better withstood the heat of the furnace.
– The pile was then heated and hammered into plates.

– Three plates were piled on top of one another and heated to the required temperature in the furnace, and were then welded together by hammering with a four-ton Nasmyth hammer.

William Clay of the Mersey Iron Co gave evidence that his company had been making armour plates for testing purposes for several years. Their preferred method was to roll three sheets of iron and then weld these together using an eight-ton hammer, adding more iron until the plate was the required length.

Samuel Beale and Co had rolled their first armour plate in December 1855[8] for the ironclad battery HMS *Terror*, which was built by Palmer's shipyard on the Tyne. The manager, George Grant Sanderson, gave evidence that they had made 1,000 tons of armour plates, including 300 tons in the last three months for *Black Prince*. He described the process by which they made the armour:

– First they hammered some 12in bars from pig iron and some from scrap iron.
– Ten layers of bars of alternating pig iron and scrap-iron were piled together; the pig iron bars were generally on the outside.
– The bars were laid out 'so that they run both left and right and all across'. This produced cross-fibres in the iron, and Mr Sanderson showed the committee a sample that had a 'uniform character' at the fracture.
– The bars were then rolled into slabs that were 5ft x 3ft x 1½in.
– The rough iron on the outside of the slab was taken off and four slabs were put together and welded by rolling into another slab 2½in thick.
– Four of these 2½in plates were then heated and rolled into the final plate.
– The rollers used by the company were 20in in diameter and 6ft 6in long. They were linked to the furnace by a railway, and the four plates, which had been heated for between four and a half and five hours, were pulled out of the furnace by a pair of tongs attached by a chain to the top roller. The plates travelled down an inclined slope to the rollers 'in about half a minute', losing little heat in the process.
– The rollers had a reverse mechanism, and the plates were rolled back and forward for about three minutes until they were the required thickness.
– The plates were then allowed to cool and the edges trimmed with a slotting machine.

Table 1: Companies Making Armour Plates

Company	Location	<4in	4in	4.5in	5.5in	>5.5in	Total
Thames Iron Works	London	13	2	13	16	0	44
John Brown & Co	Sheffield	4	1	16	10	2	33
S Beale & Co	Rotherham	12	6	9	3	0	30
Mersey Iron Co	Liverpool	10	0	3	14	0	27
Millwall Iron Works	London	1	0	3	11	5	20

Table 2: Analysis of Average Scores in Proving Tests

Manufacturer	1863 Hammered	1863 Rolled	1864 Hammered	1864 Rolled
Thames Iron Works	5.60	–	4.00	–
Mersey Iron Co	5.07	5.42	5.50	8.00
Messrs Begbie[1]	8.50	8.17	–	8.00
J Brown & Co	–	6.79	–	7.71
S Beale & Co	–	6.25	–	6.00
Millwall Iron Works	–	6.96	–	7.50
Butterley Co	–	6.00	–	–
Messrs Schencking[2]	8.00	–	–	–
Low Moor Co	3.00	–	–	–
Portsmouth Dockyard	2.00	–	–	–
Messrs Hill & Smith	7.00	–	–	–
C Cammell & Co	–	7.50	–	6.67

Notes:
1 French company   2 Probably the Compagnie des Forges et Chantiers

There was sometimes fierce debate about whether rolled or hammered armour plates were the most effective. In the Committee's early tests the hammered plates tended to perform better than the rolled plates because the early rolling mills could not exert enough pressure to weld the layers of iron together. Robert Mallet, FRS, who in 1859 had presented a paper on the coefficients of elasticity and rupture in wrought iron to the Institution of Civil Engineers, gave evidence to the Committee in 1861 as follows:

> … I do not think there are any rolling tools in the country capable of properly making plates 4in thick and upwards, and of the size required. There are none that will give sufficient compression to the iron to bring it to its best state when worked in such large masses. … If more powerful rollers existed, to give a sufficient grip at once, I believe that a rolled plate would be the best you could have, but with our present tools I think that hammered plate is the best.[9]

In March 1862 Captain John Ford, a Director and Manager of Thames Iron Works, presented a paper to the Royal Institution of Naval Architects in which he argued that first rolling bars from scrap iron made the iron fibrous and tough and that the hammering hardened the plates without making them less tough. This, he argued gave the best of both worlds and meant that his company's plates were superior to those made by other methods. Sir John Dalrymple Hay, who was present at the meeting, replied that four firms, the Low Moor Works, the Thames Iron Works, the Pontypool Works and John Brown all produced plates of equal quality by different methods.[10]

In 1863 the Committee began a series of standardised firing tests using a 68lb shot, with a 16lb charge at a range of 200yds. Each plate was scored on an eight-point 'Order of Merit' from A1 to B4.[11] In total 119 plates were scored in this way, giving a sufficiently large sample to compare manufacturers and methods of manufacture. In order to compare the scores statistically the author has converted the merit scores into a numerical value where A1 = 8 and B4 = 1 (adding or subtracting a half point where the plate was additionally described as being good or poor). Table 2 has an analysis of the scores.

The results of this analysis show that by 1863 rolled plates were scoring an average order of merit of 6.63 compared to 5.43 for the hammered plates. By 1864 rolled plates were achieving an even higher average of 7.25 compared to only 4.75 for hammered plates. The reason for this improvement was because the earlier problems experienced in welding together the layers of rolled plates appear to have been overcome through the introduction of heavier and more powerful rollers. In June 1863 Richard Hewlett, Captain of HMS *Excellent*, where the firing trials took place, commented that:

> … all our trials with hammered or rolled plates tend to show the want of uniformity in the quality of the former, not only of those in batches but also as regards different parts of the same plate, I think that there cannot be a doubt, in this respect, of the superiority of the rolled ones.[12]

The Committee's report of November 1864 confirmed this view, and in the same year the manufacturers submitted only two hammered plates for tests compared to 24 rolled plates.

The Sheffield Makers

Shortridge, Howell & Co was the first Sheffield Company to submit an armour plate for trial in April 1861. The eight plates varied in thickness from ½in to 3in. The plates were described as being of homogeneous metal, which was the company's trade name for cast Bessemer steel with a low carbon content. All the plates were badly damaged, with the thicker steel plates being completely destroyed, showing the superiority of wrought iron over steel for making armour at that time. After this the company did not submit any further plates.

John Brown & Co had begun making their first plates using a mill with 21in rollers, made by WH & G Dawes of Elsecar, near Barnsley. In May 1861 four of their plates, ranging in thickness from 2in to 3½in, were used in a test of land fortifications. After these early experiments Brown's continued to develop the manufacturing process. On 31 July 1863 Brown told a meeting of the Institution of Mechanical Engineers that in the previous fortnight he had successfully tested two plates at Portsmouth. (These were presumably the two 5½in plates tested on 18 June 1863 which were given scores of A1 and Superior A1.) He described the manufacturing process as follows:

> Bars of iron 12in by 1in, 30in long were rolled, and then five of them were rolled into a rough slab. Two of these slabs were then rolled down to a plate 1½in thick, which was trimmed to 4 feet square.
>
> Four of these plates were rolled together into an 8ft x 4ft plate, 2½in thick; and lastly, four of these were then piled and rolled to form the final entire plate. The plate therefore comprised 160 thicknesses, each of which was originally 1in thick.
>
> The final operation of welding the four plates of 8ft x 4ft x 2½in is a very critical matter. To bring a pile of four plates of these dimensions up to a perfect welding heat all through the mass, without burning the edges and ends of the plate most exposed to the fire; to drag this pile out of the furnace, convey it to the roll[er]s and force it between them, in so short a time as to avoid its losing the welding heat, is a matter of greater difficulty than those unacquainted with the work would imagine. The intensity of the heat thrown off is almost unendurable, and the loss of a few moments in the conveyance of the pile from the furnace to the roll[er]s is fatal to the success of the operation.

This description bears a remarkable resemblance to the process followed by Samuel Beale & Co at Rotherham, about 7½ miles down the Don Valley from John Brown's Atlas Works, and this may not be coincidental. In 1906 Mr Johnson claimed in the *Manchester Guardian* that:

> When chief engineer of the Park Gate Works, over 25 years ago, I replaced the original armour plate mill by a larger one for producing ordinary boiler and ship plates, and some of the old hands working for me at the time told me of the Sunday visits paid by officials of the Atlas Works to Park Gate, to make sketches of the rolling mill, the heating furnaces and the apparatus for handling heavy plates.[13]

In September 1861 John Brown's submitted two plates of different modes of manufacture for HMS *Valiant*, one of two iron-built frigates of the *Hector* class. These ships were much shorter than the *Warrior* and *Black Prince* but were better protected, carrying only slightly less armour (912 tons compared to 975 tons) on a hull that was 75% of the length. This meant that the armour covered the whole length of the ship at the level of the gun battery, but only the midship portion on the waterline.

HMS *Valiant* was laid down by Westwood, Baillie & Co on the Thames in February 1861. Due to the company's financial difficulties Thames Ironworks took over, resulting in a delay of over a year compared to her sister ship. She was the first ship armoured by John Brown's (Naval History & Heritage Command, NH71209)

The firing tests had mixed results, with one plate passing the test and the other plate failing. In November 1861 a further plate for the same ship was tested and the results were reported in *The Engineer* as being 'considered very satisfactory'.

John Brown invested heavily in a new rolling mill in order to manufacture thicker plates. The new mill was significantly larger than his first one, with 32in rollers, and it was formally opened in front of a thousand guests on 10 April 1863. Guests included the Lords of the Admiralty and other cabinet ministers, MPs, peers of the realm, and 'an array of gentlemen of scientific and inventive eminence'.[14] Special trains were laid on for the more important guests.

For John Brown the opening was clearly a major promotional opportunity, and he laid on an extravagant event for his guests, including a demonstration of a plate being rolled which was recorded in a number of etchings, some of them making the process more dramatic than others.

However, the 4½in plate that had been rolled in the presence of the Lords of the Admiralty at the opening was also tested, and was judged to be very inferior to the company's normal standard. This was attributed to the 'unusual circumstances under which it was rolled':[15] it had been kept in the furnace much longer than normal and was rolled from a much thicker (18in) piece of iron, probably to enhance the drama of the display for the audience.

The first test plate produced by John Brown's new armour mill was part of an order for HMS *Royal Sovereign*, which was being converted from an unfinished wooden three-decker into the Navy's first turret ship as a trial of Captain Cowper Coles' proposals for turret ships. Designed for costal defence duties, she had five 10½in guns housed in four centre-line turrets, four of which had a single gun and the larger forward turret two. Each turret had an iron framework to which was fastened an inner skin of ½in iron, covered with 20in of

Two drawings of armour plate being rolled at the opening of John Brown's new Armour Plate Mill. As can be seen, the upper image involves about twice as many workmen and the open furnace is much more dramatic. The lower image probably represents the actual manufacturing process more accurately. (Sheffield Industrial Museums Trust)

HMS *Bellerophon* firing at the rear turret of HMS *Royal Sovereign* in Portsmouth harbour. The sighting target on the rear turret of *Royal Sovereign* can be seen on the left of the picture (*Illustrated London News*)

EFFECTS OF FIRST SHOT—EXTERIOR OF TURRET.

EFFECTS OF FIRST SHOT—INTERIOR OF TURRET.

The interior and exterior of HMS *Royal Sovereign*'s rear turret after the trial. The thickness of the protective armour and teak backing can be seen in the image on the left. (*Illustrated London News*)

wood backing and a covering of 5½in armour plates. The armour was reinforced around the gun-ports with an extra layer of 4½in armour. The roofs were unarmoured and covered in thin iron plating.

The 5½in test plate for the *Royal Sovereign* was tested at HMS *Excellent* on 21 April 1863 and scored an order of merit of A1.[16] On 21 January 1866 the turrets underwent a further, and more severe, test when three 9in, 250lb steel shots were fired at the rear turret from 200yds by HMS *Bellerophon* to test how well the turrets stood up to gunfire. None of the three shots penetrated the turret, and although the plates were bent the turret was still operational.

However, although John Brown had submitted 33 plates for testing in total and had promoted themselves as the pre-eminent manufacturer of armour plates they were not the only Sheffield firm to invest heavily in armour plate rolling technology. Charles Cammell & Co of the Cyclops Works, next door to John Brown's Atlas Works, built an extensive new mill, and in November 1863 they submitted a 4½in and a 5½in plate, described as 'experimental', for testing. The 5½in plate gave 'such proofs of toughness and capability of withstanding repeated blows in the same vicinity as to entitle it to a figure of merit of A1 … the 4½in experimental plate … stood the test almost as well'. The 5½in plate was fired

EARLY BRITISH IRON ARMOUR

Charles Cammell's first 5½in plate, with a segment cut off for the mechanical testing, was photographed by the company and is the earliest photograph of an identifiable armour plate that the author is aware of. The impacts marks of the shots exactly match the drawing in the Committee's report. (Sheffield Industrial Museums Trust)

Although Cammell's described the plate as 'experimental' to the Committee, it was intended for HMS *Lord Clyde*, a wooden-hulled armoured frigate laid down in September 1863 at Pembroke Dock. (Naval History & Heritage Command, NH 71198)

at eleven times, considerably more than normal, and because it had stood the tests so well samples were cut from the plates for mechanical testing by William Fairbairn in Manchester. Although it was hoped that such mechanical tests would provide information to improve armour plates, the Committee concluded in their final report that the mechanical and firing tests were 'so essentially different in their nature as not to admit any fair comparison and consequently no results drawn from the one can be legitimately applied to the other.'[17]

Although Cammell's described the plate as experimental it was in fact a test plate for an order that the company had received for the armour for HMS *Lord Clyde*. The *Lord Clyde* and her sister ship *Lord Warden* were built of wood, but unlike *Royal Sovereign* they

151

were designed and built as armoured ships rather than converted from existing hulls. The battery of the *Lord Clyde* was protected by 24in of oak backing covered in a 1½in iron plating on top of which was 6in of oak and then the iron armour plates, which were 5½in thick over the battery and 4½in over the rest of the hull above the waterline. The waterline armour was protected by another layer of planking in order to stop seepage between the plates.

Like John Brown, Cammell also invested heavily in a large new armour plate mill capable of rolling 12in plates, and this was opened on 21 November 1863. The opening ceremonies show the different culture of the two Sheffield armour-making firms. Whereas John Brown used the Admiralty testing facilities to trial different types of iron and manufacturing processes, Cammell's waited until they had perfected their process before submitting plates for testing. As we have seen, John Brown saw the opening of his new mill as an opportunity to influence decision makers and opinion formers. Charles Cammell preferred to conduct business 'in a much more quiet and unobtrusive manner'[18] and there was a feeling amongst Cammell's workforce that John Brown's were inclined to 'spin' their undoubted achievements. The opening of their new armour mill was a dinner in the new works for 2,000 workers at which the local vicar was the principal guest.

From Experiments to Mass Production

The new armour plate technology rendered the traditional wooden battleships of the world's navies obsolete, and the naval building programmes provided work for the newly-built armour plate rolling mills that John Brown and Charles Cammell had installed. Records from John Brown & Co[19] show how rapid the growth in armour sales in the early years was (see Table 3). However the uncertain nature of armour plate orders impacted on the company's profits when in 1865–66 it received virtually no orders for the first six months of the year.[20]

In September 1865 John Brown's produced a price list for all of their products. This included a price for 20ft x 4ft armour plates of varying thickness:[21]

Thickness (in)	Cost per plate
4½	£28
5	£29
5½	£30
6 to 8	£32
9 & 10	£36
11	£40

As a comparison, the price quoted for cast steel ships' plates of the same size was £22 – a much smaller price differential than was found between cemented steel armour and homogeneous steel plates at the end of the 19th Century, when the cost of 4in cemented armour plate was around 20 times that of homogeneous plates.[22]

Endnotes:
[1] Lambert A, 'Politics, Technology and Policy-Making, 1859–1865: Palmerston, Gladstone and the management of the Ironclad Naval Race', *Northern Mariner* Volume VIII, No 3 (July 1998), 14.
[2] Napier J, *The Life of Robert Napier*, Blackwood & Sons (Edinburgh, 1904), 209.
[3] The National Archives (TNA), Special Committee on Iron Plates and Guns, 13/3/1860, WO33/10.
[4] TNA, Special Committee on Iron Plates and Guns, WO33/13, 155.
[5] *The Engineer*, 29/8/1862, 127.
[6] Fairbairn W, *The Life of Sir William Fairbairn*, Longman Green (London, 1877).
[7] TNA, Report of the Special Committee on Iron 1861/2, WO33/11.
[8] *Sheffield Independent*, 16/12/1863.
[9] TNA, Report of the Special Committee on Iron 1861/2, WO33/11.
[10] *Transactions of the Institution of Naval Architects*, Volume 3, 1862.
[11] *New York Times* 28/11/1863 (quoting *The Times*).
[12] TNA, Report of the Special Committee on Iron 1863, WO33/13.
[13] Grant A, *Steel & Ships: The Story of John Brown's*, Michael Joseph (London, 1950).
[14] *Sheffield Independent*, 10/4/1863.
[15] TNA, Report of the Special Committee on Iron 1863, WO33/13.
[16] *Ibid*.
[17] TNA, Report of the Special Committee on Iron 1863, WO33/14.
[18] *Sheffield Independent*, 29/9/1863.
[19] SIMT, MNBJ004, Secretary's notebook.
[20] Director's report June 1866.
[21] SIMT, MNBJ004, Secretary's notebook.
[22] Comparison taken from *King Edward VII* class, NMM, ADMB0041.

Table 3: Growth in Sales at John Brown

Month	Armour sales	Percentage of total sales
April 1862	£0	0%
May 1862	£8153	24.6%
July 1862	£7476	25.0%
April 1863	£12,811	35.2%
May 1863	£15,641	42.0%
July 1863	£32,108	54.82%

AUSTRALIA'S FIRST DESTROYERS

Mark Briggs tells the story of the six destroyers of the 'River' class built for Australia. His article deals with the political aspects of the contract, and with the modifications to the original design that were necessary to ensure that these ships conformed to Australian requirements.

In February 1909 Australia's Minister for Defence, Senator George Pearce, announced that his government intended immediately to order three torpedo boat destroyers from Britain. 'These boats', he explained, 'the first of their kind to be obtained by Australia, will serve as models for the future construction of other vessels of the same type which the Government intends shall be built in Australia.'[1] Although Pearce referred in his statement to the existing British 'River' class, the ships Australia bought were of an evolved design – larger and more heavily armed, with more advanced propulsion and a longer range – which was to influence British destroyer development down to the First World War.

Ordering the destroyers had been a long and convoluted process. Naval defence had become a controversial issue following the federation of the Australian colonies in 1901. The Commonwealth government had entered into a ten-year naval agreement with Britain in 1902 whereby Australia paid £200,000 annually to the Royal Navy to base a fixed number of warships in Australian waters. While Prime Minister Edmund Barton had defended this agreement by citing advice from the British Admiralty, which preferred cash contributions to the Royal Navy, it still provoked vocal opposition. It did not sit well with the heightened sense of national identity that had emerged in Australia after federation. To many Australian nationalists hiring naval defence from Britain was demeaning.

HMAS *Yarra*. (RAN)

Leading the chorus for an Australian Navy was Captain William Creswell, Director of the Commonwealth Naval Forces. The Commonwealth had inherited a handful of vessels, mostly gunboats and torpedo boats, that had been acquired by the colonies in the 1880s for coastal defence. If something wasn't done to replace these ageing and obsolete ships the jobs of Captain Creswell and his men would disappear. Creswell wrote profusely, churning out articles and reports with proposals to rebuild the Commonwealth naval force. The type of warship he focused on above all others to achieve this aim was the destroyer.

Torpedo boat destroyers had first appeared in the middle of the 1890s. Their purpose was to intercept and sink enemy torpedo boats before they got close enough to their targets to launch their torpedoes. The early torpedo boat destroyers, however, were small and frail, unable to operate in the open sea or far from port.

Pleas from the Mediterranean Fleet for a more robust vessel than the existing turtleback designs, with greater range and able to maintain speed in a seaway, led to a new type of destroyer for the Royal Navy.[2] The first unit of the new 'River' class, HMS *Erne*, was built by Palmers Shipbuilding at Jarrow on the River Tyne and launched on 14 January 1903. At 223ft 6in (68.1m) overall and 550 tons, she was larger than her predecessors and featured a raised forecastle for better sea keeping. *Erne* was armed with a single 12-pounder, five 6-pounder guns and two 18in (450mm) torpedo tubes. Coal fired, she had a range of 1,860 statute miles (ca 3000km).

This new breed of larger, sea-going destroyer was manna from heaven for Captain Creswell. At around £70,000 each, the 'River' class destroyers were cheap enough not to frighten budget-conscious politicians, while their relatively small crews – *Erne* embarked just seventy men – also fitted Creswell's needs, as Australia had a small population from which to recruit. Most important of all, the better sea keeping and longer range of the new designs meant they were capable of more than just the harbour and port defence role to which the earlier torpedo boats and turtleback destroyers had been limited.

Creswell was effusive in promoting the new vessels. The British Admiralty had long argued that the only threat to Australia was from one or two enemy cruisers that had temporarily escaped detection by the Royal Navy. Destroyers, Creswell declared in 1905, could deter attacks on coastal towns. He went further, claiming that the new, longer-ranged destroyers could patrol Australia's coastal shipping routes and protect the vital maritime trade between the state capital cities. As an additional inducement to wavering politicians, Creswell suggested the destroyers could be built in Australia.[3]

Australia Decides to Buy Destroyers

In his efforts to revitalise the Commonwealth Naval Forces, Creswell had a powerful ally in Alfred Deakin, who was Prime Minister for much of the period from 1903 to 1910. Deakin understood the appeal of a local navy and was suspicious of British advice on the subject, suspecting that opposition to an Australian Navy was driven more by budgetary concerns in London than an impartial assessment of Australian defence needs.

Hoping for a sympathetic hearing, Deakin wrote privately to the Secretary of the Committee of Imperial Defence (CID), Sir George Clarke, explaining that extra money could be found for any naval purpose 'visibly and concretely Australian in origin' and seeking his advice on what might be done.[3] In keeping with Admiralty policy, Clarke did his best to discourage a local navy and he urged Deakin to put the matter before the full CID. Nevertheless, if Australia insisted on going down this path, Clarke considered destroyers were a good place to start and he echoed Captain Creswell's advice, suggesting the Commonwealth could purchase eight 'River' class destroyers for about £560,000.[4]

The formal response of the CID took a much harder line. There was no 'strategic justification' for building destroyers. They were not needed for harbour defence and 'could play no effective part' in protecting Australia's coastal trade. The Committee went further: it had previously 'hesitated to recommend the extinction of the local naval forces', but now believed that the few remaining Commonwealth naval vessels should be scrapped and, presumably, Creswell and his men sacked.[5] Appalled at the lack of sensitivity to public opinion in Australia, Deakin decided to ignore the CID report and follow the advice of his Naval Director (and Sir George Clarke). His government, he announced, would purchase eight ocean-going destroyers and four torpedo boats over the next three years.[6]

The First Round of Tenders

With Australia now in the market for destroyers, the government sought proposals from British shipbuilders. All the companies that built destroyers for the Royal Navy were invited to tender: Cammell Laird at Birkenhead, Fairfield Shipbuilding at Glasgow, Denny Brothers at Dunbarton, Armstrong Whitworth at Newcastle-on-Tyne, Palmers' Shipbuilding at Jarrow, Thornycroft at Southampton, Thames Shipbuilding in London, Samuel White at Cowes and Hawthorn Leslie at Hepburn-on-Tyne. Only the response from Cammell Laird has survived but Captain Creswell, in pushing the case for Australian destroyers, had clearly overstated what the existing designs might achieve.

Based on Cammell Laird's response, Creswell's specifications called for destroyers of 230ft (70m) in length with a beam of 23ft 9in (7.2m), a draft not greater than 8ft 6in (2.6m) and a freeboard not less than 6ft 6in (2m). They were to be powered by steam turbines and the fuel was to be oil, rather than coal; they would have a top speed of 26 knots and a range of 2,500 miles (4,023km) at 14 knots. These dimensions were roughly equivalent to the most recent of the 'River' class destroyers, such as HMS *Ness*, which had completed in 1905. The coal-fired

Ness, however, while reaching 25.5 knots on her trials, had a range of just 1,870 miles (2,992km) at 11 knots.[7] 'We regret to say', Cammell Laird explained, 'we are not able to fulfil all the requirements having regard to the dimensions shown on the plan.'[8]

The Birkenhead-based shipbuilder instead provided Australia with two options. They could build a destroyer to the dimensions specified or they could build a larger destroyer that met the speed and range requirements. With a reduction in freeboard to 5ft 6in (1.7m), a destroyer with the dimensions asked for by Australia could carry up to 150 tons of oil fuel, giving a range of 1,950 miles (3,120km) at 14 knots – 2,200 miles (3,520km) if the fuel that could be carried in 'peace tanks' was included. Alternatively, Australia could buy a larger destroyer that met the 2,500-mile range requirement. Cammell Laird suggested this could be achieved in a destroyer 255ft (77.6m) in length with a beam of 26ft (7.9m). Freeboard would not be less than 6ft 6in (2m) and draft not more than 8ft 6in (2.6m) with 200 tons of oil fuel.

Cammell Laird quoted a price of £76,000 for the first unit of the smaller class and £90,500 for the first unit of the larger class; costs would fall slightly for later vessels. Neither figure included armament, which was put at £10,800 for both designs, suggesting the larger vessel was not to carry a heavier armament. The shipbuilder politely declined to respond to the second part of the Australian inquiry, which was how much it would cost if all of the ships were built in Australia, pleading they had not had enough time to make enquiries or conduct negotiations.

Before a design had been selected, Alfred Deakin travelled to London to attend the 1907 Colonial Conference. To his utter amazement the Admiralty had reversed its position on Australian defence. Destroyers, the First Lord of the Admiralty, Lord Tweedmouth, now declared were 'useful for defence against possible raids or for co-operation with a squadron.'[10] It seemed Deakin had succeeded in bringing the imperial authorities around to his way of thinking. There was a catch, however. Britain wanted Australia to agree to the termination of the 1902 naval agreement.

Strategic circumstances had changed and the naval agreement had fallen out of favour in London. Britain had originally negotiated the agreement with Australia to fit in with the (secret) provisions of the 1902 Anglo-Japanese Alliance, which obliged the Royal Navy to maintain forces in the Pacific, including Australia, equivalent to those of Russia. The Japanese victory at Tsushima in 1905 dramatically reshaped this requirement. 'Never was there such an extravagant waste of money, ships and men,' the First Sea Lord, Admiral Sir John Fisher, wrote in the margin of a letter from Alfred Deakin, 'as this agreement entails on the Admiralty.'[11]

Despite his success, Deakin, ever suspicious of British motives, had reservations. He began to fear that with no naval agreement and no close association with the Royal Navy, the proposed destroyer force would quickly become outdated and inefficient; this is what had happened to the Australian colonial navies in the 1880s. He deferred ordering the destroyers, and no decision had been made when he lost office in November 1908. The incoming Labor government had no reservations. The Labor Party had opposed the 1902 naval agreement in favour of an Australian Navy, and the new government quickly announced its intention to buy destroyers for Australia.

The Second Round of Tenders

Nearly two years had passed since the Deakin government had requested tenders from British shipbuilders to supply destroyers to Australia. In London, Australia's naval representative, Captain Robert Muirhead Collins, urged the government to update the tenders before proceeding with an order. Advances in destroyer design had taken place over this time, he argued, and Australia should have the benefit of these developments.[12] Muirhead Collins had met informally with the Director of Naval Construction, Sir Phillip Watts. When the new tender document was issued, Muirhead Collins included subtle changes that reflected the lessons learned from the first round of submissions. While the vessels were to be 'similar' to those tendered in August 1907, the shipbuilders were free to make changes to achieve the desired capabilities, most importantly, endurance. 'Should you consider that these requirements could be more easily met by a slight modification of the dimensions of the vessels as shown in the plans then forwarded to you,' Muirhead Collins explained, 'any proposals you may wish to submit, if accompanied by plans and a statement of the increased cost of such modifications, if any, will receive full consideration.'[13]

In keeping with the earlier plan the destroyers were to be powered by steam turbines with oil-fired boilers. Turbines had been trialled successfully in the 'River' class destroyer *Eden* but oil fuel was a relative novelty. It was introduced aboard the large destroyers of the 'Tribal' class, the first of which were completed in 1908. The 'Tribal' class were also turbine powered, though this was a belated demand by the Admiralty, the shipbuilders preferring conventional triple-expansion engines.[14] Armament had been set down by Australia as one 4in and three 12pdr guns and three 18in torpedo tubes.

The new request for tender called for two ships to be delivered in the UK, with a separate quotation for a third vessel to be built in Britain, taken apart and reassembled in Australia. Shipbuilders were also to supply the Commonwealth with a complete set of plans as early as possible, ostensibly to facilitate the reassembly of the third vessel in Australia, but also to enable the construction of additional ships there. Muirhead Collins was told by Melbourne that it was 'not desirable' to give this reason to the British builders.[15]

The new tenders were received in March 1909, and the main characteristics are set out in Table 1. All the shipbuilders submitted new tenders except Cammell Laird, for which only the 1907 submission survives. Cammell

Table 1: Torpedo Boat Destroyer Tenders Received by Australia 1909

Shipbuilder	Length	Beam	Depth	Draft	Radius at 14 knots	Total cost for three ships[1]
Denny Brothers & Fairfield	245ft	24ft 3in	14ft 9in	–	2,877 miles[2]	£235,500
Cammell Laird (1907)	A) 230ft	24ft	14ft	7ft 9in	920 miles	£235,610
	B) 255ft	26ft	15ft	8ft 6in	2,500 miles	£265,610
Thornycroft	243ft	24ft 6in	15ft 9in	8ft 6in	2,877 miles	£236,500
Palmers	240ft	24ft	14ft 8in	–	2,877 miles	£242,100
Armstrong	240ft	23ft 6in	14ft	–	2,877 miles	£321,220[3]
Thames Shipbuilding	235ft	23ft 6in	15ft	8ft 3in	2,877 miles	£265,423
Hawthorne	230ft	23ft 6in	14ft 10in	7ft 6in	1,553 miles	£252,300
Samuel White	235ft	23ft 6in	14ft 10in	8ft 6in	2,877 miles	£245,840

Notes:
1 Includes one vessel to be dismantled and shipped to Australia for re-assembly.
2 Radius is in statute miles.
3 Does not include cost of shipping third vessel to Australia.

Laird may not have felt the need to modify their 1907 tender, which had included a larger design, though there is no correspondence to support this. They were certainly keen to be considered for the contract because they sent a representative to Melbourne for discussions with the Australian Department of Defence, and subsequently sought information from the Commonwealth on why their bid failed.

All the shipbuilders chose to submit larger designs in order to meet Australia's demanding endurance requirements. The Armstrong Whitworth proposal, for example, was 10 feet longer than their 1907 tender and 10 inches greater in depth. At 245ft, the proposal by the Denny Brothers and Fairfield consortium was 15 feet longer than the specifications originally set down by Captain Creswell.

Choosing a Design

Recognising his lack of expertise, Captain Muirhead Collins sought permission to employ a consultant to assess the different bids. He recommended John Biles, Professor of Naval Architecture at the University of Glasgow. It was an astute move. Apprenticed at the Royal Dockyard at Portsmouth and a graduate of the Royal School of Naval Architecture, John Biles had served on the 1901 Torpedo Boat Destroyer Committee set up to investigate the loss of HMS *Cobra*, which had broken in half on her delivery voyage. He had also been a member of the influential Committee on Design established by Admiral Sir John Fisher, along with Sir Phillip Watts and such future Royal Navy luminaries as John Jellicoe and Reginald Bacon.

Table 2: Assessment of Tenders by Professor Biles

Shipbuilder	Comments
Denny Bros & Fairfield	Largest vessel in terms of length overall. Oil tanks amidships, meaning less stress on vessel than in other designs. Good stability. Scantlings greater than in other designs. Lowest price.
Cammell Laird (1907 submission)	Proposal A did not meet range requirements. Proposal B excluded due to cost.
Thornycroft	Oil tanks in the ends of vessel, which will cause greater stress when on the crest of a wave and complicates piping. Depth greater than other designs and concerns about vessel's stability. Scantlings less than in Denny & Fairfield design.
Palmers	Scantlings less than in Denny & Fairfield design. Less freeboard forward due to design of bow. Arrangement of guns allows only three to be fired on the broadside. Placement of steering gear not as accessible as in other designs.
Armstrong Whitworth	Excluded due to price.
Thames Shipbuilding	Excluded due to price.
Hawthorne Leslie	Excluded as unable to meet necessary conditions as to radius of action.
Samuel White	Smaller than other designs. Deviates from suggested arrangement by putting oil tanks in the ends of vessel which will cause greater stress when on the crest of a wave and complicates piping. Scantlings less than in Denny & Fairfield design. Higher price.

Table 3: British Ocean-Going Destroyer Classes 1903–1910

	HMS *Erne* British 'River' Class 1903	HMS *Tartar* British 'Tribal' class 1907	HMAS *Parramatta* Australian 'River' class 1909	HMS *Pincher* British *Beagle* class 1910
Length overall	233ft 6in	274ft 3in	245ft	278ft 9in
Beam	23ft 6in	26ft	24ft 3in	28ft 6in
Displacement	540 tons	870 tons	750 tons	980 tons
Horsepower	7,000ihp	14,250shp	11,500shp	14,309shp
Fuel load (max)	130 tons coal	120 tons oil	179 tons oil	224 tons coal
Range (miles)	1,735 at 11kts	1,135 at 15kts	2,850 at 15kts	1,530 at 15kts
Speed (max)	25 knots	33 knots	26 knots	27 knots
Armament	1 – 12pdr	3 – 12pdr	1 – 4in/40 cal	1 – 4in/40 cal
	5 – 6pdr	–	3 – 12pdr	3 – 12pdr
	21 – 18in TT	2 – 18in TT	3 – 18in TT	2 – 21in TT
Complement	70	74	50	96

The plans attached to the tender submissions have not survived, but Professor Biles provided a detailed assessment of each design. These are summarised in Table 2.[16]

Professor Biles recommended the design by the consortium of Denny Brothers and Fairfield, pointing out that this proposal was both the cheapest and, after Cammell Laird's expensive 'B' design, the largest vessel in the competition.

The government in Melbourne cabled Muirhead Collins accepting the Denny Brothers and Fairfield tender on 13 March. As part of the deal Denny Brothers agreed that up to twelve Australian workmen would be employed by them at British rates of pay and trained during the construction of the destroyers, the Commonwealth to pay their travel costs and expenses on the condition that upon the completion of the ships they would return to Australia to assist in the reassembly of the third destroyer and the construction of additional destroyers.[17] The Commonwealth also agreed to Professor Biles being employed to oversee the construction of the vessels in Britain.

The Australian 'River' class

The first of the new destroyers, to be named *Parramatta*, was laid down at Fairfield's Glasgow shipyard on 17 March 1909. Sister ship *Yarra* was laid down by Denny Brothers at Dunbarton a few days later. They followed the pattern of the British 'River' class with a raised forecastle extending to the rear of a small bridge above chart and radio rooms. The forecastle contained the officers' quarters, wardroom and space for most of the 50-man crew, although the accommodation for the stokers was in the stern.

The oil-fuelled Australian 'Rivers' had three Yarrow-type boilers, rather than the four coal-burning boilers of the British 'River' class. There were seven Parsons turbines driving three shafts developing 13,000shp. The layout followed HMS *Eden* with a HP turbine on the centre shaft and LP, cruising and astern turbines on each wing shaft. *Parramatta* made 29 knots on her trials, three knots more than her designed speed. Up to 179 tons of oil fuel could be carried in six tanks extending across the bottom of the ship. Two tanks in the forward end held 29 and 37 tons respectively, with a further 63 tons in the midship tanks and 36 tons in the after tank. Fifteen tons of oil could also be carried in two 'peace tanks' under the forecastle.

Parramatta and her sisters each had a single 4in/40 (105mm) Mk VIII gun on a low platform forward. Originally in open mountings, the guns were later fitted with shields. There were three 12pdrs, two in a lozenge arrangement either side of the second funnel, and one on the centre-line at the stern. Three single, rotating torpedo tubes were fitted on the ship's axis, including one right aft, with Weymouth 18in (450mm) Mk I torpedoes – the first British heated torpedo.

While like the British 'River' class in appearance, the greater size of the Australian vessels, a consequence of the extraordinary range requirement, meant they could carry a heavier armament. Questions were already being raised about the 12pdr, and the subsequent British *Beagle* class had their forward 12pdrs replaced by a 4in gun while building. The *Beagle* class, though turbine powered, burnt coal, reportedly due to fear of oil shortages in wartime.[18] Coal had many disadvantages. It took up more space than oil for the same amount of energy

The launch of *Parramatta* at Fairfield's Govan shipyard on 9 February 1910. (Royal Australian Navy)

produced, required heavier machinery and demanded a larger crew that in turn needed more space in the hull. Table 3 shows comparative data for British destroyer classes before and after *Parramatta*.

Building Destroyers in Australia

Parramatta and *Yarra* arrived in Australian waters in December 1910, calling at Freemantle and Adelaide to rapturous welcomes before docking in Melbourne where their armament, which had been shipped out separately, was fitted. A key part of the government's plan, however, was to build additional ships in Australia. The third unit, *Warrego*, building in Glasgow, was to be dismantled and reassembled in Australia as preparation for the construction of further vessels.

The government's decision to build the first three ships in Britain had attracted much criticism at the time. Senator Pearce received numerous representations from local shipyards, trade unions and professional organisations calling for the whole squadron to be built locally. He stuck to his guns, telling a deputation from the Society of Boilermakers that 'the building of the first two vessels in England will mean the commencement of building vessels in Australia at an earlier date than would otherwise be the case.'[19] Cockatoo Island Dockyard, in Sydney Harbour, was chosen as the site where *Warrego* would be reassembled and a further three destroyers built. Cockatoo was owned by the state government of New South Wales. There had been a graving dock on the site since the 1850s and a second, larger dock had been opened in 1890. Along with supporting the Royal Navy's Australian Squadron, Cockatoo Dockyard had a solid record of building commercial vessels, but there were drawbacks, not least that everything had to be lightered to the island.

The keel of *Warrego* was laid down for a second time at Cockatoo on 1 December 1910. There were problems. Some of the parts had been damaged in transit to Australia and there were complaints about the quality of some of the work completed by Fairfield, leading to an acrimonious dispute between the Commonwealth and the Glasgow shipbuilder.[20] The issues were resolved, however, and the destroyer was launched on 4 April 1911.

The first wholly Australian-built destroyer, *Derwent* (subsequently renamed *Huon* to avoid confusion with a British destroyer of the same name) was laid down on 25 January 1913. While *Derwent* was under construction, Captain Creswell investigated the possibility of modifying the ship's armament. As mentioned above, the 12pdr gun had fallen out of favour with the Royal Navy because it lacked the 'punch' now believed essential in destroyer actions and Creswell wondered if the three 12pdrs could be replaced by two 4in guns. He also wanted to know if the three 18in torpedo tubes could be replaced with two single 21in (533mm) tubes. The most recent British destroyers carried the larger 21in torpedo, and the light cruisers of the 'Town' class being built for the Royal Australian Navy would also carry this weapon. It seemed desirable to adopt a single torpedo across the

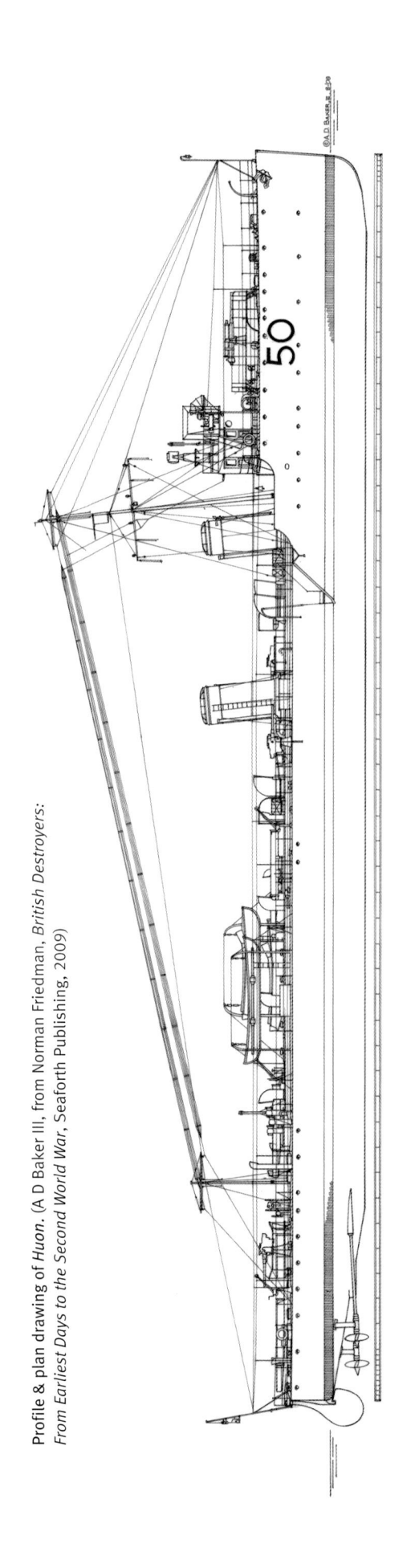

Profile & plan drawing of *Huon*. (A D Baker III, from Norman Friedman, *British Destroyers: From Earliest Days to the Second World War*, Seaforth Publishing, 2009)

AUSTRALIA'S FIRST DESTROYERS

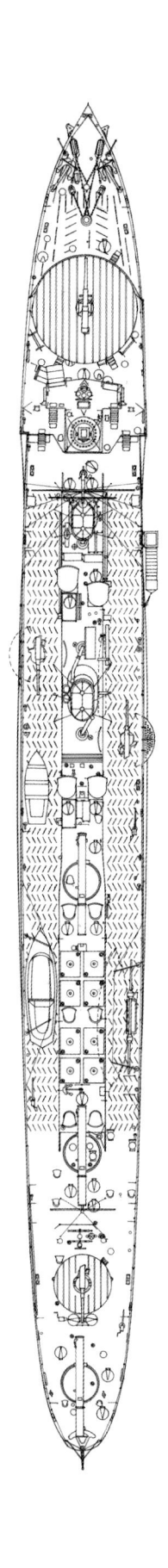

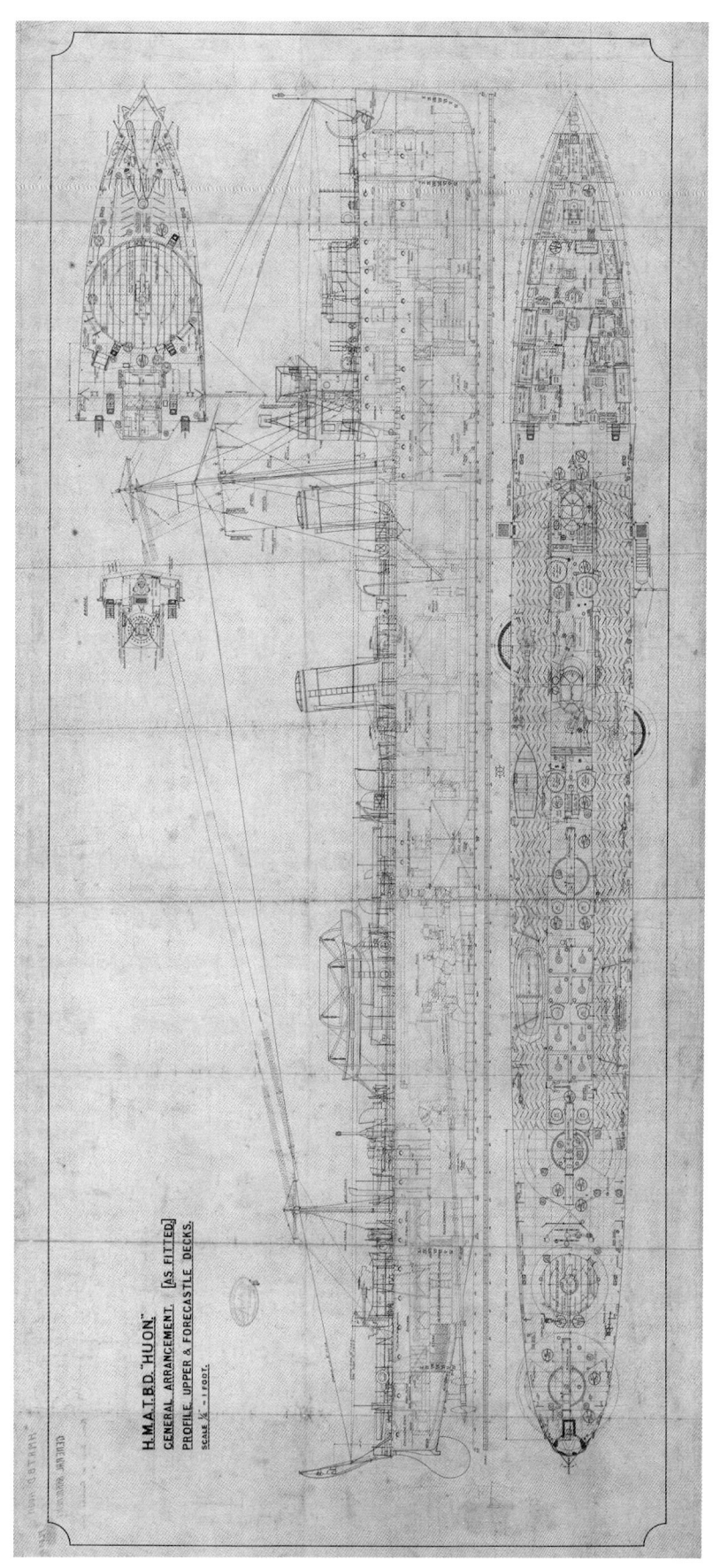

General Arrangement Plan of HMATBD (His Majesty's Australian Torpedo Boat Destroyer) *Huon*. (Australian National Archives)

Parramatta shortly after her arrival in Australian waters in December 1910. Her armament has not yet been fitted; it was shipped to Australia separately to provide space and weight for additional fuel for the long delivery voyage. *Parramatta* is flying the Australian flag, not the White Ensign, from the ensign staff at her stern. Permission to fly the White Ensign came with the formal establishment of the Royal Australian Navy in July 1911. (RAN)

Derwent (renamed *Huon* at completion) and *Torrens* under construction at Cockatoo Island. (RAN)

Derwent, later renamed *Huon*, under construction at Cockatoo Island Dockyard, Sydney, less than three weeks before her launch. (RAN)

Warrego: The absence of a shield on her 4in gun and a pennant number on her hull suggests this photograph was taken before the outbreak of the First World War. (State Library of Victoria)

Table 4: Australian 'River' class Destroyers

	Builder	Laid Down	Launched	Completed	Fate
Parramatta	Fairfield	17 Mar 1909	9 Feb 1910	10 Sep 1910	Paid off 1928 and later grounded.
Yarra	Denny	1909	9 Apr 1910	10 Sep 1910	Paid off 1920. Broken up at Cockatoo 1929.
Warrego	Fairfield	29 May 1909			
	Cockatoo	1 Dec 1910	4 Apr 1911	1 Jun 1912	Paid off 1928. Dismantled Cockatoo 1929.
Huon (ex-*Derwent*)	Cockatoo	25 Jan 1913	19 Dec 1914	14 Dec 1915	Paid off 1928. Expended as target 1930.
Torrens	Cockatoo	25 Jan 1913	28 Aug 1915	3 Jul 1916	Paid off 1926. Expended as target 1930.
Swan	Cockatoo	22 Jan 1915	11 Dec 1915	16 Aug 1916	Paid off 1928. Foundered 1934.

squadron.[21] Replacing the 12pdrs with 4in guns proved impossible: there was not enough space available to accommodate the recoil of the larger weapon. Fitting the 21in torpedo was possible but it would require some internal alterations and extra stiffening.[22] It was decided to proceed with the original design, and the outbreak of war the following year made achieving even this difficult.

With the start of construction in Australia, it was soon clear that building these technologically sophisticated vessels was much more challenging than the local shipbuilders' usual fare of tugs and dredgers. *Derwent* took twenty-three months to build with a further year for fitting out. This compares with eighteen months from the laying down of the keel to completion for *Parramatta*. The next vessel, *Torrens*, also took thirty-five months. Things improved with the final unit of the class, *Swan*, which was completed in nineteen months. To be fair, the outbreak of the First World War caused delays. Obtaining equipment from Britain became very difficult. Torpedo tubes ordered from Armstrong Whitworth in December 1914 were not received until early 1917, while 4in guns proved impossible to get hold of as Britain introduced wartime emergency construction programmes and began arming its merchant fleet. The RAN was forced to scavenge weapons from older vessels in order to complete the three Australian ships. Inevitably costs skyrocketed: the Australian-built destroyers cost about £180,000 each, twice the cost of their British-built sisters.

War Service and Fates

At the outbreak of war in August 1914 the three destroyers in commission were all in Australian waters, along with the major units of the recently-established

A fine port bow view of *Warrego*. (State Library of Victoria)

Huon during the First World War. (State Library of Victoria)

Yarra in the Mediterranean during the First World War. She is shown here operating a kite balloon to spot submarines. (RAN)

Parramatta during the First World War. (Allan C Green collection, State Library of Victoria)

Yarra out of the water for repairs at Livorno in October 1918 after a collision with sister ship *Huon*. (RAN)

HMAS *Swan* during a visit to Melbourne. (Allan C Green collection, State Library of Victoria, H91 250/1226)

Royal Australian Navy: the battle cruiser HMAS *Australia* and the light cruisers *Sydney* and *Melbourne*. The destroyers participated in the capture of German New Guinea and were then employed on patrol work in Southeast Asian waters. As the remaining three vessels entered service through 1916 they joined their sister ships in this work.

In 1917 the British government asked that the Australian destroyers be sent to the Mediterranean, where there was a growing threat from German submarines. Fitted to carry depth charges, a recent innovation, *Parramatta* sank a submarine while escorting a convoy between Alexandria and Malta. To enhance their anti-submarine capability *Parramatta*, *Yarra*, and *Huon* also landed their aftermost torpedo tube and had their mainmast moved forward to allow them to operate an observation balloon. As part of the Allied blockade of the Adriatic, *Torrens* was engaged in a skirmish with Austrian destroyers.

In 1918 the Australian destroyers became part of the Royal Navy's 5th Destroyer Flotilla. They saw service in the Sea of Marmora and the Black Sea, HMAS *Swan* being sent into the Sea of Azov to support anti-Bolshevik forces fighting in the Crimea. All six destroyers returned to Australia in 1919. *Parramatta*, *Warrego* and *Yarra* were refitted at Cockatoo where the middle 18in torpedo tube was replaced with a twin 21in (533mm) mounting, and a 24in (609mm) searchlight was fitted on a platform immediately forward of it.[23]

These modifications seem to have been something of a waste of money, as the ships saw relatively little service after the war. The *Marksman*-class destroyer leader *Anzac* was transferred to the RAN in March 1919 followed by five 'S' class destroyers in June. These war-built ships replaced the 'River' class in active service, with the old destroyers being laid up. All six vessels were brought back into service for the visit of the Prince of Wales in May 1920. *Torrens*, *Warrego*, *Parramatta* and *Huon* undertook stints as training ships for naval reservists at ports around Australia, but by 1928 all were slated for disposal. *Huon* and *Torrens* were used as gunnery targets, while *Yarra* and *Warrego* were broken up. *Parramatta* and *Swan* were sold for use as fishermen's cabins on Sydney Harbour but the plan was abandoned. *Swan* foundered while under tow to the breakers in 1934. *Parramatta* had the most protracted fate: breaking loose from her moorings, she ran aground near the mouth of the Hawkesbury River in Sydney Harbour, her hulk remaining there until 1973. Hopes of restoring the old destroyer as a museum ship came to nothing, but her bow and stern were removed as a naval memorial in Parramatta.

Huon moored alongside two of her sisters in Sydney Harbour after the end of the First World War. The battle cruiser HMAS *Australia* is in the background. The old cruiser to the left of the picture is HMAS *Encounter*, used by the RAN as a training ship. (RAN)

Endnotes:

1. Statement by Senator George Pearce, Minister for Defence, 5 February 1909, National Archives of Australia (NAA), MP1049/14, 1911/3652.
2. Norman Friedman, *British Destroyers: From Earliest Days to the Second World War*, US Naval Institute Press (Annapolis 2009), 86–87.
3. 'Naval Defence: Report of the Director of the Naval Forces for the Year 1905', *Commonwealth of Australia Parliamentary Papers*, 1906 session, Vol II, No 44.
4. Alfred Deakin to Sir George Clarke, 3 October 1905, National Library of Australia (NLA), Deakin Papers, Ms 1540/15/3522.
5. Sir George Clarke to Alfred Deakin, 6 October 1905, NLA, Deakin Papers, Ms 1540/15/3526.
6. 'Report of the Committee of Imperial Defence upon a General Scheme of Defence for Australia', *Commonwealth of Australia Parliamentary Papers*, 1906 session, Vol II, No 62.
7. *Commonwealth of Australia Parliamentary Debates*, 1906 session, Vol XXV, 5577–78, 26 September 1906.
8. Friedman, *op cit*, 293.
9. RR Bevis, Director, Cammell Laird & Co, to Australian Commonwealth Offices, London, 13 August 1907, NAA, MP1049/14, 1911/3652.
10. 'Colonial Conference 1907: Minutes of the Proceedings of the Colonial Conference', *Great Britain Parliamentary Papers*, 1907 session, Vol LV, Cd 3523, 130.
11. Marginal note by Admiral Sir John Fisher on a letter from Alfred Deakin. Deakin to Fisher, 12 August 1907, The National Archives (London), Adm 116/1241B.
12. Muirhead Collins to Department of Defence, Melbourne, 19 February 1909, NAA, MP1049/14, 1911/3652.
13. Muirhead Collins, 6 February 1909, NAA, MP1049/14, 1911/3652. All the shipbuilders who had submitted tenders in 1907 were sent this request to submit a revised tender.
14. Friedman, *op cit*, 107.
15. Department of Defence, Melbourne, to Muirhead Collins, 12 February 1909, NAA, MP1049/14, 1911/3652.
16. Assessment of tenders by Professor John Biles, 3 March 1909, NAA, MP1049/14, 1911/3652.
17. SA Pethebridge, Acting Secretary, Department of Defence, Melbourne, to Muirhead Collins, 5 March 1909, NAA, MP1049/14, 1911/3652.
18. Edgar Marsh, *British Destroyers: A History of Development 1892–1953*, Seeley Service (London 1966), 80.
19. 'Record of Meeting with Deputation from the Society of Boilermakers, NSW', 12 February 1909, NAA, MP1049/14, 1911/3652.
20. 'HMAS *Warrego* – Reconstruction in Australia and Australian facilities for completing torpedo boat destroyers', NAA, MP104/14, 1912/4646.
21. Memorandum by Rear Admiral Sir William Creswell, First Naval Member of the Australian Naval Board, October 1913, NAA, MP472/1, 16/16/6917.
22. Memorandum by Engineer Captain William Clarkson, Third Naval Member of the Australian Naval Board, 26 November 1913, NAA, MP472/1, 16/16/6917.
23. Published descriptions of these vessels make no mention of these alterations, but there is a set of 'as fitted' plans from Cockatoo Dockyard dated 26 May 1919 that show the changes, suggesting the destroyers were modified in this way. NAA, C3598.

NORTH SEA PARTNERS:
The British and Dutch Navies in the Cold War Era

Jon Wise takes a close look at the successes and failures of Anglo-Dutch collaboration on a series of postwar naval programmes and initiatives.

The British Ambassador to the Netherlands wrote to the Foreign Secretary Ernest Bevin on 25 September 1945 praising a Naval Exhibition and Navy Week which had just taken place in Rotterdam with Royal Navy (RN), US Navy and Royal Netherlands Navy (RNlN) vessels in attendance. He remarked:

> Showing the Flag is always good commercial propaganda and under these circumstances where the people are particularly well disposed to us and admire us it comes at an opportune time.[1]

War-torn Rotterdam had been liberated just five months earlier, and the RN, with hardly any vessels left in the Home Fleet, had managed to scrape together enough ships to represent the country abroad.

Close wartime Anglo-Dutch naval co-operation had commenced with the fall of the Netherlands in 1940, when the Royal Family and the Cabinet had fled to Britain along with a considerable portion of the Netherlands Navy. Thereafter, for the duration of the war, the RNlN integrated operationally with the RN. Likewise, in the Far East, the RNlN was to play a significant role firstly in the losing battle against Japan and, some three years later, in ultimate victory over that opponent and the reclamation of its overseas territories.

Regeneration

Foreign Office documents of the immediate post-war period illustrate the commitment of both the British Government and the Admiralty in rendering assistance to the Dutch Navy.[2] This was in recognition for the part played by a faithful ally, to perpetuate an historic alliance and to acknowledge the on-going importance of co-operation in the face of a growing threat from the Soviet Union. This article traces the course of that partnership during the Cold War years, as the UK helped materially in the re-building of the RNlN and later collaborated in aspects of the design and equipping of their respective navies. Although this is not a narrative of unalloyed success, throughout the period relations both at diplomatic and service level remained positive and predominantly close.

An ambitious plan was drawn up by the Dutch Naval Staff at the end of the war for the construction of a so-called 'Seagoing Fleet' which combined the latest theories of naval warfare with practical lessons learned from the recent conflict. In broad terms this envisaged the creation of two 'Light Naval Task Forces', each comprising a fixed-wing aircraft carrier and two cruisers.[3] Before indigenous new construction could be completed, the Dutch relied heavily on the purchase or loan of ex-British warships, as shown in Table 1.

In January 1946, the British Admiralty agreed to loan

Table 1: RN Vessels Sold/Loaned to the Royal Netherlands Navy Following the Second World War

RNlN Name	RN Name	Type & Class	Sale/Loan	Commissioned in RNlN Service
Karel Doorman	*Nairana*	Escort Carrier/*Vindex*	Loan	20 Mar 1946 to 9 Mar 1948
Karel Doorman	*Venerable*	Light Fleet Carrier/*Colossus*	Sale	28 May 1948
Evertsen	*Scourge*	Destroyer/'S' Class	Sale	1 Feb 1946
Kortenaer	*Scorpion*	Destroyer/'S' Class	Sale	1 Oct 1945
Piet Hein	*Serapis*	Destroyer/'S' Class	Sale	5 Oct 1945
Banckert	*Quilliam*	Destroyer/'Q' Class	Sale	21 Nov 1945
Marnix	*Garland*	Destroyer/'G' Class	Sale	16 Jan 1950
Dolfijn	*Taurus*	Submarine/'T' Class	Loan	4 Jun 1948 to 8 Dec 1953
Zeehond	*Tapir*	Submarine/'T' Class	Loan	18 Jun 1948 to 15 Jul 1953
Pelikaan	*Thruster*	LST (1)/*Boxer*	Sale	3 Jul 1948
Vulkaan	*Beachy Head*	Maintenance/Depot/'Cape' Class	Loan	15 Apr 1947 to 1950

In addition, eight Australian-built *Bathurst* class minesweepers were sold in 1946 for service in the Dutch East Indies. Other minor vessels sold included 17 motor minesweepers, three open launches, three river minesweepers, six LSTs, a high speed patrol boat, one MFV, two refuelling tenders, two seaplane tenders and two LCMs.

Source: FO371/90037

The postcard shows the first *Karel Doorman* in Dutch service. Pre-dating any NATO nomenclature, the prominent 'QH' visible on her side denotes 'Auxiliary Aircraft Carrier'. (Foto Stevens Hilversum)

an escort carrier, either the *Nairana* or *Campania*, for two years on the understanding that the Dutch would purchase a *Colossus*-class light fleet carrier at the end of that period. The RNlN had already gained experience of operating naval aircraft at sea in the Arctic, Atlantic and Pacific Oceans.[4] In the event HMS *Nairana* was loaned, becoming HrMs *Karel Doorman*, and in 1948 HMS *Venerable* was purchased and given the same name as her predecessor. Initially, she operated the Hawker Sea Fury FB Mk 50–60 and the Fairey Firefly FR Mk IV.

The Netherlands already possessed two other major surface warships which had remained partially constructed throughout the Second World War. The 11,850-ton cruiser *De Zeven Provinciën*, (later re-named *De Ruyter*) had been launched by the Germans with the intention of employing her as a block-ship at Rotterdam, while *Eendracht* (later *De Zeven Provinciën*) was still on the building slip. The armour and decks were complete and the boilers and some of the auxiliary machinery in place. However, the majority of the equipment had been removed by the Germans; after nearly a decade of inactivity most of it was in any case obsolescent. In 1945 two Dutch senior officers travelled to London to enquire whether the *De Zeven Provinciën* could be towed to the UK to be fitted with a British main armament and a modern superstructure. More presciently, it was suggested that the cruiser could be despatched to the Pacific to act as an HQ ship before the end of the war.[5]

An investigation examined the possibility of the cruisers being fitted with the triple 6in gun mountings of the *Fiji* class. Although the displacements of the two classes were similar, it was found that the *Fiji* class was designed to carry a considerably greater weight of armament: 1,183 tons compared with 689 tons in the Dutch cruisers. Despite this initial setback, the Dutch were afforded full cooperation, and their Chief Naval Constructor and five draughtsmen subsequently worked at the Royal Corps of Naval Constructors' headquarters at Bath during the latter part of 1946 on the drawings and calculations required to recast the design of the cruisers.[6]

When they commissioned in 1953, the two cruisers were very different from those which had been envisaged

The second carrier to be named *Karel Doorman* in Netherlands Navy service. The *Colossus*-class ship was purchased in 1948 and sold to Argentina in 1969, becoming ARA *Veinticinco de Mayo*. (Author's collection)

The Maintenance Ship HMS *Beachy Head* was employed as a Minesweeper Depot Ship and re-named HrMs *Vulkaan* for the loan period. (Author's collection)

The submarine chaser *Freyr*, built at Den Helder 1953–54, was one of several classes of minor Dutch vessels paid for with US funds under the Mutual Defense Assistance Program. (GEBR Spanjerberg NV, Rotterdam)

in the 1930s, not least in their main armament. To counter the aerial threat, eight dual-purpose 152mm guns, manufactured under licence from the Swedish Bofors company, were fitted. These had an elevation of 60°, were fully automatic and radar controlled, with a rate of fire of fifteen rounds per minute. Major internal modifications included different arrangements for magazines, hoists and boiler trunking, and a unit system for the propulsion machinery which required a second funnel. An Action Information (AIO) Centre, new fire control systems and the prominent LW-01 air-search radar were further evidence of a 'new-age' design. The LW-01 set, which formed a very narrow 1-degree beam, was manufactured by NV Hollandse Signaalapparaten (HSA), part of the excellent Dutch electronics industry which had been rapidly resurrected post-1945.[7]

The above account infers an untrammelled transition from war to peace for the Netherlands and its navy, but this was not the case. The immediate future of its vast and troubled Eastern Empire and the role its small but ambitious navy was to play in the developing Cold War were two issues which needed to be settled, and were causes of friction both internationally and domestically.

Overseas Commitments

At the end of the war, the Netherlands East Indies (NEI) was riven with unrest due to the irresistible drive towards independence led by the *Partai Nasional Indonesia* (PKI). Moreover, there was still a huge garrison of Japanese military *in situ*, while an often-reviled colonial ruling class was emerging from the prison camps demanding a return to pre-1942 *Rust en Orde* – tranquillity and order. This situation presented problems for the Navy which, among other tasks, was required to counter smuggling, which was seriously undermining the economy. This task required a constabulary role previously undertaken by the civilian-manned *Gouvernementsmarine*, which had been organised and paid for by the local NEI Government.[8]

In 1945 the Netherlands Admiralty enquired about the transfer of two destroyers and a cruiser with immediate effect in order to provide a naval presence in the NEI region. Although the destroyers were not a problem, the issue of the transfer of a cruiser proved contentious. Despite an acceptance by the British that 'the Dutch constitute a special case in the matter of transferring naval warships to foreign powers other than the Dominions', the RNlN made it clear that a modern cruiser such as a *Fiji* was required. There was division of opinion within the British Admiralty, and protracted debate before the much older HMS *Leander* was offered. The Dutch declined.[9]

A more serious international political row erupted in 1947 over the interception, boarding and impounding of British and American merchant ships within a self-

This quarter view of *De Zeven Provinciën* taken in 1961 shows how the threat from the air influenced naval designers in the immediate post-war years. Both the main and secondary armaments had dual-purpose, rapid-fire guns. (Koninklijke Marine)

NORTH SEA PARTNERS: THE BRITISH AND DUTCH NAVIES IN THE COLD WAR ERA

HrMs *Rotterdam* at sea sometime between 1970 and 1976. The after Bofors twin 120mm gun mounting is prominent in this photograph, likewise the LW-02 air surveillance radar. The covering over the forward funnel indicates that the after boiler only is in operation. (Author's collection)

The *Friesland* class destroyer HrMs *Drenthe* driving through choppy seas. These were purpose-built, ocean-going ASW destroyers, evidence of their country's determination to be recognised as a 'blue water' navy in the Cold War era. (GEBR Spanjersberg NV, Rotterdam)

171

proclaimed maritime exclusion zone around the NEI. The Dutch claimed that the ships might be carrying contraband cargoes obtained illegally by the nationalists from Dutch-owned plantations. The British countered that the ships were bringing much-needed supplies such as palm oil, tea and sugar to the austerity-hit UK. The Chancellor, Hugh Dalton, stated that it was up to the Dutch 'to see we get, promptly and urgently, the Indonesian supplies we need so urgently' in partial reparation for seven years of support. He added that the matter was contributing to a 'very grave foreign exchange crisis'. The tension subsided once the Dutch Navy officially withdrew from NEI waters at the end of 1950, having transferred some coastal patrol and anti-smuggling vessels to the nascent Indonesian Navy.[10]

Meanwhile, as the Dutch overseas empire was crumbling, NATO's military command wanted the Netherlands to prioritise the build-up of their land and air forces in the face of the threat from the Soviet Red Army. Jan Brouwer asserts that it was reasonable 'to ask the Dutch to limit their maritime efforts to coastal defence and mine-sweeping in the North Sea, while leaving defence of the Atlantic to the larger navies'.[11]

The British added their voice to this apparent call for realism. A senior civil servant wrote in 1950 that in 'the modern world' the Dutch could no longer expect to 'range the seas with their ships'. However, it was an alleged public remark by Lord Tedder, at a NATO Chiefs of Staff Meeting in October, which threatened seriously to upset relations between the North Sea partners. The *New York Times* reported that the Air Marshal had remarked that the Netherlands would 'have to reduce its navy to light vessels only'. This prompted a somewhat hysterical dressing down for the British Ambassador in Washington by his Dutch counterpart who, according to the British diplomat, 'had got the impression that the United Kingdom was behind the whole business and that this was yet another example of the centuries-old British jealousy of the Dutch fleet'.[12]

Meanwhile, at home, the majority of the Dutch Cabinet opposed the naval expansion plan, claiming that it was far too expensive. They were supported by the opposition party, the Social Democrats, who agreed that in time of crisis NATO allies would protect the sea lanes and that the Navy should be sacrificed in order to build up the army and air force. However, the majority in Parliament, the right-of-centre parties and the general public rallied in support of the Naval Staff, and in the autumn of 1950 the government bowed to the wishes of both Parliament and the people and accepted the plans for the reconstruction of the Royal Netherlands Navy.[13]

Innovation

The RNlN's 'blue water' naval ambitions received official recognition in 1950 when, following intensive lobbying, NATO handed the RNlN the task of helping to defend sea lines of communication in the Northeast

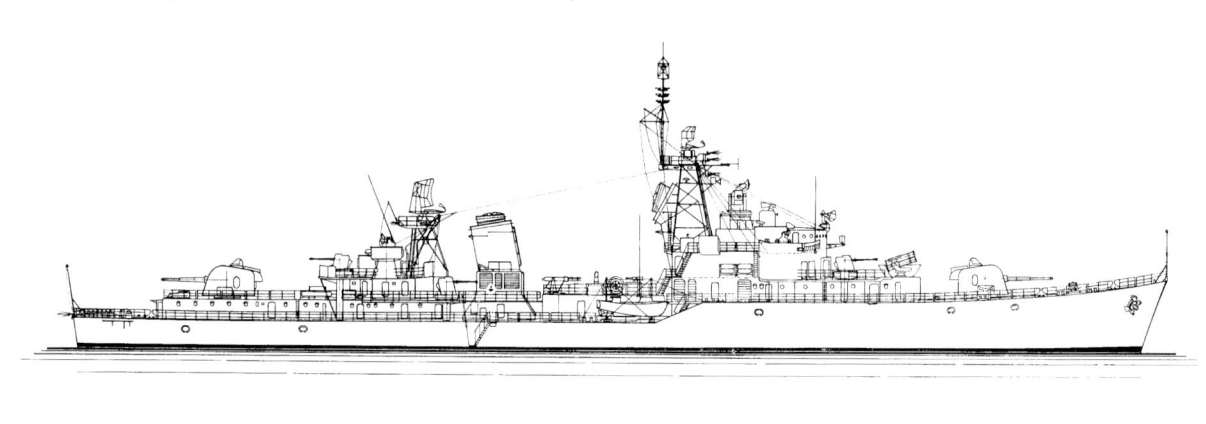

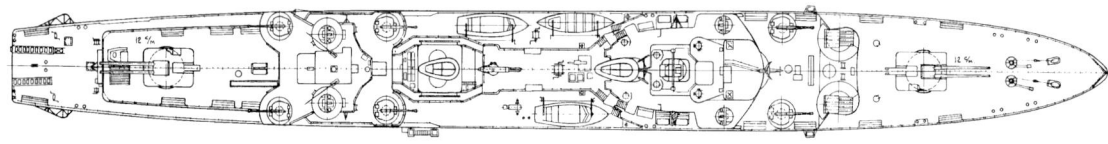

PRINCIPAL CHARACTERISTICS H. N. Ms ANTI-SUBMARINE DESTROYERS "FRIESLAND" CLASS

The eight destroyers of this class as well as the four destroyers of the *Holland* class were built by The Rotterdam Dockyard Co. Ltd., Dock- and Shipbuilding Co. Wilton-Fijenoord Ltd., Netherlands Dock- and Shipbuilding Co. Ltd. and Royal Company "De Schelde" Ltd.

Commissioned 1954-1958.

Length over all .... 380' 6''	Displacement full load .... 3,070 tons	Armament: guns .... 4 × 4,7 inch
Length between p.p. ...... 370'	Maximum S.H.P. 60,000	6 × 40 mm.
Breadth moulded on w.l. .. 38' 6''	Maximum speed .... 36 kts.	A.S. rocket throwers (mortars) .. 2 × 4
Displacement standard .... 2,476 tons	Complement 283	depth charge racks 2
		flare rocket launcher 1

The silhouette of the slightly smaller Holland class destroyers resembles in general the Friesland class.

Plans prepared by the Netherlands United Shipbuilding Bureaux Ltd, from Hubert V Quispel, *The Job and the Tools* (Rotterdam 1960).

The distinctive profile of HrMs *Karel Doorman* following her 1955–58 reconstruction, which included the construction of an angled flight deck. (Koninklijke Marine)

Atlantic. Later, NATO's Standing Group, which was charged with organising allied defence, specifically included HrMs *Karel Doorman* in its plans. Up until then, the carrier had been considered the most superfluous of the Dutch warships.[14] Assurances in the form of financial backing by the Dutch Government, coupled with a specific maritime role within NATO, allowed the Dutch Navy to forge ahead during the 1950s with its modernisation and improvement plans.

The construction of twelve destroyers of the *Holland* and *Friesland* classes, which were all completed in Dutch shipyards within a four-year period between 1954 and 1958, was a remarkable achievement in the face of initial delays due to the devastation inflicted on the country's defence industry by the war. Although they bore a passing visual resemblance to contemporary British destroyers, that was all. These warships were specifically designed for ocean ASW work and, significantly, were fitted with indigenous sonar sets.[15]

The extensive modernisation of *Karel Doorman*, undertaken by Wilton-Fijenoord between 1955 and 1958, borrowed extensively from three British innovations in postwar carrier design: the steam catapult, the angled flight deck and the stabilised optical deck-landing

The 'island' following HrMs *Karel Doorman*'s 1955–58 reconstruction. Her radar sets include, from left to right: VI-01 height-finder, LW-02 and LW-01 air surveillance, the smaller DA-01 combined air/surface surveillance antenna and a second VI-01 array. (Leo van Ginderen collection)

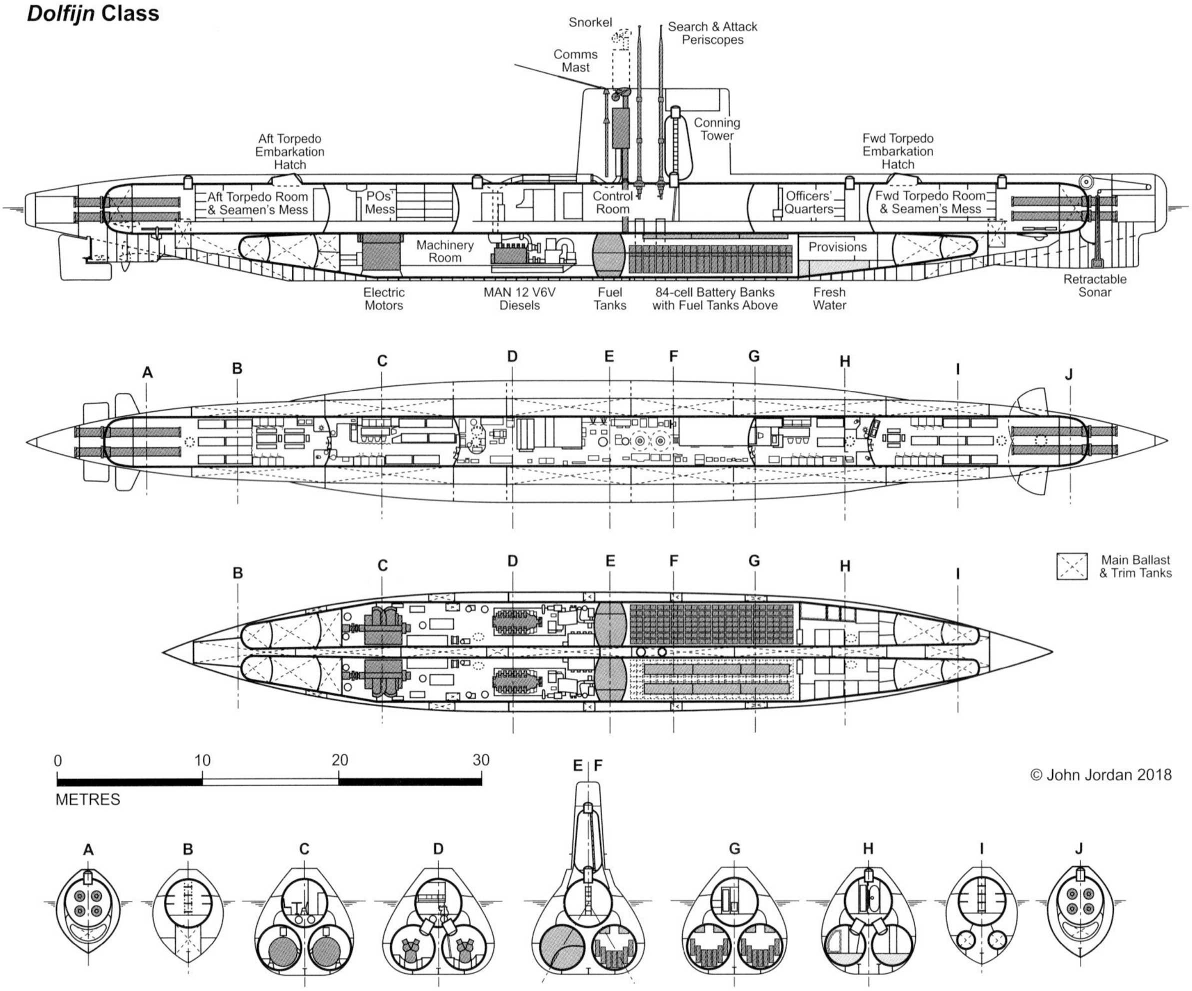

Dolfijn Class
Snorkel
Search & Attack Periscopes
Comms Mast
Conning Tower
Aft Torpedo Embarkation Hatch
Fwd Torpedo Embarkation Hatch
Aft Torpedo Room & Seamen's Mess
POs' Mess
Control Room
Officers' Quarters
Fwd Torpedo Room & Seamen's Mess
Machinery Room
Provisions
Electric Motors
MAN 12 V6V Diesels
Fuel Tanks
84-cell Battery Banks with Fuel Tanks Above
Fresh Water
Retractable Sonar
A
B
C
D
E
F
G
H
I
J
Main Ballast & Trim Tanks
0
10
20
30
METRES
© John Jordan 2018
E|F

S805

The *Dolfijn*-class submarine *Tonijn* underway on the surface. Her prominent pennant number (visual call sign) indicates that the photograph was taken before 1972. (Henk Visser)

sight. These allowed the ship to operate a squadron of Sea Hawk jet fighters.[16] Post-refit, advances in the Dutch electronics industry were evident in the impressive radar fit, which included both the LW-01 and LW-02 air search radars. The latter was claimed to have an aircraft detection range of 100 nautical miles and a ceiling of 59,000 feet. A further air/surface search capability was provided by the S-Band DA-01 set plus a pair of V1-01 'nodding beam' height-finding radars which were fitted at either end of the island structure.[17]

The most interesting Dutch innovation of the period, one which has received scant attention hitherto, was the design and build of four three-cylinder submarines of the *Dolfijn* class. The pressure hulls were arranged in a pyramid shape. The upper cylinder accommodated the crew plus navigational equipment and armament, while each of the lower two housed an independent propulsion unit comprising a diesel engine and an electric motor.[18] The steel used in the construction of the cylinders and the small diameter of the latter allowed the *Dolfijn* class to achieve a remarkable safe diving depth of 200 metres and a maximum depth of 300 metres.

The British Admiralty was privy to the details of this design throughout its maturation process. Max Gunning, the Dutch shipbuilding engineer and Director of the Netherlands Shipbuilding Bureaux, came to Britain from Portugal during the early part of the war.[19] While working for the RN in London, he offered the design to the Admiralty as a transport submarine, specifically with the siege of Malta in mind. The island was relieved before the concept could possibly have been realised, but it caught the attention of Rear Admiral Max Horton, Flag Officer Submarines (FOSM) at the time. Horton called the design 'revolutionary' and considered it adaptable for the attack submarine role. Subsequently, he ordered the design to be trialled at the Admiralty Experiment Works at Haslar.[20]

After the war, the Dutch required a new class of submarine to replace the numerous but much-depleted 'O' and 'K' classes.[21] By that time Gunning had retired to Britain, but was retained as a consultant by the Dutch Navy between 1949 and 1954, during which time the three-cylinder boats went through four design phases. At each stage the boat had to be progressively reduced in size, and displacement declined from 2,500 tons to 1,200 tons due to cost. In 1954, Project 2076 was accepted as 'Onderzeeboot 1949', the necessary finance was approved and the four boats were officially named the *Dolfijn* class. Along with the destroyers of the *Holland* class, they were the first major naval vessels of local design and construction to be commissioned in the Netherlands after the war, and became important symbols of national pride.[22]

In 1953, when the design of these submarines was almost complete, the RN ordered a technical appraisal of the type in the light of its own current and near-future ASW requirements. The report that followed stated that the triple hull would be much stronger at all depths and more capable of withstanding damage than a single-hull configuration. Against the latest Type 170 sonar, which would provide data for the Limbo three-barrelled mortar system, it was calculated that survivability would be enhanced at these greater depths. 'Up to five times as many salvoes of Limbo may be needed for a submarine at 1,000 feet (300 metres) as would for a submarine at periscope depth'. The Navy's report added that although the Type 170 had been designed to measure depths down to 1,200 feet (365 metres), an improved successor would be required in order to produce effective data at that depth. However, it was concluded that for the most part the deep-diving submarine would operate in a conventional manner at shallower depths unless it was aware that it was being hunted.[23] Thus, while recognising the advantages associated with the Dutch design and the detection problems it raised even for its most advanced sonar, the Royal Navy chose not to collaborate over this project.

During the 1950s there were several exchanges of information and visits made to naval research establishments and electronics companies by both British and Dutch personnel. There continued to be expressions of goodwill on the part of the RN regarding the advantages of partnerships and of the sharing of equipment; however, this apparent openness masked concerns about security, even with a close NATO partner. There was also evidence of an NIH (Not Invented Here) culture. For example, following the observation of a demonstration of what was described as a 'Dutch Asdic set' and a Bofors-designed 'rocket propelled depth charge', the Director Torpedo and Anti-Submarine Warfare (DTASW) remarked that the Dutch were 'technically substantially advanced as we are'. The Director Underwater Weapons (DUTW) disagreed: in his opinion, Bofors had produced a 'satisfactory weapon' and the Dutch 'a good Asdic set', but there was no evidence as yet that the components could work together as an efficient asdic attack unit. On the matter of sharing the latest Limbo mortar and Type 170 sonar, it was agreed that there was 'no chance of Dutch ships being fitted [with] and operating' the equipment in the foreseeable future.[24]

Collaboration

In contrast to the previous decade, the period 1960 to 1970 produced four major Anglo-Dutch collaboration projects. Disappointingly, despite genuine commitment on both sides, only one was brought to fruition.

During a succession of meetings at the Admiralty in London in late April 1961 regarding a new-design guided weapons frigate eventually to succeed the *Leander* class, referred to as CF 299, there was considerable pressure from Board and Naval Staff members for a 'packaged deal' with another Western European nation, and a surveillance radar was singled out as an appropriate project to pursue. This decision carried with it powerful political undertones. It led to a seven-year-long debate about whether or not the RN should collaborate with the Netherlands Multi-Target Tracking radar (MTTR)

HrMs *Van Speijk* under tow. The absence of a pennant number suggests the photo was taken during her sea trials prior to February 1967. (Henk Visser)

Two failed attempts at Anglo/Dutch naval equipment collaboration are seen in this photograph of the missile frigate HrMs *De Ruyter*, seen here entering Portsmouth on 6 April 1991. The HAS 3-D radar beneath the giant radome was rejected by the British while the single-arm Tartar SAM launcher, which the Dutch adopted instead of Sea Dart, is sited on the after superstructure. (John Jordan)

project currently underway – referred to in British circles as 'Broomstick'.[25] At the time the RN's requirements not only covered the CF 299 frigates but also the proposed new aircraft carrier CVA-01, its Type 82 destroyer escorts and a 'large cruiser' class. The RNlN, in turn, wanted to fit the radar in two missile frigates intended as replacements for the cruisers of the *De Ruyter* class.

At the same time, the Admiralty Surface Weapons Establishment (ASWE) was also embarking on its own advanced radar project. Initially, this was aimed at major warships in excess of 6,000 tons. The necessary shift in priorities to include the needs of the much smaller CF 299 meant that the Dutch radar was at a more advanced stage and therefore attractive in terms of collaboration.

Technical discussions with the Dutch later in 1961 drew attention to the limitations of their design regarding weight, size, power demands and the apparent lack of tracking capacity which made it fundamentally unsuitable for fitting as part of a CF 299 suite in a 3,000-ton frigate. There was also concern expressed about the use of the S band, which the British Admiralty considered overcrowded; instead it favoured the C band, which was also less susceptible to jamming. However, despite the differences in performance expectation, a compromise was reached. This was prompted by the political pressures at play, and in particular the timescale involved for production equipment to be available for ship fitting: for the Netherlands it was 1966, the UK 1967.[26]

By early 1962, the British considered Broomstick to be a 'necessary' interim solution. The ASWE project, thought in Admiralty circles to be 'substantially better', could be retro-fitted during a first major refit. Understandably, the need to acquire adequate air defence for the fleet by the 1970s was the driving force as far as the RN was concerned. This was particularly the case with regard to CVA-01, which had an ambitious 'alongside date' of 1968. Other than the Type 984 radar, there was no alternative set available. In the event, the Controller proposed that a case for Broomstick should be put to the Treasury for authorisation, a move which at worst might involve paying a proportion of the development cost if no UK order finally materialised.[27]

The collaboration proceeded for a short while after the cancellation of CVA-01 in 1966 and the reduction of the Type 82 class to a single vessel, HMS *Bristol*. Central to the future planning for maritime operations in the next decade was the concept of 'relatively lightly armed frigates' supported by 'some ships of high quality to give credibility to the force as a whole'. These 'cruisers' would have a radar fit capable of dealing with the 'increasing sophistication of enemy missiles and aircraft … up to the twenty-first century'. It was anticipated that these task forces in the post-carrier era would be capable of establishing a 'maritime presence', perhaps independently, both East and West of Suez. To this end, what was by now referred to as the Type 988 radar was still perceived as the only viable option capable of offering the necessary air and surface surveillance and three-dimensional information to a data handling system for control of gun/missile systems and aircraft.[28]

However, the voices of dissent within the UK MoD

HrMs *Kortenaer*, the first ship of the 'Standard' class frigates to complete, leaving Portsmouth Harbour on 29 August 1989. The 40mm mounting on the hangar roof would later be replaced by a Goalkeeper CIWS. (John Jordan)

A Sea Wolf GWS 25 launcher aboard HMS *Chatham* in 1990. The Dutch Navy's decision not to acquire Sea Wolf contributed to the failure of the 'integrated' frigate project. (Author's collection)

remained, and continued to focus on the shortcomings of the Dutch radar. Britain's decision in 1968 to withdraw from East of Suez, which would effectively see its naval forces restricted to NATO duties in the Northeast Atlantic, seemed to put an end to the requirement for cruiser-led independent task groups of the kind outlined in the previous paragraph. The decision to withdraw was made in Autumn 1968, leaving the thorny question of when and how to inform the Dutch. This matter was diplomatically sensitive owing to the acrimony which had been caused in the recent past over the cancellation of the Sea Dart order (below). In the end the Dutch were informed in early October.

Possible collaboration over the sale of the Hawker Siddeley GWS 30 surface-to-air missile (SAM) system, or Sea Dart, was far more short-lived than Broomstick, but the failure of the Dutch to agree to its purchase was, at the time, just as damaging to the goodwill which existed between the two countries. The issues at stake were quite simple. Sea Dart was new and therefore technologically advanced, but largely untried and more expensive than its rival, the General Dynamics RIM-24 Tartar. Tartar enjoyed the benefit of having a proven track record in combat and, for a small nation such as The Netherlands, it was more affordable. A capable SAM system was required to enable the RNlN's two new *Tromp* class frigates to undertake their primary function of protecting a task force or convoy against aircraft and guided weapons.

Matters came to a head during 1966. Early in the year there was a belief within the Admiralty in London that the RNlN preferred Sea Dart, as it was considered superior in capability to Tartar. Unfortunately, it was also understood that the Americans considered this a key contract to win, and would offer terms so financially attractive it would be hard for the Dutch to resist. Dutch and British representatives met in The Hague in October of that year. Admiral H Bake, representing the RNlN, agreed that Sea Dart had a superior range and ECCM performance, flew under continuous thrust and had a faster reaction time. However, Tartar was already in use with several NATO navies. Crucially, it was considered only marginally inferior to Sea Dart, whose performance data were based primarily on computer simulation; only two live firings had been carried out to date, and one of those had been unsuccessful.[29]

The Dutch remained unmoved by the arguments made by the British, and opted for Tartar; despite every effort to cut financial corners, affordability was the dominating factor. Privately, the British estimated that the cost per ship of GWS 30 was £4.28 million and likely to rise, compared with £3.1 million per ship for the General Dynamics Tartar system. Without doubt, this was a grave failure which damaged relationships at a diplomatically sensitive time. Sea Dart and Broomstick were certainly linked as a 'package deal', and there was annoyance in London as to what was seen as a betrayal on the part of their North Sea neighbour. Naturally, there was commensurate frustration and disappointment in The Hague two years later when the decision over Broomstick was announced.[30]

In contrast, earlier in the decade, the collaboration over the Dutch acquisition of the *Leander* class design had been a simple and striking success. In November 1959, Vice Admiral L Brouwer, the Dutch Chief of Naval Staff, informed the Secretary of Defence of the need to replace their six, ageing US-built destroyer escorts of the *Van Amstel* class in the light of increasing NATO responsibilities and the likelihood of a forthcoming conflict in New Guinea. Subsequently, a constructor team visited the RN Director General Ships at Bath in August of that year. A decision to adopt the general-purpose *Leander* class design, the FSA 25 (HMS *Arethusa* model) was approved by the Netherlands Government for the first four of six frigates in June 1961, and a contract was signed with the British the following November.

In order to facilitate rapid construction in Netherlands shipyards, it was decided to accept the overall hull and superstructure design, although there would be differences in the internal compartment layout. The surface weapons equipment, identical to that fitted in the RN

A HSA Goalkeeper CIWS aboard HMS *Cornwall* opposite the Tower of London in 1989. Its acquisition was opposed by the Royal Navy at the last minute, ostensibly on grounds of cost. (Author's collection)

frigates and purchased from the UK, comprised a 4.5 inch (114mm) Mk 6 twin gun mounting, two Seacat SAM missile launchers, and a Mk.10 Limbo ASW mortar.[31] The design drawings were prepared by J Samuel White at Cowes and the contracts, worth 117.66m guilders, were awarded to NDSM of Amsterdam and KM Schelde, Vlissingen. Work began in October 1963 and, despite delays, the last of the six warships completed in December 1967.

The Dutch shipbuilders installed two-shaft Werkspoor-English Electric Y-136 geared steam turbines and an entirely indigenous HSA electronics suite. Specifically, this comprised the LW-02 long-range air-surveillance radar, the DA-02 medium-range air/surface search radar and the Mk 45 combined radar and optical fire control system for the 4.5 in guns. The Mk 44 radar/visual director for Seacat could automatically follow the target in elevation and bearing; its compact design allowed two launchers to be fitted to the hangar roof instead of the one on the RN *Leanders*.

When the contract had been signed Peter Carrington, First Lord of the Admiralty, sent a warm message to his Dutch counterpart. It ended:

It is good of you to mention the help we were able to give; we were of course, very glad of the opportunity to co-operate with you.[32]

Here, after nearly two decades of effort, was the proof that the two navies could bring a major project to fruition in a comparatively short space of time. It was therefore unsurprising that in 1968 the Netherlands Government proposed that the RN and RNlN should collaborate in the construction of a new class of what was termed 'integrated', *ie* identical, frigates for introduction into service with both navies in the mid-1970s. The proposals were developed at a series of informal meetings in 1968 and 1969 between Admiralty Board Members and the Dutch State Secretary and Chief of the Defence. The Netherlands representatives pointed out at an early stage the importance of recognising discrepancies in size and therefore affordability between the two navies, an issue that had bedevilled the Sea Dart collaboration. Despite this warning, basic differences in staff requirements and design philosophy soon became apparent.

The RN wanted to base the design around the Sea Wolf Close-in Weapons System (CIWS), which would be

A Limbo Mk 10 triple A/S mortar being lifted from a frigate of the *Van Speijk* class berthed at No 18 Jetty, Den Helder Naval Base, as a prelude to her midlife modernisation. The ship's two Seacat launchers and her Mk 44 director have already been removed. The modernisations of this class took place between 1976 and 1982, and were intended to bring the ships up to the same standard as the *Kortenaers*. They would be fitted with an Italian OTO Melara 76mm gun forward in place of the British Mk 6 4.5in twin mounting, and the well for the Limbo mortar was to be plated over to enable the ships to operate a Lynx helicopter; new radars of local design and manufacture were fitted. (Courtesy of John Jordan)

Table 2: Anglo–Netherlands Collaboration Projects & Committees

Radar fire control equipment	1950–52
Project 'Dimple': intercept receiver	1957–64
Establishment of Anglo–Netherlands naval collaboration in R & D	1957
Anglo–Netherlands Joint Naval Committee (ANJNC)	1957–64
ANJNC Communications	1957–60
ANJNC Fighter Direction Radar & Navigation	1957–60
ANJNC Gunnery & Guided Weapons	1957–64
ANJNC Mines & Mine Countermeasures	1957–64
ANJNC ASW	1957–64
ANJNC Anti-submarine Weapons	1957 onwards
ANJNC Communications & Radar	1959–64
High Speed Sonar Dome	1960
Shock trials using two *Van Straelen* class MCMV	1963
Anglo/Netherlands/French ASW projects: sonar domes	1963
Anglo/Netherlands/French ASW projects: beam pressure field	1963
Anglo/Netherlands/French ASW projects: bow sonar dome trials	1965
Anglo/Netherlands/French Mine Countermeasures Committee	1965–68
Anglo/Netherlands/French ASW projects: long range sonar	1966
Anglo/Netherlands/French ASW projects: sonar signal processors	1966
Anglo/Netherlands/French ASW projects: sonar signal processors & data processing and displays	1969

Source: ADM 1/31054

served by the GWS 25 back-to-back aerial array for surveillance and target indication. By contrast, the Dutch were unconvinced of the efficacy of the guided missile on grounds of efficiency and cost; they preferred a modernised version of the Bofors 120mm gun which would be available shortly.

Nevertheless, in June 1969, the Dutch reaffirmed that they were keen to pursue the 'integrated' design route, as the possibilities of reducing overhead costs were attractive. The British saw what they now referred to as the Type 22 as the long-term frigate replacement. In the interim they were planning to discontinue the *Leanders*, which were expensive to man and which had also reached the end of their 'design life', in favour of a stop-gap, commercial design, the Type 21.[33]

By April 1970 discussions were ongoing but there remained areas of disagreement. The Dutch were still unhappy about the choice and cost of the Sea Wolf CIWS, and in particular the overall length of the proposed frigate, which could not be comfortably accommodated in the dock at their main base, Den Helder.[34] However, it had been agreed that as a collaborative venture unit cost would be the responsibility of the individual nation. The Dutch were pressing for closer cooperation on such matters as standardisation of equipment, harmonisation of personnel structure and organisation, training, joint exercises and research & development. The British were promising to consider those matters.

By the middle of the year the overall length of the frigate had become the dominant factor, as the RN pressed for it to be increased in order to meet their latest staff requirement. Although the RNlN partially acquiesced on that point, there was still no agreement about internal layout and other technical differences. Most importantly for this joint venture, the Dutch were demanding a sizeable, guaranteed Dutch element to be included in the construction before any competition for contracts between private firms was introduced. Fearful that this would lead to increased costs, re-design and thus delays, there was pressure from within the Admiralty for the British to withdraw.

By the end of November, both the Controller and the Vice Chief of Naval Staff were supporting abandonment, citing particularly the issue of length. The 117 metres overall favoured by the Dutch would mean that the Type 22 would not drive through the seas as well as the *Leander* and would thus fall short of requirements for full speed and endurance. There was a good deal of political pressure, particularly from the British Foreign Office, to continue negotiations before both sides agreed there were irreconcilable differences.[35]

Close Partners Perhaps, But ...

Although the 1970s had begun inauspiciously, there were important agreements reached during the first half of the decade which breathed life again into the partnership. In 1972, agreement was reached which saw Dutch submarines use Faslane as a peacetime operational base, forming part of the RN's 3rd Submarine Squadron under British command. Dutch warships also used the Sea Training facilities at Portland for post-refit work-up, and participated in operational training during the annual Joint Maritime Courses. In June 1973, a Memorandum of Understanding was signed which led to the formation of the United Kingdom/Netherlands Amphibious Force. This was timely for the Dutch, who had recently decided to relinquish the western half of New Guinea, leaving their marine corps without a major role. In December they signed a £6 million contract for four sets of Rolls-

Royce Olympus TM3B Gas Turbine engines with an option for a further four.[36] The same year, a destroyer, frigate and helicopter unit joined one of the RN's 'East of Suez' deployments designed to show a presence in the region following official military withdrawal some five years earlier.[37]

This article demonstrates that, despite many positive outcomes, it often proved stubbornly difficult to achieve satisfactory collaboration on major projects; the *Leander/Van Speijk* agreement proved to be the exception. Both navies have suffered severe cutbacks in numbers and resources since the end of the Cold War. If the lessons of their recent, shared history are to be learned, their best chances of survival as front-rank naval powers must lie in persisting with a policy of technology exchange and the sharing of resources and expertise, despite the differences in national objectives which will inevitably occur.

Acknowledgement:

I would like to extend my sincere thanks to Warrant Officer Henk Visser RNlN (retired) whose extensive knowledge and experience of the operating practices both of the Royal Netherlands Navy and of the Royal Navy was extremely useful in writing this article.

Endnotes:

1 The National Archives (TNA) FO371/49466 Naval Exhibition held in Rotterdam 1945.
2 TNA FO371/60230 Rebuilding the Netherlands Navy 1946. An Admiralty delegation went to The Hague in July 1946 to elicit details of the logistical support required by the Dutch pending the rebuilding of their defence industries. Post-war, the RNlN also benefited from US support through the Mutual Defense Assistance Program (MDAP). This enabled six *Cannon* class destroyer escorts to be loaned, while six *Roofdier* class coastal escorts were built in the USA for the Dutch Navy. Five *Balder* class submarine chasers, built in the Netherlands in the 1950s, were also paid for with MDAP funds. Eighteen of the 32 *Dokkum* class coastal minesweepers, built 1954–56, were again MDAP funded and based on the RN 'Ton' class, to which they bore a strong visual resemblance.
3 Robert Gardiner (Ed), *Conway's All the World's Fighting Ships 1947–95*, Conway Maritime Press (London 1995), 269. See also TNA FO371/60230.
4 860 Squadron Netherlands Naval Air Squadron (NAS) operated Swordfish from the MAC ships *Gadila* and *Macoma*, which were manned by Dutch mixed merchant and naval crews, on escort duties in the North Atlantic, and also from HMS *Indefatigable* during the *Tirpitz* raid in July 1944. Netherlands 1839 NAS flew Hellcats from HMS *Indomitable* between February and April 1945 as part of British Pacific Fleet operations.
5 TNA ADM1/18316 Dutch Cruisers, completing building in the Netherlands 1945-47. See also *CAWFS 1947–1995*, 269.
6 TNA ADM1/18316. See also *CAWFS 1947–1995*, 272; Norman Friedman, *Naval Radar*, Naval Institute Press (Annapolis 1981), 212.
8 Christopher Bayley & Tim Harper, *Forgotten Wars: The End of Britain's Asian Empire*, Penguin (London 2008). *Forgotten Wars* examines the short but bloody conflict which led to Indonesian independence in December 1949 within the context of colonial unrest across the South-East Asia region in the postwar years.
9 TNA ADM1/18272 Request for the allocation of two destroyers and one cruiser to the Royal Netherlands Navy 1945; TNA ADM1/18545 Sale of eight Bathurst Class minesweepers to the Dutch Navy 1945-47. The sale of minesweepers proved uncontentious. These ex-Australian vessels were used mainly on anti-smuggling duties, but proved to be a maintenance nightmare as they were worn out after lengthy and arduous wartime service.
10 TNA PREM8/596 Blockade by Netherlands Navy of Dutch East Indies 1947; G Teitler, 'Decolonisation & Coastal Operations in the East Indies, 1945–50', in Nicholas Rodger (Ed), *Naval Power in the Twentieth Century*, Palgrave Macmillan (London 1996), 178–180; TNA FO371/88770 Withdrawal of the Dutch Navy from Indonesia 1950. A scathing report on the dire situation in the NEI, as he saw it, was written by Vice Admiral JN Edelsten, Commanding Officer of the 4th Cruiser Squadron, following an official visit in June 1946 – see TNA ADM1/20582, Review of Situation in Netherlands East Indies 1946.
11 Jan Willem L Brouwer, 'Dutch Naval Policy in the Cold War Period', in Jaap R Bruijn et al (Eds), *Strategy and Response in the Twentieth Century Maritime World: papers presented to the fourth British-Dutch Maritime History Conference*, Batavian Lion International, (Amsterdam 2011), 43.
12 FO371/90037 Denial by Lord Tedder that Dutch Navy should have only light forces 1950.
13 Jan Brouwer, *op cit*, 44–5; G Teitler, 'Anglo-Dutch Relations, 1936–1988. Colonial and European Trends', in GJA Raven & NAM Rodger (Eds), *Navies and Armies: The Anglo Dutch Relationship in War and Peace 1688–1988*, John Donald (Edinburgh 1990), 76–79. Another vociferous opponent of the Dutch 'Task Force' plan was Vice Admiral AS Pinke, Flag Officer Commanding NEI. His views on the way in which his government handled the transfer of power to the nationalists, coupled with his accusations of the Dutch Admiralty 'thinking too big and building castles in the air' with its blue-water navy plans, cost him his expected promotion to C-in-C Netherlands Navy and left him a disillusioned man – see G Teitler, *op cit*, 177–182.
14 Jan Brouwer, *op cit*, 46.
15 *CAWFS 1947–1995*, 273. The attack sonar, PAE-01, and search sonar, CWE-010, were developed and manufactured by Van der Heem, a Dutch company specialising in radio equipment. The attack sonar provided fire control for the principal ASW weapon on the destroyers, the Swedish Bofors 375mm rocket launchers (*Raketdieptebomwerpers*).

In 1946–47, the RNlN asked the British for assistance with the design of the *Holland* class. The Chief of the Imperial General Staff at the time, Bernard Montgomery, opined that the role of the RNlN should be to 'sweep the Channel' in support of the BAOR. The Dutch, understandably angered, turned to other nations for assistance in the design and equipment of the destroyers. One former commanding officer remarked that the ships' engine/boiler room layout made them 'half a *Gearing* class'.

The classification of the *Holland/Friesland* classes as 'ocean-going anti-submarine destroyers' dated from 1939 rather than the 1950s when several navies were constructing specialist classes of destroyers or frigates. See Hubert V Quispel, *The Job and the Tools*, The Netherlands United Shipbuilding Bureaux Ltd (Rotterdam 1960), 157.

[16] In 1965–66 the *Karel Doorman* was re-boilered with units removed from the incomplete and redundant *Colossus*-class carrier HMS *Leviathan* (*CAWFS 1947–1995*, 272).

[17] Friedman, *Naval Radar*, 212–214. The cruiser *De Zeven Provinciën* was also extensively modernised in the 1960s: her after main guns were replaced with a US twin Terrier SAM to counter the threat of the new wave of Soviet bombers in the North Atlantic. Cost precluded her sistership from receiving a similar conversion.

[18] Raymond Blackburn (Ed), *Jane's Fighting Ships 1961–62*, Sampson Low, Marston & Co Ltd (London 1961), 162.

[19] Bob Roetering & Henk Visser, *HrMs Tonijn: Van onderzeeboot tot museumschip*, Lanasta (Emmen 2011), 29; KHL Gerretse & JJA Wijn, *Drie-Cylinders Duiken Dieper*, Uitgeverij de Bataafsche Leeuw BV (Amsterdam 2011), 33. The advantages of triple-hull design are explained in Quispel, *op cit*, 168–69.

[20] The Netherlands United Shipbuilding Bureaux Ltd. had been formed in 1934 of the four biggest shipyards in The Netherlands together with a large engineering company. The Bureaux had a central organisation which produced the designs and drawings for ships to be constructed to the requirements of the Netherlands Navy – see Quispel, *op cit*, p.28.

[21] Submarines destined for service in home waters were prefixed with the letter 'O' for the Dutch word *Onderzeeboot*, while those allocated for service in the East Indies carried the 'K' prefix for *Kolonien*.

[22] Gerretse, *Drie-Cylinders*, 34–40. Orders for the submarines were subject to constant delays owing to the stringent financial conditions caused by the recent war – see Quispel, *op cit*, 165.

[23] TNA ADM 1/25154 Design of Deep Diving Submarine by the Dutch 1953. Willem Hackmann, *Seek and Strike: Sonar, anti-submarine warfare and the Royal Navy 1951–54*, HMSO (London 1984), 346–48.

[24] TNA ADM1/24119 Collaboration with the Netherlands Navy on research and development 1952. One major fear at the time concerned the proximity of the Netherlands to the Soviet border. It was thought likely that a swift Red Army invasion would put Dutch naval vessels carrying the latest British equipment at risk of capture. A report on a visit by a British Admiralty Mission to Holland in April 1952 concluded that while HSA was reliable security-wise, the same could not be said of what became the electronics giant Philips at Eindhoven – see TNA ADM1/26864 Dutch Assistance in the Development of Radar Fire Control Equipment 1951–52. However, it would be misleading to allege there was no naval collaboration during the 1950s (see Table 2).

[25] ADM1/31054 Anglo/Netherlands Collaboration on Project Broomstick, 3-D Radar 1961–64. Later, the RN gave the radar the official designation Type 988; the Dutch designation for this radar was SPS-01. It combined search and target tracking in a single, multiple antenna – see Friedman, *Naval Radar*, 214–15.

[26] ADM1/31054, *ibid*.

[27] ADM1/28043 Dutch 3-D Surveillance Radar: initial purchase 1962. Development of the Type 984 radar had commenced in the late 1940s; it was subsequently installed in the carriers *Victorious*, *Hermes* and *Eagle*.

[28] ADM1/31054. Much earlier, in 1964, the increased weight of Broomstick had caused the planners to doubt if it could be carried by the Type 82. Alternatives were considered, including a modified Type 992, but that would result in the ship's ability to deal with low, fast targets halved, and no defence picket or air defence coordination capability. The post-CVA-01 maritime environment in which the Type 988 radar would play a central part is set out in DEFE10/532 Operational Requirement Committee Memoranda 1–24 (9) 1967.

[29] DEFE 69/482 Meeting between British and Dutch Ministers at The Hague, 1966.

[30] John Jordan, *Tromp*, in John Roberts (Ed), *Warship No 9*, Conway Maritime Press (London 1979), 28-29. The cancellations of Sea Dart and Broomstick also need to be viewed alongside other politically charged decisions affecting relationships between The Netherlands and Britain at the time, namely landing rights for KLM in Hong Kong and the purchase of the Chieftain tank.

[31] The total cost of the weapons fit was £469,000 – see ADM1/28895 *Leander* Class Frigates: proposal to construct in India 1960–63.

[32] Quoted in SG Nooteboom, *Deugdelijke Schepen-Marinescheepsbouw 1945–1995*, Europese Bibliotheek (Zaltbommel 2001), 55.

[33] This was acknowledged when the comparatively inexpensive 'broad beam' design modification was suggested in 1964 – see ADM1/31008 *Leander* Class Frigates – widening of beam 1964.

[34] Dock 6 at Den Helder measured 153.2 x 23 x 10 metres. The frigates of the 'Standard' class, when eventually built, were 130.1 metres overall.

[35] T225/3604 Design and Construction of the Standard Frigate Type 21 & 22 1970; FCO46/513 Collaboration between the United Kingdom and The Netherlands on naval matters 1970. Revealingly, a senior Foreign Office civil servant found the two British senior officers most closely involved in the integrated frigate project, Admirals Ashmore and Pollock, 'most disappointingly rigid and narrow-minded' in their attitude. An anecdotal view, perhaps, but the importance of the 'human factor' is not to be underestimated in understanding how and why complex and costly ventures succeed - or fail. See also Norman Friedman, *British Destroyers & Frigates: The Second World War and After*, Chatham Publishing (London, 2006), 297–299.

[36] G Teitler, *op cit*, 80; FCO46/1086 Collaboration between the UK and The Netherlands 1973. The Dutch NAS 860 took delivery in 1967 of the Westland Wasp HAS, which was also purchased by other Western European navies.

[37] On a sourer note, the Netherlands' decision not to purchase the Abbeyhill naval interception system drew this sobering comment from a Foreign Office official, a reminder that, fundamentally, attitudes had not changed: '... since the breakdown of the joint frigate project in 1970, a feeling seems to persist in Dutch naval circles ... that naval cooperation always comes secondary to our commercial interests' (see FCO46/1086).

USS *LEBANON* (AG-2): A JACK OF SEVERAL TRADES

In the latest of his drawing features, **A D Baker III** focuses on the USS *Lebanon*, taken up from trade as a collier and subsequently used for a variety of tasks.

With the impending onset of the Spanish-American War, the US Navy found itself woefully short of auxiliary vessels to carry coal and supplies in support of its still-nascent 'Steel Navy' combatant fleet. It was estimated that 20 colliers were required to support the fleet should it need to operate far from home waters, but only three were readily available. Of these, the Reading & Philadelphia Railroad's 1,468GRT SS *Lebanon* was purchased for $225,000 and quickly adapted for naval service at the Chelsea Annex of the Boston Navy Yard, commissioning on 6 April 1898 under the command of Lt-Cdr CT Forse. The ship had been launched by William Cramp & Sons at Philadelphia on 29 September 1894, and the Navy retained her name, which commemorated both a county and its county seat in Pennsylvania coal country.

USS *Lebanon* completing outfitting for US Navy service at the Chelsea Annex, Boston Navy Yard in 1898. Note the open steel hatch to Hold No 2 and the shields to the 6-pounder guns mounted forward. Although two more guns were authorised for fitting aft on the main deck, none were available, and she never received them. (US Navy)

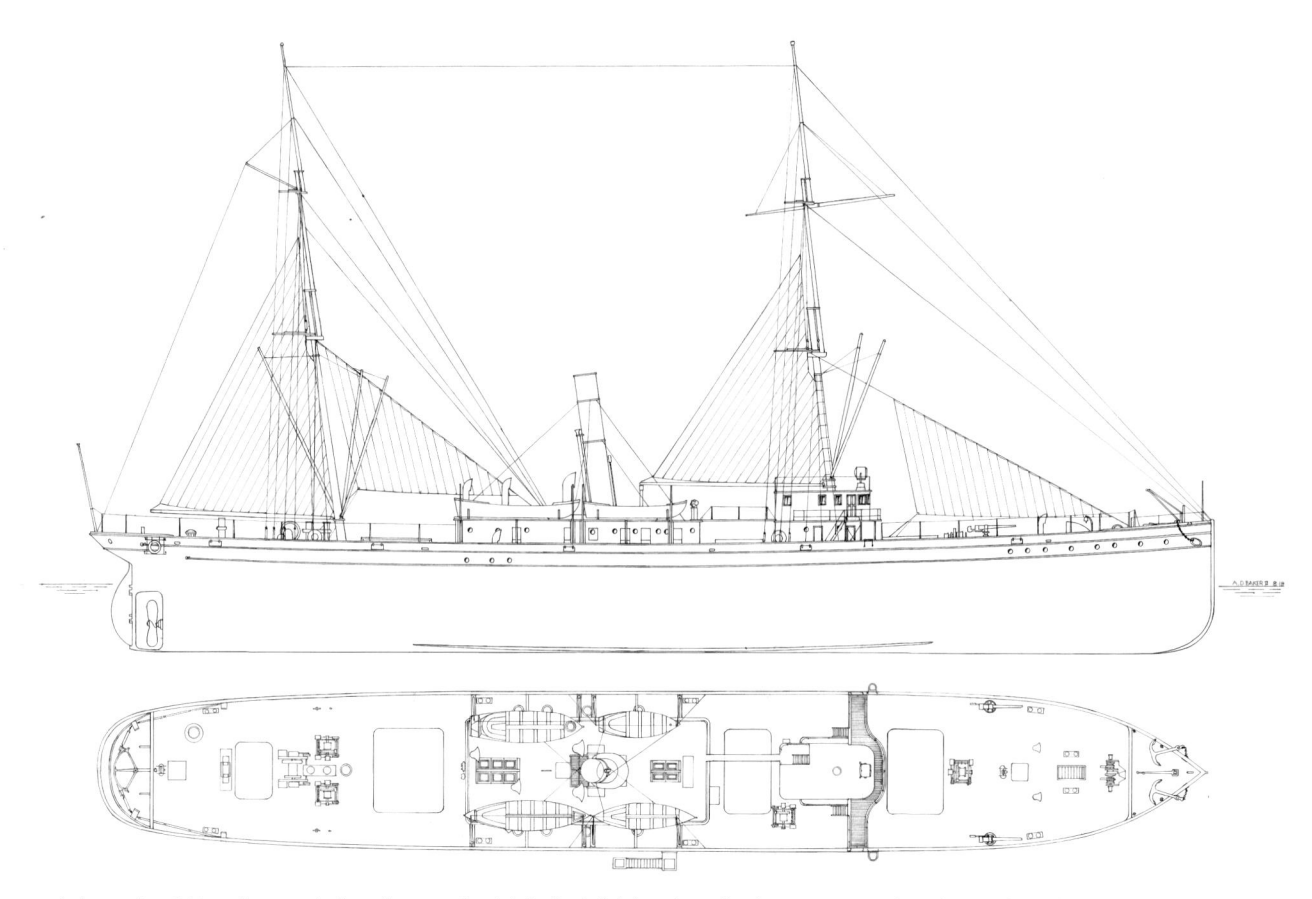

Lebanon in 1901 at the completion of an overhaul. It is doubtful that the sails shown were employed very often, if ever, but the ship had been rigged to carry them since completion and still carried them in 1910. A 1917 Booklet of General Plans for the ship omits them. (Drawn by the author)

USS *Lebanon* (AG-2): Characteristics

Displacement:	
normal	3,235 tons (18.5 tons/inch immersion)
full load	3,375 tons
Tonnage	1,468grt
Dimensions:	
length	249ft pp; 259ft 6in overall
beam	37ft 4.5in
draft (mean)	17ft 3in at normal load
Armament:	2 x 6pdr QF (not always mounted; see text)
	1919: 2 x 3in 50-cal HA
Engineering	two single-ended boilers
	one VTE engine (16in stroke HP, 50in LP, 30in LP)
Speed	10 knots light / 8.5 knots loaded
Fuel:	
1912	188 tons coal
1917	332.3 tons coal
Electrical (1898)	2 x 5kW 80V General Electric dynamos
Complement:	
1908 proposal	189 total (ship: 10 officers, 8 warrant officers, 6 CPO, 45 enlisted, 69 total; submarine crew: facilities for 8 officers, 24 CPO, 88 enlisted, 120 total)
1920	102 total as target towing and service ship

Lebanon was one of a fleet of seven colliers operated by Reading & Philadelphia Railroad, and was equipped not only to carry coal in her four holds but also had a robust capability to tow one or more of her former owner's three dozen seagoing coal barges to ports up and down the US East Coast. The towing capability was ultimately to provide the ship with the majority of her service assignments during the second half of her Navy career.

USS *Lebanon's* initial service was to supply Navy ships with coal, and she arrived at Cardenas, Cuba, on 4 June 1898. No longer required with the conclusion of the conflict, the ship was laid up at the Norfolk Navy Yard on 15 April of the following year. She was refitted and reactivated at the Portsmouth, New Hampshire Navy Yard in 11 August 1905 and remained operational, with a Naval Auxiliary Service civilian crew operating her in support of the fleet as far south as Nicaragua, until she was again placed in reserve at Norfolk, Virginia, on 2 October 1909.

During 1908, plans were drawn up to convert the ship to act as a tender for a flotilla of the Navy's early submarines, a dozen of which were in service at that time. The new configuration was illustrated in a new general arrangement plan dated 25 November 1908 and titled 'Rearrangement of Parent Ship for Submarines'.

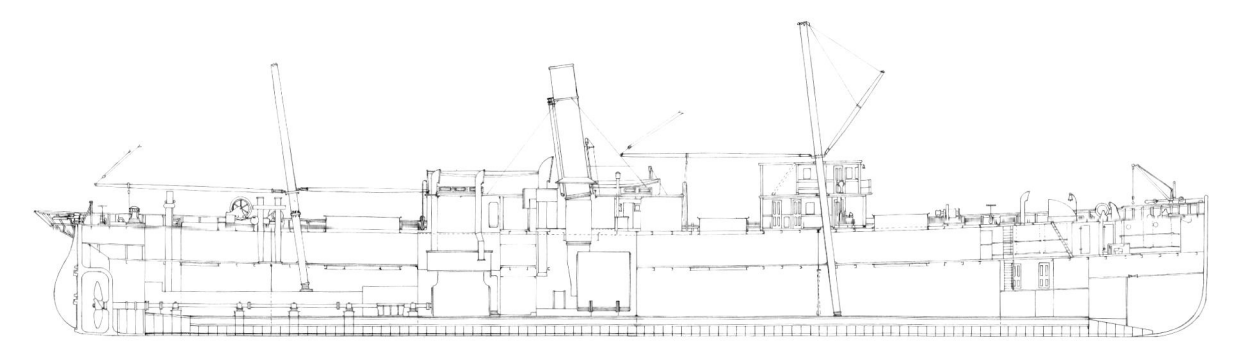

Internal elevation of *Lebanon* as of January 1906, showing the four coal holds and the towing equipment aft. (Drawn by the author)

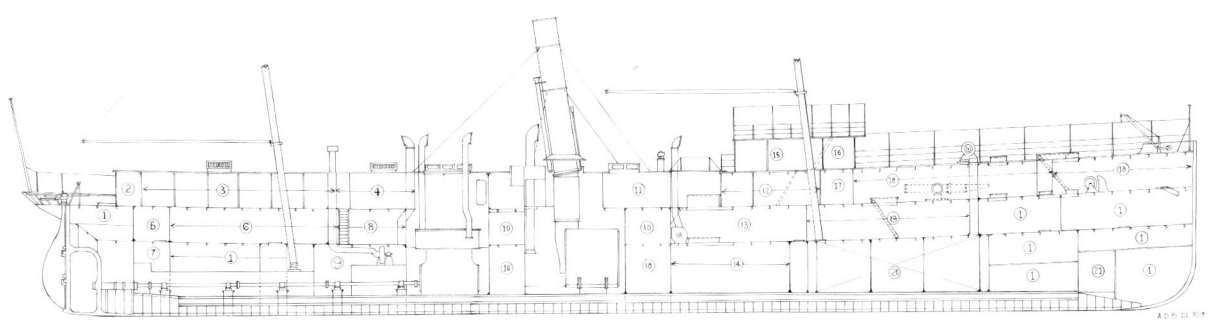

Lebanon as a submarine tender. The drawings are based on plans drawn up by the Boston Navy Yard and dated 24 November 1908. (Drawn by the author)

1 Dry Stores
2 Wireless Office
3 Officers' Cabins
4 Wardroom
5 Ship's Magazine
6 Petty Officers' Quarters
7 Lower Magazine
8 Offices
9 Dynamo Room (2 x 75kW to port, 2 x 50kW to starboard)
10 Coal Bunkers
11 Officers' and Crew's Galley
12 Warrant Officers' Quarters
13 Cold Storage & Paymaster's Stores
14 Submarine Stores
15 Captain's Quarters
16 Pilot House
17 Steering Gear
18 Crew's Space (stbd), trainable 18in TT (port)
19 Crew's Space & Stowage for 20 torpedoes
20 100,000 gallons gasoline (swash bulkhead athwartships between the two original coal holds)
21 Cable Locker

A new forcastle deck was to be added, stretching aft to the pilothouse, with the former foredeck fitted to berth enlisted crewmembers to starboard, while a single trainable 18in torpedo tube would be installed behind shutters to port for testing torpedoes. Below decks, the upper level of the two former cargo coal bunkers was to be fitted with a torpedo workshop and space for 20 torpedoes to starboard, with additional enlisted berthing facilities to port. The lower levels of the two coal bunkers were to be reworked to carry up to 100,000 US gallons of gasoline for use by attached submarines, nearly all of which continued to be equipped with gasoline-fuelled engines up to and including the 'G' class, the last of which was completed in 1913. Abaft the new forecastle, a deckhouse extending nearly to the stern would house submarine officers, while the after coal holds were to have been rebuilt to accommodate submarine warrant officers and chief petty officers, as well as providing a number of compartments to carry submarine-related stores. The conversion, however, was not carried out.

In the event, *Lebanon* did not remain inactive for long. Recommissioned on 1 July 1911 as a 'firing range tender', she was assigned the dual roles of towing and repairing gunnery target barges and transporting ammunition to Navy facilities along the US East Coast. In 1912 *Lebanon* deployed to the Caribbean to support Navy fleet manoeuvres. While both roles were of equal importance, it is worth noting that a booklet of general plans for the ship dated 6 November 1906 and updated on 3 April 1917 refers to *Lebanon* as an 'Ammunition Ship' on its title sheet, and lists her ammunition capacity as including space for 1,400 12in, 1,000 7in, 800 6in, and 297 5in projectiles, and additional hundreds of rounds of smaller projectiles plus small arms ammunition. The upper portions of the holds were partitioned to carry powder bags.

The ship remained in use for a decade towing and repairing target barges and occasionally carrying munitions, until newer and more economically-operated ships could replace her. Designated as Auxiliary, General, Number 2 (AG-2) under the new, comprehensive fleet ship typing and hull numbering system signed by the Secretary of the Navy on 17 July 1920, *Lebanon* was relieved as fleet target support ship by the much larger

Lebanon as a dual-purpose gunnery range support ship/collier in October 1912, showing heightened masts to carry radio antennae. No armament is fitted here, and it was not until 1917 that two 6-pounder guns were re-installed, this time without shields. (US Navy)

A towed gunnery target barge is being repaired alongside *Lebanon* in this photo dating from 1920–21. The US Navy's gunnery targets at that time consisted of vertical metal racks to support wooden slats that could quickly be replaced at sea by the towing ship's crew if hit. A cinecamera, used to record the results of the firings, has been added on a platform above the towing winch area. (US Navy)

Lebanon with four 'R'-class submarines alongside, including *R3* (SS-80) and *R9* (SS-86), during 1919. *R3* was transferred to the Royal Navy in November 1941 as *P511*. Note the portside enclosed latrine structure added *c*1918 to serve *Lebanon*'s greatly enlarged crew. (US Navy)

Procyon (AG-13) and was put up for sale on 21 November 1921, with decommissioning scheduled for 6 February 1922.

In the event, *Lebanon* was sold to an NF Dillon of New York City, to whom she was turned over on 14 June 1922. On re-entering commercial service, she became *Taboga*, being renamed again in 1926 as the *Homestead*. Transporting a variety of cargoes, the ageing vessel came to an end on 26 July 1932 when she foundered in the Humber Arm, the southernmost arm of the Bay of Islands on the west coast of Newfoundland, near the port of Corner Brook. While she was hardly a combatant, it is highly probable that *Lebanon* saw more naval gunfire during the Great War than any other US Navy vessel – by a large margin.

Acknowledgements:

Many thanks to: naval architect and expert ship modeller John F Cheevers, who got me interested in *Lebanon*; Christopher C Wright, editor of *Warship International*, who obtained copies of the plans of the ship from the US National Archives; and Dr Stephen S Roberts, author and custodian of the invaluable shipscribe.com website, who provided the best possible copies of the photo illustrations.

WARSHIP NOTES

This section comprises a number of short articles and notes, generally highlighting little known aspects of warship history.

VICKERS CRUISER DESIGN NO 866: AN ALTERNATIVE TO THE 'COUNTY' – WITH A JAPANESE CONNECTION?

Some four years ago plans of a 10,000-ton 'treaty cruiser' designated 'Design No 866' were posted online as part of Hiraga archive at the University of Tokyo. **John Jordan** suggests an explanation for the plans, and compares the design with the contemporary British 'County' class.

In the wake of the Washington Treaty of February 1922, the Chief Naval Architect of Vickers, Sir George Thurston, drew up a series of possible designs for a cruiser of 10,000 tons standard displacement armed with 8in guns. Sketches of three of these designs, dating from 1923 and identified as Types 'A', 'B' and 'C', accompanied a chapter on light cruiser design by Thurston published in the 1925 edition of *Brassey's* (see Sources).

The Thurston notebooks in the NMM archive that record these designs generally focus on general characteristics, distribution of weights, etc, and the sketches that illustrated the *Brassey's* article were based on a selection of these. It was unusual for detailed plans to be produced by Vickers except where a design was put forward as a serious proposal for export (see the plans of Design No 924, which became the Spanish *Canarias*, in Friedman, *British Cruisers*, 114). When such plans for Design No 866 were recently made available online as part of the Hiraga Archive by the University of Tokyo, they therefore elicited a considerable amount of interest (and no little puzzlement). One possible explanation is that the plans were drawn up prior to Hiraga's visit in April 1923 (see Appendix below), in an attempt to interest the IJN in post-Washington Vickers designs for a battleship similar to the British *Nelson* armed with 16in guns (Design No 873, which likewise features in the Hiraga Archive), and for a 10,000-ton 'treaty' cruiser (Design No 866).

The original plans, which were drawn to a scale of 1/200, had been scanned by the University at a much-reduced size, and the annotations, which included figures for armour weights (in pounds not inches) were barely legible. Moreover, interpreting the plans was complicated by the standard Admiralty practice of using the profile to show not only external features, but also inboard compartmentation and armoured decks.

These plans are of interest not only because of the questions surrounding their rationale and context, but because they show a fully-worked British treaty cruiser design that presents a serious alternative to the 'County' being developed concurrently by the RCNC, with a

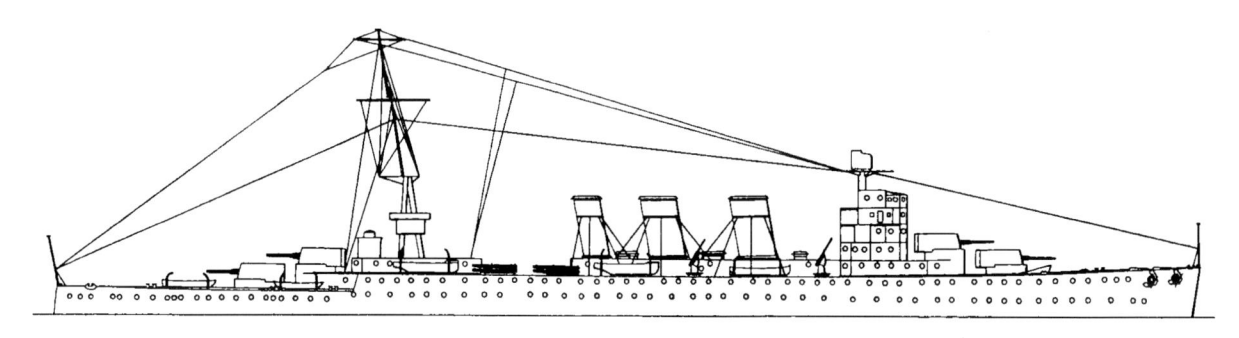

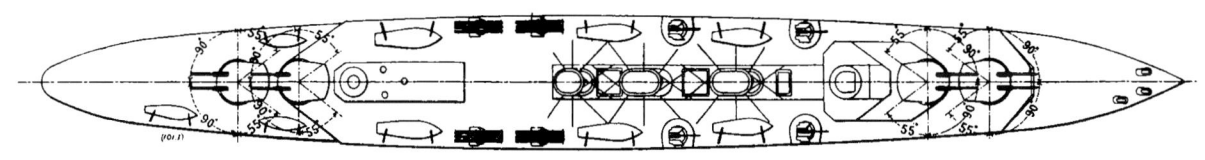

DESIGN OF LIGHT CRUISER, TYPE "B."
Length, 585 ft.; displacement, 10,000 tons; speed, 34½ knots; armament, eight 8-in., four 4·7 in. h.a., twelve 24-in. torpedo tubes. Protection, 1 in. to 3 in. on barbettes and decks.

The sketch of Light Cruiser Type 'B' published in Brassey's 1925. Design No 867, which it represents, differed in only a few small respects from No 866: the hull was 5ft shorter and had 2ft more beam, and there were only four HA guns; 867 was even faster (34.5 knots on trials) but slightly less well protected. (*Brassey's* 1925, page 69)

WARSHIP NOTES

CRUISER OF ABOUT 10,000 TONS DISPLACEMENT AND 33.5 KNOTS SPEED

Design Nº 866

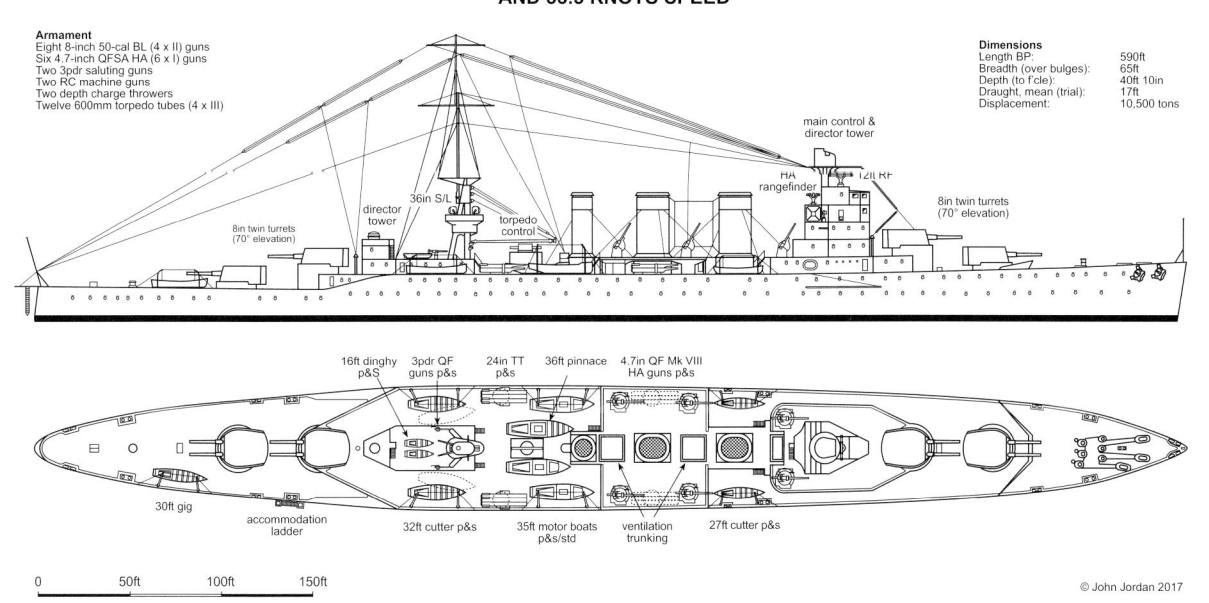

Armament
Eight 8-inch 50-cal BL (4 x II) guns
Six 4.7-inch QFSA HA (6 x I) guns
Two 3pdr saluting guns
Two RC machine guns
Two depth charge throwers
Twelve 600mm torpedo tubes (4 x III)

Dimensions
Length BP: 590ft
Breadth (over bulges): 65ft
Depth (to f'cle): 40ft 10in
Draught, mean (trial): 17ft
Displacement: 10,500 tons

© John Jordan 2017

different hull-form, a different scheme of protection, more powerful propulsion machinery, and alternative weaponry. They have been redrawn here by the author with a view to separating the external profile and plan from the protection scheme and improving the clarity of the annotations.

General framework for the design

In his article Thurston makes it clear that for an 8in-gun cruiser of 10,000 tons displacement only light protection for the ship's vitals were either feasible or worthwhile, thus the twin priorities should be firepower and speed. Apart from its tactical value, Thurston sees high speed as key to this lightly-armoured vessel's survival, both against enemy battleships and submarines. He further opines that no distinction need be made between a cruiser built for fleet duties and one intended to protect trade, and he opposes the embarkation of aircraft because current stowage, handling and launch arrangements interfere with the fighting qualities of the ship.

Hull-form

The hull of Design No 866 appears to be derived from that of the *Hawkins* class, with a long forecastle that extends aft as far as 'X' turret, and a 'rombhoid' hull section with inclined sides and a shallow bulge. Hull depth is recorded on the plans as 40ft 10in, significantly less than that of the famously deep-hulled 'County' design, which had a completely flush deck without the cut-down aft; the original legend for *London* gives a draught amidships of 17ft and freeboard of 26ft 6in for a total of 43ft 6in. This gives a depth to length ratio of approximately 1:13.7 for the 'County' and 1:14.5 for No 866. The saving in terms of hull weight, according to the Thurston notebook, would have been about 65 tons, although the type of steel is not specified.

Armament

In the Thurston plans the main gun is an 8in 50-calibre model on a mounting with a maximum elevation of 70 degrees, which would make long-range barrage fire against aircraft a theoretical possibility. The eight guns are conventionally disposed in four twin turrets fore and aft with arcs of 310 degrees for the forward turrets and 300 degrees for the after pair, the superstructure sides being suitably angled to permit fire on forward and after bearings. Other Thurston designs in the series featured three triple turrets: either two forward and one aft (No 848 – Type 'A' in *Brassey's*) or all-forward in a *Nelson*-style pyramid (No 847 – Type 'C').

For the primary anti-aircraft gun Thurston favoured a 4.7in gun of 45 or 50 calibres, which could also be used to repel enemy destroyers. The HA gun depicted in the plans appears to be the 4.7in/50 Mk VIII (on Mk XII mounting) developed post-war for the new generation of capital ships and fitted in the contemporary *Nelsons*. This would have incurred a significant weight penalty compared to the 4in Mk V gun selected for the 'Counties', each mounting weighing 12 tons *vice* 7 tons for the Mk V. Moreover, there would have been six guns rather than four, so the total weight of the mountings (without taking account of ammunition) would have been 72 tons for Design No 866 *vice* only 28 tons in the 'Counties'.

Similar considerations apply to the torpedo armament. In his *Brassey's* article Thurston cites the contemporary RN 'E' class as the new standard, and all the designs in the series feature twelve torpedoes in four triple mountings. Of the major navies, most gave serious consideration at the end of the First World War to a larger model of torpedo capable of delivering increased range in a fleet action. The Germans adopted a 60cm (23.6in) model, while the French opted for 550mm

189

(21.65in) and the Japanese 61cm (24in); Thurston himself favoured a new 24in model, which would be allied to improved torpedo fire control. The two British new-build battleships of the *Nelson* class would be armed with a newly-developed 24.5in Mk I torpedo; however, for its new treaty cruisers the Royal Navy retained the standard 21in torpedo, which was fitted in two quadruple mountings in the 'County' class (and, retrospectively, in the two 'E's). Unusually, on the No 866 plans the 24in torpedo is designated by its metric diameter (600mm), all other measurements on the plans being in imperial. The weight of twelve 24in torpedoes, assuming an individual torpedo weight of 2.3 tons, would have been 27.6 tons, compared with little more than half that figure (14.7 tons) for the eight 21in torpedoes embarked in the 'Counties'. The weight of each of the triple torpedo mountings would probably have been *ca* 11.5–12 tons; the weight of the quadruple mounting in the 'County' would have been similar, but there were only two.

Propulsion

Design No 866 was for a fast cruiser, designed for a speed at normal displacement of 33.5 knots with 105,000shp; with forced draught the ship would have been capable of 34.5 knots with 120,000shp. She was to be powered by four-shaft turbines in two consecutive engine rooms with the turbines in the forward engine room driving the wing shafts. Steam was supplied by ten Yarrow-type small-tube boilers disposed in three adjacent boiler rooms – see Protection Plan for probable layout. The funnels in the Thurston plans correspond in dimensions to those in the earlier sketches, but are now vertical, as is the tripod mainmast. The cost in terms of weight over the propulsion machinery in the 'Counties', which was rated at 80,000shp (eight boilers in two boiler rooms) for a designed speed of 31.5 knots, would have been just under 300 tons (see table).

Protection

Arguably the most interesting feature of Design No 866 as it appears in the Thurston plans is the protection system, which dispensed with a side belt in favour of a combination of a protective 'turtle' deck over the propulsion machinery and armoured boxes over the magazines fore and aft. The protective deck extended from the forward turret to the after part of the ship; it was flat at the centre-line with slopes that met the side plating just above the load waterline. It comprised 60lb (1.5in) NVNC plates over the machinery spaces, reducing to 40lb (1in) in way of the magazine boxes (see plan) with a 40lb carapace over the steering gear. The boxes for the 8in shell rooms, magazines and handing rooms had 120lb (3in) NVNC crowns and 80lb (2in) sides. A shell passing through the slope of the deck at this point would therefore have to penetrate 40lb plating, then 80lb plating on the side of the box. The 4.7in magazine and lobby, the transmitting station, the ring bulkheads for the 8in ammunition hoists and the gunhouses themselves had only 40lb plating, sufficient to keep out splinters but not much else.

Design No 866 differed from the 'Counties' and from later Admiralty cruiser designs in having the shell rooms above the magazines. This was more economical in terms of armour coverage and enabled the shell rooms to be given the same level of protection as the magazines, but it meant that the crowns of the shell rooms were above the waterline. In contrast, the 'Counties' had all these key spaces well below the waterline, which meant that they were less exposed to enemy fire.

Like the *Kent* class, Design No 866 would have had underwater protection in the form of a shallow (5ft) bulge outboard of the machinery spaces to minimise the ingress of water into the engine and boiler rooms in the event of the ship being struck by a torpedo or mine. The plans show the bulges divided into 14 watertight compartments (WTC) on either side of the ship.

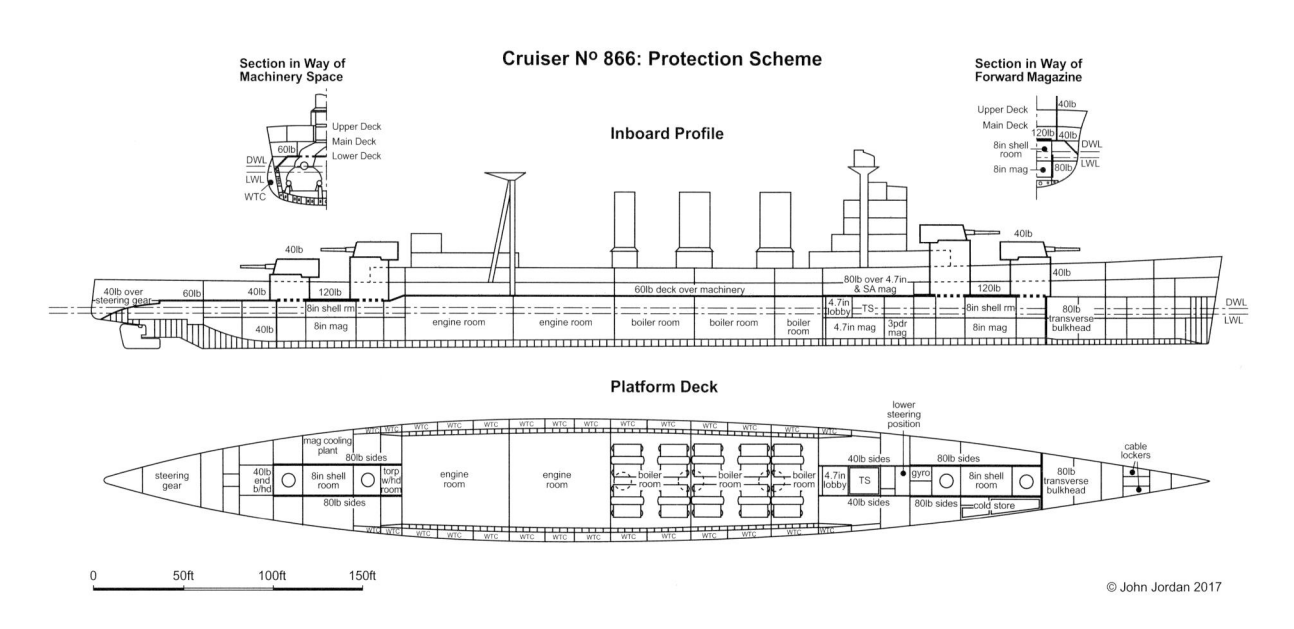

Some final observations

The plans describe this ship as a 'Cruiser of *about* 10,000 tons' (author's italics). The Thurston notebook gives a trial displacement of 10,500 tons, including 500 tons of oil fuel that would not be counted under Washington. There is no 'board margin', and the disparity in the allowances made for 'General equipment' between No 866 and the *Kents* (see table) suggest that the fully-worked design might have emerged slightly over treaty limits, which would have been unacceptable to the British government of the day. The obvious candidates for savings in the design are two of the six 4.7in HA guns (*ca* 40 tons including hoists and ammunition) and two of the four triple torpedo tubes (*ca* 50 tons). These changes would have resulted in a ship whose military capabilities would have been comparable to a 'County', with greater speed but less protection for the waterline and the magazines. If the plans of Design No 866 reveal anything, it is just how tight the 10,000-ton limit was for a cruiser armed with eight 8in guns.

Appendix: Hiraga's Visits to Europe and the USA 1923–24

The following brief summary of Hiraga's movements has been kindly provided by regular *Warship* contributor **Hans Lengerer**:

Hiraga departed Yokosuka on 22 November 1923 on board the *Katori Maru* for London. The ship docked on 12 January 1924, and Hiraga visited the Admiralty and the Greenwich Naval College over the course of the next few weeks. On 18 February he left for Paris, visiting all of the French naval dockyards, eleven shipyards and three research establishments. There were then similar fact-finding visits to Italy, Germany and Austria.

Hiraga returned to London on 6 April, where he was shown drawings of 35,000-ton battleships (15 April) and 10,000-ton treaty cruisers (16 April); it is reported that

Table of Weights

	Kent Class	Design No 866	Comparison
Hull	5,400	5,335	-65
Protection	1,025	915	-110
Armament	1,050	1,125	+75
Machinery	1,850	2,125	+275
General equipment	675	500	-175
Margin	0	0	–
Total	10,000 tons	10,000 tons	–

not only was he shown plans but he also received copies, including sketch plans of the new 16in-gun battleships and the *Kent* class. He left London on 21 April and visited two naval dockyards, eleven shipyards, 16 factories, a research establishment and a training establishment over the next five weeks. On 4 June he was at Vickers, and was given detailed data for the new 16in gun the following day. *(Editor's note: It is possible that he was also given information about the latest Vickers designs, including the as-fitted plans of Design Nos 873 and 866, at this point in his visit).*

Hiraga left London on 7 June for the USA and arrived at New York on 11 June, subsequently visiting two US navy yards and four shipyards. During his visit he made a request for IJN naval architects to study at MIT (a similar request had previously been made in France). He departed San Francisco on 18 July and arrived at Yokohama on 3 August.

Sources:

Plans of Design No 866 posted online as part of the Hiraga Archive.

Data from Sir George Thurston's Notebooks (courtesy of David Murfin).

Brassey's Naval Annual 1925 Edition, Sir George Thurston, Chapter IV: Light Cruisers, 59–73.

THE HELICOPTER CRUISER HMS *BELFAST*

Growing recognition of the utility of naval helicopters during the 1950s saw various methods adopted to deploy them at sea in meaningful numbers. The hybrid helicopter cruiser was one possible approach. The Royal Navy considered converting existing gun-armed cruisers to this configuration on several occasions from the late 1950s onwards. Here **Conrad Waters** examines a little-known proposal to adapt the 'Town'-class cruiser *Belfast* to an amphibious role.

The development of effective helicopter designs towards the end of the Second World War had a significant impact on the future direction of naval aviation. Early operational use tended to focus on search and rescue and casualty evacuation missions. Notably, the helicopter's suitability for performing the carrier 'plane guard' function previously allocated to an escorting destroyer gained early acceptance in both the Royal Navy and USN. However, the helicopter's potential value in anti-submarine warfare had been recognised from the outset, and evaluation of its use as an amphibious assault vehicle soon followed. This was demonstrated to good effect during the 1956 Suez operations when Royal Marines were flown ashore from the extemporised helicopter carriers *Ocean* and *Theseus* during the assault on Port Said.

The evident utility of naval helicopters resulted in much thought about how best to deploy them at sea in effective numbers. Availability of war-built aircraft carriers provided an obvious route forward for the United States and its British ally, which adapted a number of these vessels as helicopter platforms. Other major fleets had to adopt more innovative solutions. One result was the hybrid helicopter cruiser, which combined traditional cruiser functions with the ability to embark varying numbers of anti-submarine or transport helicopters. Among the earliest were the Italian Navy's two *Andrea Dorias* and France's sole *Jeanne d'Arc*, all ordered during the latter half of the 1950s. These developments may have influenced the RN's interest in a new type of escort cruiser, combining area air defence missile capabilities with a squadron of anti-submarine helicopters, around the turn of the decade. Although this project ultimately proved abortive, the escort cruiser concept ultimately formed the starting point for the 'through-deck cruisers' (TDC – later 'support carriers') of the *Invincible* class.

Roughly contemporaneous with the escort cruiser project were a number of studies that examined converting some of the RN's existing gun-armed cruisers to carry meaningful numbers of helicopters. The first of these appears to have been conducted in late 1959 at the request of the Third Sea Lord (Controller) when 'face saving' efforts were underway to find an alternative use for *Swiftsure* following cancellation of her major refit two years after work had commenced. Three different schemes to convert her into an amphibious commando carrier capable of carrying 200–350 troops, two landing craft assault (LCA) and between eight and twelve helicopters were sketched out. The most substantial involved replacement of all superstructure with a hangar and flight deck. The schemes were estimated to cost between £6m and £7.5m compared with the £2.5m spent on adapting the aircraft carriers *Albion* and *Bulwark* to the commando carrier role. Unsurprisingly, it was concluded that none of the *Swiftsure* schemes represented value for money.

Despite this setback, the conversion of a cruiser as an amphibious ship continued to attract interest. At this time, the United Kingdom relied heavily on mobile carrier and amphibious task forces to maintain a credible naval presence 'East of Suez'. This approach demonstrated its value in deterring an Iraqi invasion of Kuwait in July 1961. It appears that this resulted in consideration being given to the most effective and timely means of bolstering these capabilities. *Belfast's* conversion to operate as an amphibious transport and landing ship was seen as a key element in the resulting plan. Although extensively refitted between 1956 and 1959, the cruiser was due to reduce to operational reserve after the delivery of the three new *Tiger*-class ships. While not explicitly stated, it is likely reconfiguration was seen as attractive in extending the active life of a recently updated vessel. Manpower was to be obtained by placing two escorts in the training fleet in reserve.

The basic requirements of the rather austere conversion proposed by the Director of Plans in late 1961 were:

– removal of 'Y' turret and as much as possible of the aft superstructure as could be undertaken in a six-month conversion period
– provision for operating as many 'Wessex'-type helicopters as possible, each being provided with limited maintenance facilities commensurate with those in a frigate
– the addition of at least four LCAs on davits
– accommodation for at least two companies of infantry (*c*260 men).

Belfast pictured in Plymouth Sound at the end of her extended refit in 1959. Her proposed conversion to operate 'Wessex' helicopters in the amphibious transport and assault role would have seen the after turrets and structure replaced with a helicopter landing area and hangars. (Crown Copyright 1959)

WARSHIP NOTES

Belfast as Amphibious Ship

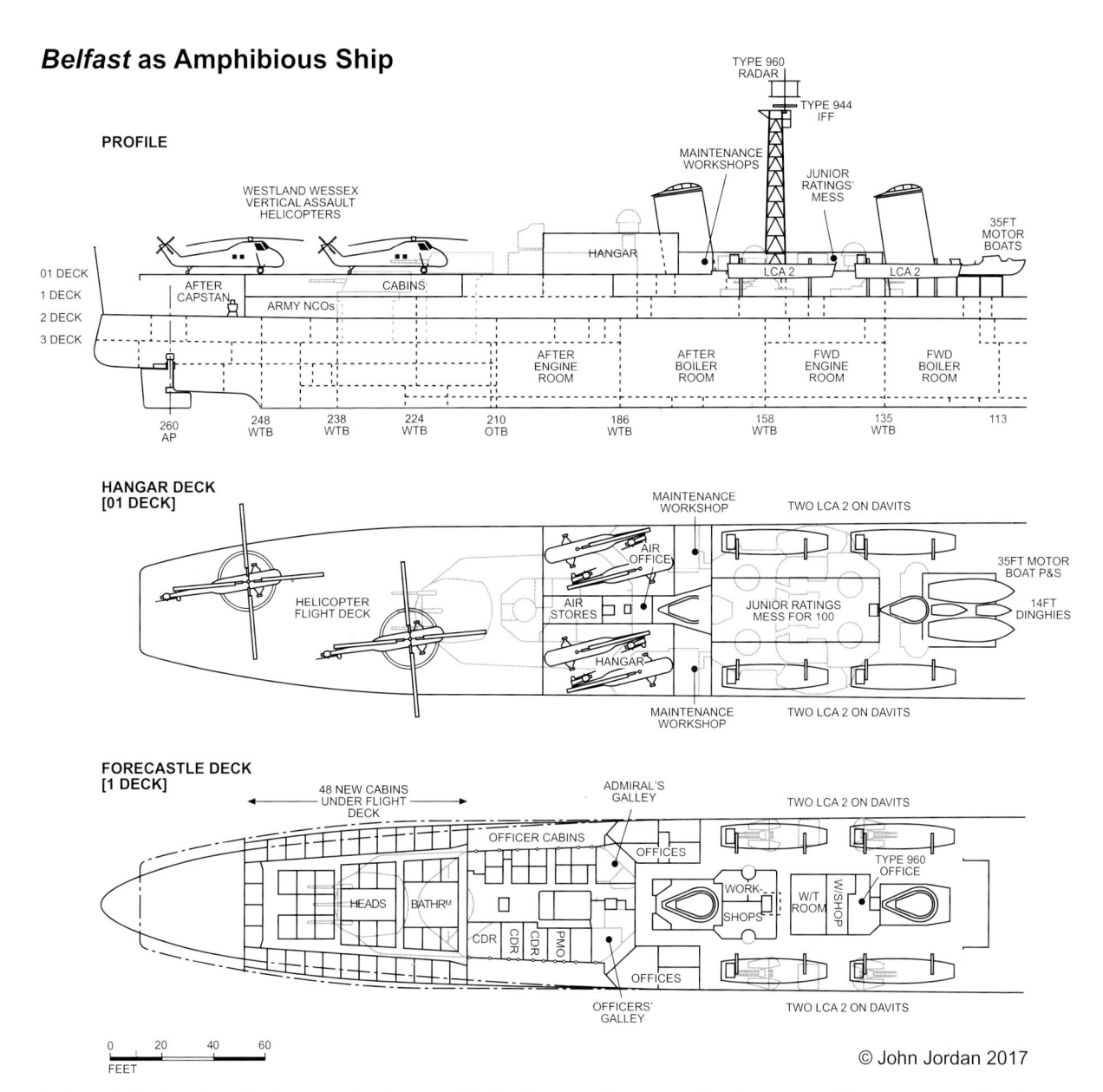

Profile and plan drawings showing the main elements of *Belfast's* proposed reconstruction into an amphibious transport and landing ship. The helicopter flight deck arrangements and hangar stowage arrangements are based on earlier drawings for a helicopter support ship filed in the *Belfast* cover and which were probably used to speed completion of the *Belfast* conversion sketches.

It was accepted that manning would have to be kept to a minimum to compensate for the embarked infantry and additional air crew. This would be achieved by operating only one boiler room and confining armament to the bombardment capability of just one 6in turret. Equipment not needed for the new role would be removed or preserved. It was hoped that the conversion might be carried out at Singapore in the second half of 1962 and that the converted ship would remain in service until 'about 1967'. The latter timing may have been influenced by estimated commissioning dates for the planned *Fearless* class LPDs.

Based on these requirements, a rough sketch of the proposed conversion was drawn up during December 1961 for further discussion. Key elements of the scheme are depicted in the drawings and can be summarised as:

Structural alterations: 'X' and 'Y' 6in gun turrets and supports were to be removed down to 2 Deck (Upper Deck) level. 01 Deck (Superstructure Deck) would be plated in from Station 168 aft to provide a 160ft-long helicopter landing area, with two hangars replacing the after superstructure, Bofors mounts and directors at its forward end. The space between this extension and 2 Deck was to be reconstructed to provide accommodation for the naval and army/marine officers embarked under the ship's new role. Additional accommodation was to be provided by construction of a

Although *Belfast's* planned conversion into an amphibious helicopter cruiser never progressed beyond preliminary sketches, two of the *Tiger* class were later converted as anti-submarine helicopter cruisers – *Blake* is seen here at Chatham. Their reconstruction shared some similarities with that proposed for *Belfast*. (Author's Collection)

Table: Accommodation Requirements

	Officers	Men/Ratings
Additions		
Army: 2 infantry companies (130 maximum)	30	230
Navy: 6 Wessex crews (each 3 officers + 1 rating)	18	6
Navy: maintenance personnel (10 per helicopter)	–	60
Total	48	296
Reductions		
Navy: 1 boiler room watch	–	12
Navy: 2 6in turret crews	2	100
Navy: 4 4in gun crews	2	68
Navy : aft DCT crew	–	6
Total	4	186
Net Addition	44	110

Provided by:
For officers: 48 new cabins (including 4 for deleted cabins) on 1 Deck cabin flat.
For army senior NCOs: Mess for *c*20 on 2 Deck.
For army soldiers: 205 spaces in former junior rates' messes on 3 Deck forward.
For navy ratings: 100 spaces in new mess on 01 Deck.

ratings mess at 01 Deck level between the two funnels. This replaced the four close range blind fire directors (CRBFDs) used to control the 4in armament.

Helicopters: The structural alterations would be sufficient to stow four 'Wessex' helicopters in the twin hangars, with space for two more on the flight deck. Maintenance facilities would be provided at the forward ends of the hangars but it was noted that space would be insufficient to support all three types of 'Wessex' helicopter then in service. The landing area was to be provided with a magnetic loop and landing lights.

LCA provision: The four twin 4in gun mountings and shelters were to be removed and replaced by davits for four LCAs, two on each side.

Accommodation: As detailed in the accompanying table, it was calculated that the conversion would require a net increase in accommodation sufficient to house around 44 officers and 110 men and ratings. The officers were to be housed in 48 cabins in the new cabin flat under the helicopter landing area at 1 Deck (Forecastle Deck) level. A further 20 senior army non-commissioned officers were provided with their own mess one deck below. Embarked troops would be housed in former ratings messes in the forward end of the ship, with the displaced crew being transferred to the new mess – able to hold 100 crew – between the funnels. Dining hall sittings would be increased to feed the additional personnel.

In line with the Director of Plans' guidance, it was assumed that only No 1 boiler room would be operated. However, the conversion went further than his requirements in that both 'A' and 'B' 6in turrets were retained. It also seems that the four forward Bofors mounts and their twin CRBFDs would have remained *in situ*, providing a residual anti-aircraft capability.

The sketch conversion was accompanied by calculations of the effort required to undertake the work involved. These suggested that the task would be too much for Singapore, as access to UK-based design departments would be needed. It would also require between nine and twelve months to complete, much longer than anticipated. Cost would be in the region of £1m to £1.5m.

The results of the study were presented to the Admiralty on 15 December 1961. It was considered the time required to undertake the work involved was unacceptable. Alternative, more limited conversions involving fewer helicopters and other deletions were discussed but rejected as either requiring too much work or providing negligible benefit. The scheme was abandoned.

The design effort expended on *Belfast* may not have been entirely wasted. The RN was to get its helicopter cruisers – albeit focused on anti-submarine operations – with the conversions of *Blake* and *Tiger* in the 1960s. These had some similarities with the *Belfast* adaptation and it is possible the earlier proposal influenced the new scheme. More broadly, the abandonment of *Belfast*'s conversion must be welcomed as helping to ensure the cruiser's survival into preservation. The adaptation would have radically altered the ship's appearance and she would have been a less attractive candidate for retention in her new guise.

Sources:
Ships' Cover 839A, *Belfast* Extended Refit, Folios 16 and 18 (*Belfast* Conversion).
Ships' Cover 567F, *Fiji* Class, Volume 7, Folio 26 (*Swiftsure* Conversion).

BERTHING HMS *VICTORIOUS* AFTER RECONSTRUCTION

In the course of researching his new book on HMS *Victorious*, **David Hobbs** came across a plan showing how the ship was to be berthed at Portsmouth following her major 1950s reconstruction. Reasons of space meant the plan had to be omitted from the book, so instead the author discusses here what it shows.

Modernisation of HMS *Victorious*

When the majority of the fleet carriers on order in 1945 were cancelled, the Admiralty was forced to consider a modernisation programme for the *Illustrious* group, and in 1947 the Assistant Chief of the Naval Staff (Air) was tasked to study the question. He recommended that only a full modernisation could be justified; this would give a further two decades of useful life for what, the Director of Naval Construction advised, would be about half the cost of a new carrier. Admiralty approval for a modernisation programme was given in January 1948 and detailed design work began in February. Drawings were completed in June 1950, when it was decided that the first carrier to be modernised would be *Victorious*.

The task involved taking *Victorious* into 'D' Lock in Portsmouth Dockyard on 23 October 1950, dismantling the hull down to hangar deck level and making structural changes to many compartments in the decks below to provide new auxiliary machinery, magazines and bulk aviation fuel stowage. The largest project of its kind ever undertaken in a Royal Dockyard, work was originally expected to be completed in 1954 with modernisation of *Implacable* and *Indefatigable* to follow. However, contemporary advances in aircraft operating technology, including the angled deck, led to major design changes which both extended the time required to complete the work and increased cost. *Victorious* was finally taken out of dry dock in January 1958, by which time a virtually new ship had been built onto the skeleton of the original lower hull, retaining only the frames, plating, refurbished steam turbines, anchors and steering gear.

The initial design was for a straight flight deck which would have had a minimal impact on dockyard infrastructure, although there was to have been a side lift on the port side forward. However, following the success of flying trials in HMS *Triumph* and USS *Antietam*, in 1953 the Admiralty Board approved the design and fitting of a full 8.5-degree angled deck to *Victorious*, a change that required a large sponson to be fitted on the port side and brought the ship's deep load displacement to 35,500 tons. This lowered the hull deeper in the water and put the hangar deck only 14 feet above the waterline. The sponson overhung the tracks on which a dock-side crane ran and these had to be re-routed to clear it. This would be the first of several dockyard changes required to accommodate the 'new' carrier, which had significant flight deck overhang around the entire hull, the intention being to put the largest possible aircraft operating area onto the original hull's water-plane area.

Internally *Victorious* became a completely new and innovative ship, with changes in the design introducing new armament, the first computerised action information system, three-dimensional Type 984 radar, the latest facilities to operate transonic fighters armed with nuclear weapons and guided missiles, helicopters and accommodation to the RN's latest standards. On completion she was arguably the most advanced aircraft carrier in the world, with many of the advantages of a new ship but without the same expectation of hull life. The result of the modernisation was a well-equipped and effective

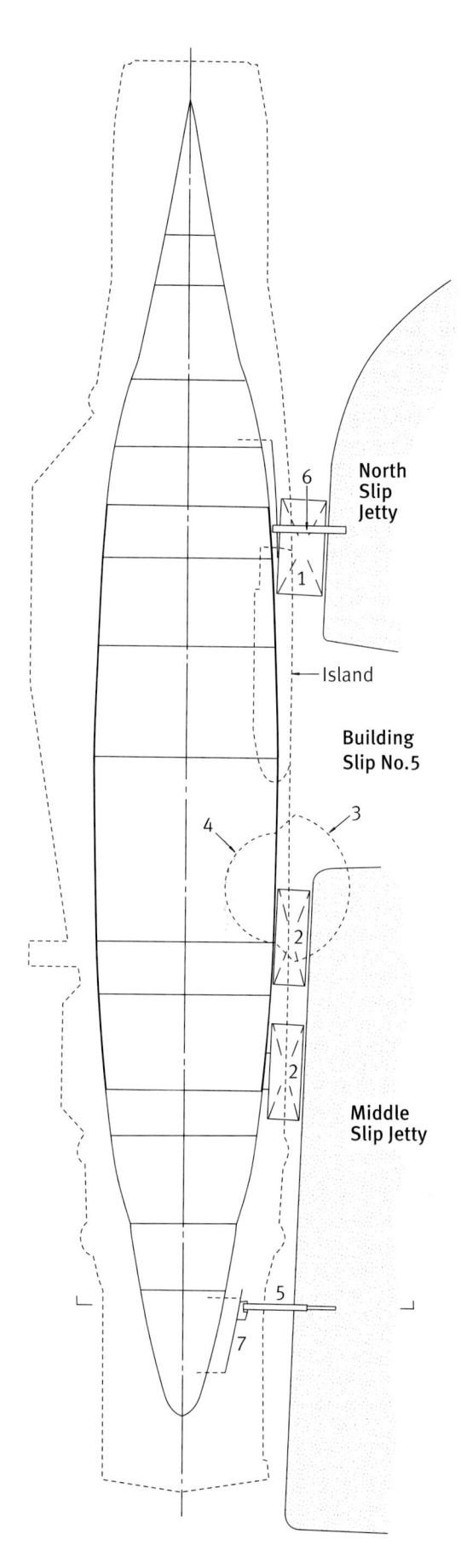

Key:
1 forward fender
2 after fenders (52ft x 20ft)
3 crane radius at maximum extension over the jetty
4 maximum crane radius over the flight deck
5 after brow (36ft, attached to 16ft light brow)
6 forward brow (42ft)
7 quarterdeck opening and brow platform
8 position of the hull at Station 162 relative to the jetty at low tide (L)
9 position of the hull at Station 162 relative to the jetty at high tide (H)

On the ship's position shown on the plan drawing, the broken line shows the outboard edge of the flight deck, which can be compared with the solid line which shows the hull at the normal waterline; the outline of the island is shown as a separate broken line. The overhang to starboard outboard of the island was increased in 1963 by the fitting of an 'Alaskan Highway', on which flight deck tractors could be parked when not in use and ready-use ammunition prepared ready to be loaded onto aircraft. This also had to be factored into berthing arrangements in due course.

(Drawing by Stephen Dent, based on NMM M0623 with additional information from the author)

Section at Station 162 looking aft

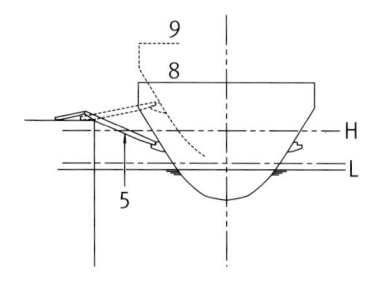

WARSHIP NOTES

aircraft carrier, but one which cost more and had taken longer to complete then a new ship. Further carrier modernisations were cancelled once the cost and complexity of such projects became clear.

Berthing Arrangements After Modernisation

There was insufficient capacity within Portsmouth Dockyard's Construction Department to fabricate the new hull structure above the hangar deck. The new hull above 4 deck, including the island, was therefore built in 30-ton blocks in Devonport Dockyard, which at that time had the capacity to do so, and transported to Portsmouth by sea. Once there they were craned into position and welded into the new structure using the latest techniques. As the new hull began to grow above the residual structure below the hangar deck it became clear that the new *Victorious* had a far greater sheer and flight deck overhang than previous British aircraft carriers or, indeed, any other type of warship.

The first requirement after 1954 was to move the track for the mobile crane on 'D' Lock to clear the angled-deck sponson on the ship's port side. However, there were also longer-term implications, as the ship was to be based in Portsmouth and new berthing arrangements had to be planned to accommodate her. When she first emerged from 'D' Lock she was actually berthed alongside South Railway Jetty port side to, so that LST 3522 *Tracker* could be moored outboard on her starboard side for use as a Dockyard workers' canteen, storage area and workshop – the angled deck had prevented *Tracker* from being secured on *Victorious*' port side. Normally, however, carriers were berthed starboard side to, so that the Captain and Navigating Officer on the compass platform would have the best view of the jetty, and this is the way she is drawn on the accompanying diagram, which is based on contemporary plans prepared by the design staff at Portsmouth Dockyard. Once operational, *Victorious* was to be secured alongside Middle and

This photograph shows *Victorious* in the King George VI Dry Dock during a refit in HM Dockyard Singapore. It clearly shows the effect of the angled deck sponson on dock-side arrangements and gives some idea of the material that was lifted onto the flight deck while the ship was undergoing maintenance. She would have had to be manoeuvred into the dock at an angle at first so that the small travelling crane on her port quarter could be eased past the overhanging sponson before she was centred up and docked down. Fortunately this dock was sufficiently large to achieve this but it would not have been possible with a larger carrier.

Note the work in progress on various parts of the flight deck including the port steam catapult, and that both jet blast deflectors are in the raised position. 'Jumbo', the mobile flight deck crane, was invaluable for moving stores and heavy machinery about the deck and it can be seen parked on the angled deck. Apart from the narrow underwater part of the bow, every part of the ship visible in this photograph was assembled using 30-ton blocks prefabricated in Devonport Dockyard and welded together in Portsmouth Dockyard between 1954 and 1958.
(David Hobbs Collection)

North Slip Jetties, which were specially prepared for her.

Note how a single fender 30 feet wide had to be used forward (1) where the hull began to taper towards the bow, and two fenders each 20 feet wide were used farther aft (2) to give adequate clearance between the overhanging flight deck and the jetty. Great care had to be taken to position the ship in exactly the right place so that the boat crane could plumb the northern edge of Middle Slip Jetty in order to hoist in aircraft, boats, stores and vehicles. The maximum radii of the crane jib (3) over the jetty and the flight deck (4) are represented as broken lines. The crane was on number 4 deck, and for the outer part of the jib to plumb the flight deck the rear section had to be raised to the vertical, reducing the area of the deck on which heavy loads could be deposited. With no restriction over the jetty, the jib could be extended farther for lighter loads. During storing and de-storing, lorries could be craned on board so that their cargoes could be off-loaded at points nearest to the access routes for various armament, naval, general and victualling stores complexes, reducing the need for the working parties known as 'human chains'.

Note also the length of the after brow (5) required to span the distance between Middle Slip Jetty and the specially-designed after brow platform on its sponson off the quarterdeck (7). The position of the hull relative to the jetty at various states of the tide meant that the after brow had to be flexible; it was therefore constructed in two sections, one of 36 feet and the other of 16 feet, with a hinge between them that allowed the brow to remain usable at any angle between the lowest (8) and highest (9) tides ever recorded. Its length was due to the way the hull tapered aft at quarterdeck level with a considerable overhang that had to be kept clear of the jetty. The hull forward did not taper to the same degree where the forward brow (6) was attached and a single, unhinged brow 42 feet long over the forward fender position was adequate at all states of the tide.

The wide fenders were custom-made to cater for the unique way in which *Victorious*' hull tapered fore and aft; they had to be positioned accurately if they were to be effective, as shown in the berthing plan. Although not shown, securing tank cleaning vessels and other auxiliaries to the ship's port side while she was alongside for routine maintenance also posed significant problems, and several vessels had to be modified so that they could be secured under the angled deck sponson.

Arriving alongside starboard side-to from sea was a straightforward evolution, but leaving harbour required the ship to turn through 180 degrees immediately after coming away from the jetty in what were confined pilotage waters. There was only just sufficient room to do so, and constant bearings of conspicuous objects had to be taken to fix the ship's position accurately. As a Midshipman, the author was one of the navigation team that took these bearings from the compass platform on 8 July 1966 when *Victorious* sailed for what was to be her last deployment to the Far East.

As explained in the author's new book *Aircraft Carrier Victorious* (Seaforth Publishing, 2018), she was arguably one of the best designed and most advanced warships of her era, and it is particularly interesting to compare the preparations made to get ready an operational berth for her in Portsmouth Naval Base with those recently carried out to berth HM Ships *Queen Elizabeth* and *Prince of Wales*.

A's & A's

FOG OF WAR (*WARSHIP* 2018)

David Murfin provides some additional comment relating to his article on cruiser designs for trade protection

On page 163 of the article there is a reference to the 500-ton difference in fuel allocation and displacement between the B4 design with a mix of coal-fired and oil-fired boilers, and the B4 variant which burned only oil. The following detailed discussion of the likely arrangements in the two variants was omitted due to lack of space. Although by David's own admission it is largely speculative, he thought readers might like to see what he had to say about this:

Without drawings, internal arrangements for B4 with mixed fuel are speculative, but the coal has to go somewhere – most likely behind the armour alongside the machinery, increasing beam. There the coal aids the ship's protection, with ready access to stoke the boilers – harder than pumping oil. Space for 500 tons can be estimated assuming that B4 'mixed' carried 1,100 tons of oil located exactly as in B4 'oil'. The bulk density of anthracite is 50–58lb/ft^3. Spaces 20ft high along 250ft of both sides of the ship, each at least 2ft deep, would provide the 20,000ft^3 required. Eight hundred tons would require 32,000ft^3, with 11,000 ft^3 from 300 tons less oil – plus some rearrangements to accommodate stokers. A beam of least 57ft would seem to be required, 4ft more than the legend for B4 'oil'. However, the legend beam of B4 'mixed' was only 54ft.

Also, if B4 'oil' *and* B4 'mixed' have the same length and draught, their beams must be roughly in proportion to their displacements with B4 'mixed' again 57ft (ie 53 x 7000/6500), not the legend 54ft. However, the ¹⁄₁₆in = 1ft plan of B4 'oil' shows her beam as only 50ft. This would make beam of B4 'mixed' 54ft (50 x 7000/6500 = 53ft 7in) with draughts equal. The speed differences between B3, B4 'oil' and B4 'mixed' are then easier to understand, as is the 30,000shp of B4 'oil', giving a speed only half a knot less than Design 'A' with a beam of 49ft 6in and 10,000shp greater power. Dimensions, speeds and tonnages all fit better. B4 'oil' hardly needs greater beam than B3 (whose plan does match her legend), so

53ft may be an error, or a late note of 'insurance' for stability as oil (all of which was stowed low in the hull) was burned, with the detailed plans not redrawn.

THE ARMOURED CRUISER *JEANNE D'ARC* (*WARSHIP* 2018)

Aidan Dodson has pointed out that the photograph on page 84 showing *Jeanne d'Arc* in the Landévennec 'ships' graveyard' also includes, at bottom left, the light cruiser *Metz* (ex-German *Königsberg*), which featured in his article 'After the Kaiser: The IGN's Light Cruisers after 1918', in *Warship* 2017.

WARSHIP GALLERY: JAPAN'S U-BOATS, SASEBO, 1921 (*WARSHIP* 2018)

Aidan Dodson comments on the issue of the distribution of ex-German submarines after the First World War

Stephen Dent's Warship Gallery text includes the statement that 'it was agreed that submarines be divided up, roughly in proportion to the number of ships each of the allied countries had lost to enemy action, between Britain (105 boats), France (43), Italy (10), the United States (6), Japan (7) and Belgium (2)'. While the principle stated is indeed that which was ultimately applied, the numbers quoted are not those finally agreed, with no boats actually being transferred to Belgium.

All other powers listed had accepted boats in the relatively informal distribution undertaken during the winter of 1919/20, to be used for 'propaganda' purposes – tests, public exhibition, etc – for a strictly limited period of time. The UK had been allotted 22, France 15, Italy 10, the USA 6, Japan 7, and Belgium 2. The latter did not, however, accept her two boats (*U-90* and *UC-92*), on grounds of a lack of suitable personnel and excessive running costs, although one (not specified in the extant data) was taken on loan for a few weeks only. Belgium's warship spoils were thus restricted to a few torpedo boats (see the author's article in the present issue of *Warship*). In addition, the UK, France and Italy had further boats laid up in their ports.

The numbers quoted in the article for the UK, the USA and Japan reflect the aggregate numbers of both the latter boats and those provisionally allocated for 'propaganda' still in their respective hands in January 1920, when the formal determination of entitlements based on wartime losses were made. However, the numbers for France and Italy do not fit in with these, the January 1920 numbers (recorded in ADM 116/1992) being 37 and 22 respectively (the latter, however, including ex-Austro-Hungarian boats) – for a total of 177 ex-enemy submarines in Allied hands.

The January 1920 'proportionate' allocation awarded the UK 124 boats (70%), France 34 (19%) and Italy 19 (11%) – with the USA and Japan receiving none at all. Uniquely, France was to be permitted to commission for long-term service ten of the boats in French waters; all other boats were to be destroyed. Thus, the UK was 'owed' 19 boats, the other powers all being in surplus by 3, 3, 6 and 7 vessels, respectively. However, no actual return of vessels to the UK was contemplated, it being decided that boats 'shall be broken up in countries where they are now, under the same conditions as other ex-enemy submarines, in order that the proceeds may assist in compensation for the expense incurred in towing the propaganda submarines to the countries concerned'.

THE PEN & THE SWORD (*WARSHIP* 2017)

Reader **Sergey Trybitcyn** has written in with the names of two ships serving with the present-day Russian fleet named after well-known writers:

The minesweeper **Valentin Pikul** is named in honour of Valentin Savvich Pikul (1928–1990), who served on destroyers of the Northern Fleet 1943–45. After the war, he studied at the 1st Baltic Higher Naval School, but failed the course. He began writing historical novels, taking as his theme Russian 18th century history. The books proved so popular that their price on the black market reached 25 rubles – for reference, my first salary in 1989 was 45 rubles. He also wrote several books with a naval theme: *Moonzund* (war in the Baltic 1914–17), *Cruiser* (Vladivostok detachment of cruisers in the Russo–Japanese war), *The Three Ages of Okini-San* (life of a Russian naval officer from the end of 19th century to the 1920s), and *Requiem for Convoy PQ17*. The border ship of the Caspian Flotilla is also named after Pikul.

The rescue vessel **Viktor Konetsky** is named after Viktor Konetsky (1929–2002), who took the same course as Valentin Pikul, which was intended primarily for officers of the submarine fleet. Graduates of the school were distributed among the rescue ships of the Northern Fleet. Some years later, Konetsky resigned and joined the merchant navy. He wrote novels in a very unusual genre: travel books with a lyric hero (the author himself). Against the background of a voyage on board a merchant ship he philosophises and reflects on various problems of the universe; every so often, events in the life of the crew bring him back to reality. The books are a strange mixture of philosophical and humorous prose. He was the first in the USSR to reveal to the general public the tragedy of the submarine *M 200*, sunk with the loss of 28 of her crew on 21 November 1956 after colliding with the destroyer *Statny*, which he recounted as fiction. His major work was the eight-novel cycle *Behind Kind Hope*. Konetsky also wrote scripts for two feature films: *The Way to the Pier* and *Striped Trip*. The second of these films, a comedy about the escape of tigers on the ship which was transporting them, was a huge success in the USSR.

THE PEN & THE SWORD (*WARSHIP* 2017)

Kenneth Fraser has written in with some additional information about ship names.

It now appears likely that the destroyer *Tipperary* (purchased on the stocks from Chile in 1914) was named after the famous song, which is known to have been at the height of its popularity in that year. The name is at first sight that of the Irish county and town; but no destroyer up to this time had ever been named after a county, and only one *(Larne)* after a town. On the other hand, another of the ex-Chilean destroyers was named for General Botha, the South African Prime Minister who had been instrumental in bringing his country into the War, and it was unprecedented to name a warship after a living Commonwealth statesman. In 1915 we find two 'M'-class destroyers given names of Great War battles, *Marne* and *Mons*. These examples suggest that *Tipperary* may also have been intended to commemorate the War.

'HIGHBALL' (*WARSHIP* 2016)

Stephen Dent provides an update on his article on the French battleship *Courbet* and the 'bouncing bomb' trials.

At the end of the article it was mentioned that a diving expedition in 2010 had located a number of the dummy weapons on the bed of Loch Striven, and that plans existed to return at some point and attempt to salvage at least one of them.

In July 2017 this took place, when a team comprising members of the Royal Navy's Northern Diving Group and the East Cheshire branch of the British Sub-Aqua Club, based on the mooring support vessel *Moorfowl*, conducted a week-long operation in the loch, during which they succeeded in raising a pair of 'Highballs' from depths of around 100 feet (35 metres). From the deck of *Moorfowl* the two 'Highballs' were transferred to tanks ashore filled with a saline solution, prior to conservation

The second 'Highball' after recovery, partly cleared of encrustation. (© Lindsay Brown, Phil Grigg and BSAC)

The first 'Highball' breaks surface, after 74 years on the bed of Loch Striven. (Dr Iain Murray)

work. In fact, despite heavy encrustation resulting from 74 years on the seabed, both turned out to be in extremely good condition, and are now on display at the de Havilland Museum in Hertfordshire (http://www.dehavillandmuseum.co.uk/) and at the Brooklands Museum in Surrey (https://www.brooklandsmuseum.com/).

In addition to salvaging the 'Highballs', a sonar survey of the loch was carried out, while the BSAC team have produced an underwater trail guide to allow experienced recreational divers to explore the site and see a number of the other 'Highballs' *in situ* (as well as a side charge from an X-craft); this contains directions both to and around the site as well as diving instructions, and is available for download at http://www.sirbarneswallis.com/files/HighballsDivingTrail.pdf

The project was initiated by research conducted by Dr Iain Murray of the University of Dundee, who is also a trustee of the Barnes Wallis Foundation. Incidentally, Dr Murray suggests that *Courbet*, having been subject to attack by a German mini-submarine during her time off the Normandy beachhead, may hold the unique distinction of being the only ship to have been the target of both Allied and Axis secret weapons during the course of the Second World War. *Warship* would be interested to hear from any readers who can expand upon this.

Acknowledgements:

Thanks to Dr Iain Murray and to John Jordan for information used in compiling this update.

NAVAL BOOKS OF THE YEAR

Robert Brown
Battleship *Warspite*: detailed in the original builders' plans
Seaforth Publishing in association with the National Maritime Museum; hardback, 144 pages, complete sets of full colour plans; price £30.00.
ISBN 978-1-5267-1937-9

One of the glorious but less familiar treasures in the National Maritime Museum is the collection of original ship plans ('draughts') held in its Brass Foundry outstation in Woolwich. The set of plans for each ship or class of ships generally comprises a profile (in effect a multi-coloured labelled X-ray representation showing internal and external details), a plan for each deck, and cross sections. The profile, almost a work of art in itself, is complemented by a simplified rigging plan showing masts, rigging and (later) aerials and antennae. Details of armour and protection often appear separately on simplified plans. Covering a ship with a very distinguished record in two World Wars, this book launches a new series based entirely on reproductions of these plans, scanned at high resolution and printed in full colour.

The 'as fitted' originals, at scale 1:96 (⅛in = one foot), are about seven feet long for *Warspite* and fragile; until recently, the Museum's policy for reproducing draughts for researchers was by full-size black and white photocopying (a tedious and potentially destructive process); visitors were free to take digital photographs of the originals. From late 2015, photocopying was replaced by very expensive digital scanning; the publishers are therefore to be congratulated on making plans available at a very reasonable price – the complete set for *Warspite* scanned *ab initio* would cost a researcher in excess of £2,000.

The 'as fitted' drawings for *Warspite* as built and as refitted in an interim modernisation with a single trunked funnel in 1924–26 were apparently destroyed when she underwent major reconstruction between 1934 and 1937, so for the early ship drawings of the very similar *Malaya* have been substituted. Two-thirds of the book show the post-reconstruction *Warspite* in great detail. Additional drawings from the Gun Manual show the 15in Mark 1 turrets; from Puget Sound Navy Yard come drawings dated November 1941 showing post-refit aircraft and antenna arrangements of 'Request S-126', *Warspite*'s designation in the then-neutral United States.

The treatment of plans is remarkable in conception and brilliantly executed. The 'as fitted' profile from 1937 is first reproduced at slightly under half size on a four-page gatefold, measuring approximately 95x29cm. The reader is then taken from stern to stem in labelled double-page spreads, each spread taking a slightly enlarged part of the reduced original, with neighbouring areas faded and with the appropriate cross-section alongside. Extended captions guide the reader to important features. The eight successive decks are then reproduced at about one quarter size before being similarly examined one by one in larger labelled double-page spreads, each including the corresponding part of the profile. The result is just about the best possible representation in two dimensions of a three-dimensional structure, almost an MRI scan of a ship set out ready for the researcher or model maker.

This book closely follows its mission statement; it is meant to supplement, not supplant, the standard battleship books by Raven & Roberts, RA Burt and recently by Norman Friedman. There are no photos, few construction details, little information on genesis or on wartime operations except when action damage was involved; even then, there are no drawings or sketches to show the damage. No search for original plans at the private shipbuilders that contracted for *Valiant*, *Barham* and *Malaya* is mentioned. It seems churlish to criticise excellence, but it should also be noted that not all scales are clearly shown. Such minor reservations aside, this is an incomparable reference work, absolutely ideal for serious enthusiasts and model makers.

Ian Sturton

Conrad Waters
Cruiser *Birmingham*: detailed in the original builders' plans
Seaforth Publishing in association with the National Maritime Museum, 2018; hardback, 144 pages, complete sets of full colour plans; price £30.00.
ISBN 978-1-5267-2497-7

The second of this new series by Seaforth drawing on the collection of original plans held by the National Maritime Museum focuses on HMS *Birmingham*, the fifth of ten 'Town' class cruisers authorised under the RN's 1933–36 construction programmes. The builders' plans for *Birmingham* were selected because they are unusually well-preserved and comprehensive; all were produced by draughtsmen at the Royal Dockyards.

The principal problem faced by a book publisher in reproducing these plans is the size of the originals, which are not only extraordinarily detailed but are fully annotated. Even if reproduced over a double-page spread the labelling is generally illegible, which considerably reduces the value of the plans. The usual solution adopted is to use multi-page gatefolds, and there are three/four-page gatefolds here for the three profiles (as-fitted 1937, 1943 and 1952). However, where this book really scores is in the centre section, in which sections of the 1943 and 1952 profiles have been enlarged and paired with the station section drawings, which have been reproduced at a matching scale. The highlighted sections have been reproduced in the centre of the page, permitting a brief introduction together with more detailed annotations to draw the reader's attention to points of interest. This works extraordinarily well; it is like having a digital copy of the plans on screen which you can zoom in on, but with the added benefit of a detailed running commentary by an author who, in this case, has really done his research, to the extent that on occasions he is able to highlight detail differences between *Birmingham* and her sisters.

One unusual feature is the reproduction of the 'Incident Board' plans of the decks that were used for damage control. These perspective views highlight aspects of the design which are less apparent in conventional two-dimensional profile, plan and section drawings, in particular the armoured 'boxes' which enclosed the magazines and shell rooms fore and aft, and the separate handing rooms inside the cordite magazines, the propellant charges being passed through via flashtight scuttles.

The book is printed on thick, good-quality paper and the plans are beautifully reproduced throughout. Page layouts are well-spaced and attractive. Printing in colour has made possible the use of colour for the text and headings. One particular advantage of this is that the modifications to the ship up to 1943, which were traditionally marked in green on the original plans, are described using green blocks of text with detailed comments in a matching colour. There was a small problem with the binding of my review copy, which after every block of twelve pages is very 'tight' and does not allow the book to be fully opened; however, this is only really an issue where plans have been reproduced across the double-page spread (pages 12–13, 36–37 and 72–73).

This is a stunning book, and a 'game-changer' in terms of the way original ship plans are used. The companion volume on *Warspite* is reviewed above by Ian Sturton, and the series is scheduled to continue with HMS *Victorious* and SMS *Helgoland*.

John Jordan

Anthony R Wells
A Tale of Two Navies: Geopolitics, Technology and Strategy in the United States Navy and the Royal Navy, 1960–2015
US Naval Institute Press, Annapolis 2017; hardback, 264 pages, 23 B&W images; price £33.95/$35.00.
ISBN 978-1-68247-120-3

Written by a former naval intelligence officer who holds the unusual distinction of having worked for both the Royal Navy and the US Navy while a citizen of both countries, *A Tale of Two Navies* looks at the evolving relationship between the two fleets over the past 55 years. Given the author's background, it is not surprising that much of the book's focus is on intelligence links between the two forces. However, many other significant areas of collaboration are also discussed.

The book comprises twelve chapters which are arranged in broadly chronological order but focus on a number of distinct themes. The lead chapter looks at the governmental structures that underpin the two navies and how this impacts on their respective political influence – in the author's view, much to the US Navy's advantage. Many of the subsequent chapters look at cooperation to counter the threat posed by the Soviet Union's powerful submarine fleet at different stages of the Cold War, particularly from an intelligence perspective. Other chapters examine areas such as Anglo–US collaboration during the Falklands conflict and the changed maritime security environment after the end of the Cold War. The concluding chapter is essentially a call for an increased emphasis on a maritime-centric strategy as part of a strengthened partnership between the two navies.

The author's direct involvement with both navies throughout much of the period in question provides some valuable insights, although there is often an impression that ongoing security constraints leave some areas unexplored. The emphasis on the 'big picture' also results in minor errors of detail: for example, the number of Type 209 submarines available to the Argentine Navy in 1982. However, these are minor quibbles. *A Tale of Two Navies* provides an interesting and readable account of a lengthy naval partnership that continues to play a vital role in ensuring global maritime security.

Conrad Waters

David Kohnen
21st Century Knox: Influence, Sea Power, and History for the Modern Era
Naval Institute Press, Annapolis 2016; paperback, 176 pages; price $24.95/£20.95.
ISBN 978-1-61251-980-7

Few readers will be familiar with the name Knox beyond its association with a class of postwar frigates. Commodore DW Knox served in the US Navy during the Spanish–American and First World Wars. He commanded a variety of warships and served in flag headquarters before becoming a prominent naval historian, overseeing the US Navy Department's historical office. In the latter role he was instrumental in assisting President Franklin D Roosevelt to educate the American public in the role of naval power and the relationship between war and peace. Knox was both a historian and a public influencer.

This is the fourth book in USNIP's 21st Century Foundation Series and follows the now familiar format, with the editor providing an extensive introduction which, in this case, provides a biography on Knox's life – especially useful as no biography currently exists. The main body of the book is a selection of seven essays which span Knox's extensive service. Knox was keen to use his knowledge to help shape the navy that he believed the nation needed as it faced a time of great uncertainty; the essays cover issues such as strategy, doctrine, and leadership that remain relevant today. A key thread running through his writing is 'influence': that of individuals, and also the unique collective role of naval forces not just in high-end warfighting but also in conflict prevention.

The essays also highlight the importance of history in support of contemporary analysis. Knox's 1926 essay 'Our Vanishing History and Traditions' inspired the formation of the Naval Historical Foundation. However, he was also aware of history's limitations, writing in 1950: 'In these days of guided missiles, jet propulsion, atomic power, and a great variety of other new weapons and devices we cannot, of course, rely too precisely upon recent wartime experience for future guidance.'

This volume offers insights into leadership and officer professionalism as well as a well-crafted case for history and sea power. Knox knew that only by preserving the historical record can the appropriate lessons be drawn, thus feeding the development of future doctrine in a changing world. This is history with purpose, and Kohnen's book deserves as broad a readership as possible.

Philip Russell

Malcolm Wright
British and Commonwealth Warship Camouflage of WWII, Volume 3: Cruisers, Minelayers and Armed Merchant Cruisers
Seaforth Publishing, Barnsley 2016; hardback, 192 pages, numerous colour line drawings; price £30.00.
ISBN 978-1-84832-420-6

This book is the third and final volume of a series by Australian war game designer and maritime artist Malcolm Wright. Aiming to provide a quick reference source to the wide range of paint schemes adopted by British and Commonwealth warships during the Second World War, the series has gained both admirers and detractors. The former typically refer to the comprehensive nature of the coverage and the clarity with which the various schemes are presented; the latter tend to focus on the accuracy of the information presented.

Following much the same format as its predecessors, the book commences with a short, four-page introduction outlining some of the practical factors that impacted on the application of official guidance on camouflage in operational conditions, and the consequent difficulty in accurately reflecting the actual schemes used. The limitations of photographic evidence are also discussed. This is followed by a page of references, three summarising the various camouflage schemes used and two pages listing the various colours and symbols used in the drawings.

The remainder of the book, encompassing some 177 pages over five chapters, is taken up with around 800 colour drawings of the ships themselves. These are typically arranged four to a page, often depicting their subjects at different stages of their careers. All of the variations are illustrated by either a port or starboard profile, frequently supplemented by the opposite profile and/or a plan view where appropriate (depending on the availability of the information available). In a change from previous volumes, each chapter commences with an alphabetical list of its contents, facilitating reference to specific ships. This reflects the well-structured and clearly presented layout evident throughout the book.

The author has clearly undertaken a very significant amount of work in producing both this book and the wider series. It is apparent that he has a good general understanding of the ships presented, while the breadth of information provided can be particularly useful. However, the reliability is suspect: there are often errors in the colours of the schemes depicted, incorrect descriptions of the paints themselves and confusion over some of the major modifications made to ship appearance. It should also be noted that none of the books in the series take account of the most recent research on Royal Navy camouflage of the period, which has significantly advanced understanding of the subject.

In conclusion, all three volumes represent an ambitious, but flawed attempt to provide a comprehensive overview of the camouflage used by Royal Navy vessels in the Second World War. While providing a useful aid to ship recognition during this period, they cannot consistently be relied upon for accurate representations of the vessels portrayed.

Conrad Waters

Andrew S Erickson (editor)
Chinese Naval Shipbuilding: An Ambitious and Uncertain Course
China Maritime Studies Institute & Naval Institute Press, 2016; hardback, 376 pages, 35 maps, plans & tables; price US$39.95.
ISBN 978-1-68247-081-7

This is the sixth in a series of books examining current Chinese naval capabilities jointly published by the US Naval War College's China Maritime Studies Institute (CMSI) and the Naval Institute Press. In essence, it collates the views of more than thirty experts in their subject drawn from participants in the CSMI's 'China's Naval Shipbuilding: Progress and Challenges' conference held in May 2015. The main aims of both were to assess the strengths and weaknesses of China's current shipbuilding infrastructure, the resulting consequences for the Chinese People's Liberation Army Navy and, in turn, the likely implications for the US fleet.

The book is structured into five main sections, each comprising three or four chapters authored by one or more experts. The initial section provides a strategic framework by assessing China's evolving warship building requirements and the resources available to the sector. The subsequent three sections provide more granularity on the industry by assessing shipbuilding infrastructure, warship design and the major challenges that remain to be overcome. The final section returns to a more strategic level by assessing the potential future direction of Chinese naval shipbuilding and the possible consequences for the US Navy's own requirements.

The expert analysis can be somewhat heavy going at times, while overall readability is not assisted by NIP's typically austere approach to the inclusion of illustrative material. The inevitable US-centric approach seen in the book's conclusions means that the opportunity to provide lessons for other regional naval powers such as

Japan and India is largely missed. Nevertheless the book provides authoritative, in-depth analysis of an area crucial to current and future global naval developments that makes it essential reading for any serious student of contemporary maritime affairs.

Conrad Waters

Admiral Raoul Castex (edited & translated by Eugenia C Kiesling)
Strategic Theories

US Naval Institute Press, Annapolis 2017; paperback,
494 pages; price US$29.95.
ISBN 978-1-59114-594-3

First published as a hardback in 1994, this edited and translated edition of *Théories Stratégiques* has now been published in paperback format by the US Naval Institute. *Théories Stratégiques* was first published as five volumes spanning 2,493 pages between 1929 and 1935, and its publication confirmed Castex as France's foremost naval strategist of the day. Compressing Castex's writing into a 500-page single-volume work was clearly a major undertaking, and Kiesling has opted to focus on his method of strategic analysis while omitting most of the historical narrative. Included are chapters defining strategy and relating it to policy and geography, analysing the role of maritime forces and the significance of command at sea, prescribing a theory of conduct of operations, and introducing Castex's favourite themes: strategic *manoeuvre*, *stratégie générale*, and the theory of 'perturbation'.

Although Castex accepts fleet action as the goal of naval warfare, his interest is in the plight of the second-rank navy (such as that of France) that must assert itself in other ways, accepting battle only after having used strategic *manoeuvre* to create a favourable shift in the naval balance. This involves forms of naval warfare – attacks on commerce, blockade, naval raids, mine warfare and amphibious operations – dismissed by Mahan as distractions from the battle fleet's struggle for mastery of the sea. In the chapter 'Second Theory for the Conduct of Operations', the map shows two opposing powers ('red' and 'blue') with a common land border; between them is a gulf dotted with islands, ports and naval bases. Naval power is seen as an adjunct to land power, with naval forces used for fire support, for denial of the enemy's use of his bases, for the protection of one's own maritime communications and for attacks on those of the enemy. The narrow waters bring submarines and minefields into play, and the proximity of land enables both powers to deploy aerial forces for reconnaissance and attack. In the view of Castex control of the sea is a 'local' concept, and a goal only in so far as it enables the fleet to carry out specific missions that directly or indirectly impact on the progress of the war on land. Unlike territory, the sea cannot be usefully 'occupied', nor can control of the sea be as absolute as it was during the 19th century, given the advent of the submarine and maritime aviation.

The concept of *manoeuvre* is so particular to Castex's thinking that the word has been left untranslated. In essence it derives from the military theories of Napoleon Bonaparte: given forces of roughly equal size, the aim is to pin down a large part of the enemy's strength using a minimum of one's own in order to concentrate forces at a key point (either in the centre or on the flank) where they can overwhelm the enemy forces facing them. It is about keeping your own 'balance' while upsetting the enemy's by using decoys and feints. Castex also promotes the case for 'combined arms' operations using a combination of air power and naval forces – a concept which had a major influence on Soviet postwar naval strategy.

Castex was very much of his age, and some of the political analysis jars with modern sensibilities: his view that war is an inevitable, even desirable, way of resolving conflicting political ambitions, that peace in Europe will always be at the mercy of an ambitious *perturbateur* – Germany, and probably the Italy of Mussolini are in Castex's sights – and that Europe needs to come together to defend the white race and European civilisation against the Asiatic hordes, are a hangover from the *faux*-Darwinism of the late 19th century.

Despite the limitations of Castex's *magnum opus*, it is good that USNI have made *Strategic Theories* available to a wider audience at a reasonable price. Kiesling supplements Castex's own pompous and self-regarding footnotes with her own, more grounded commentary, and the translation cannot be faulted.

John Jordan

Barry Gough
Churchill and Fisher: Titans at the Admiralty

Seaforth Publishing, Barnsley 2017; hardback,
600 pages, illustrated with 35 B&W photographs and 2 maps; price £35.00.
ISBN 978-1-5267-0356-9

The papers of Admiral Lord Fisher and Sir Winston Churchill, together with several other related collections, are held at Churchill College, Cambridge. This book on these two titans of the Admiralty contains the fruits of two years' research by the distinguished naval historian Professor Barry Gough as resident archives fellow. It begins with a fairly brief account of Fisher's early career, followed by a longer treatment of his appointments as Second and First Sea Lords. Following the public controversy which attended his later years as First Sea Lord, Fisher was pushed into retirement in January 1910 while Churchill became First Lord of the Admiralty in October 1911. Especially at the beginning of his appointment, he sought Fisher's advice, notably on such technical advances as oil fuel and the increase in heavy gun calibre to 15in. The war began badly for Churchill and the Admiralty with the escape of the *Goeben*, while the decisive outcome of the Battle of Heligoland was marred by poor staff work. Churchill indulged his taste for action in the Dunkirk 'circus' and his Antwerp adventure. Further disasters at sea culminated in defeat at the battle of Coronel. Battenberg was replaced as First Sea Lord by

Fisher, who immediately took energetic action to despatch two battlecruisers to the Falklands. Both Churchill and Fisher were attracted to operations against Borkum and into the Baltic. However, when Churchill became committed to an attack on the Dardanelles, initially by ships alone, Fisher remained aloof but unwilling openly to oppose the First Lord's plans. It was not until May 1915 that Fisher finally carried out his often-repeated threat to resign and walked away from his post at the Admiralty; Churchill was soon replaced by Arthur Balfour. Fisher was appointed chairman of the Board of Invention and Research but his naval career was over. Churchill became an effective Minister of Munitions and in 1939 famously returned as First Lord, though this second time at the Admiralty is mentioned only very briefly in Gough's final chapter.

The preface declares that: 'My theme, different from recent historical preoccupations [with technical issues], plays on aspects of character and personality'. However, it is soon apparent that, in writing about Fisher, technicalities have to be addressed, although the author's grasp of these matters is at best tenuous. Gough seems unable to decide whether Fisher lacked 'technical or engineering competence' (page 12) or that his technical knowledge 'marked him out' (page 50). He is also unsure of Fisher's technical appointments; he does not mention that Fisher held the post of Director of Naval Ordnance and appears to believe that he was Controller (the principal technical position in the Admiralty) not in the 1890s but before he was appointed Captain of *Excellent* in 1883. There is no mention of Fisher's construction policy once he became First Sea Lord, let alone that much of it had to be reversed towards the end of his time in office. His responsibility for the first diesel-powered 'overseas' submarine (the 'D' class) is acknowledged, as is his advice to Churchill that substantially influenced the adoption of oil firing for the *Queen Elizabeth* class battleships and *Arethusa* class light cruisers; however, Fisher's unheeded demands that the new capital ships should be battlecruisers are overlooked.

Gough also claims in his preface that this book represents the first full use of the papers in the Churchill archive and that it will disclose the secrets they hold. However, the end-notes show that the majority of citations are to already-published primary sources (including those at Churchill) and to secondary sources; for example, of the 147 notes to the key Chapter 9 on the Dardanelles, only 26 refer directly to documents at Churchill. It is also striking that Gough makes almost no direct use of the Admiralty and other state papers in the UK's National Archives; this reviewer found only eight such references in the whole book. This substantial omission is not made up by references to recent authors who have used TNA material, notably Nicholas Black in his book on the naval staff. For example, Gough mentions serious reservations expressed by Admiral Sir Henry Jackson about the projected Dardanelles operation but not, as Black has established, that Jackson's memo was dated 5 January 1915, a day before Churchill's telegram to Admiral Carden stating that 'High Authorities here concur in your opinion' that an attack could succeed.

The cover of *Titans at the Admiralty* promises a 'revelatory saga', but Gough does not explicitly identify these revelations nor indeed the secrets that he has uncovered. His study may well be valuable to specialist naval historians of the Fisher era in pointing out what can be found in the papers in the Churchill archive – and, if largely by omission, what cannot. However, for the general reader of naval history, this account of Fisher and Churchill at the Admiralty provides too limited a perspective on these tumultuous events. *Warship* readers will find the technical errors more than a little irritating.

John Brooks

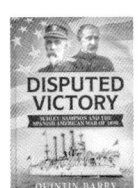

Quintin Barry
Disputed Victory: Schley, Sampson and the Spanish–American War of 1898

Helion & Co, Solihull 2018; hardback, 253 pages, 42 B&W photographs, 5 maps; price £35.00.
ISBN 978-1-912-1749-11

For the US Navy, the victory in the Spanish-American War was marred by a bitter dispute over who was responsible for the American success in the most spectacular sea battle of the war when the best ships of the Spanish fleet, four armoured cruisers and two destroyers, were sunk at Santiago de Cuba. At the moment of battle the US Commander-in-Chief, Rear Admiral William T Sampson, in the armoured cruiser *New York*, was *en route* to a meeting with the American land commander, so that during the action the senior officer was Commodore Winfield S Schley in the armoured cruiser *Brooklyn*. Thus an increasingly acrimonious debate grew up over who was 'really' in charge and thus deserving of the ultimate credit for the victory. Introduced into the debate were criticisms of Schley's conduct in the earlier phases of the war, culminating in a Court of Inquiry, an account of which bookends Barry's narrative.

However, this book is far more than a 'courtroom drama'. Rather, it is a fine history of the US Navy during the later 19th century and the Spanish–American War, seen through the lens of the key personalities of the era. It begins with the Navy in its dark years following the Civil War, then traces its climb from wood and sail towards becoming a first-class modern navy. This historical account is followed by sketches of the careers of the key naval protagonists in the main story: not only Schley and Sampson, but also Admiral George Dewey, victor of the Battle of Manila Bay and later President of the Court of Inquiry, and Theodore Roosevelt, Assistant Secretary of the Navy (and later President). A series of chapters then outline the situation in Cuba (the ultimate cause of the conflict), Hispano–American relations, the events surrounding the explosion of the battleship *Maine* in Havana harbour, the state of the Spanish Navy, and American war plans, before moving on to the war itself. This is covered in detail over eleven chapters, before a final two (plus an epilogue) cover the postwar controversy, the conduct of the Court of Inquiry and its aftermath.

The story is well told in an even-handed way, and gives the reader all the material necessary to understand the way in which the Court of Inquiry came about, and the inevitability of the result (a verdict of two to one against Schley).

The book is well-produced, although it is a pity that the numerous photographs are concentrated in a plate section and are nowhere referenced in the text, even when of direct relevance (*eg* the sinking of the Spanish cruisers). The author also seems less comfortable with ships than with events, with a number of errors on both technical matters (the Spanish cruisers had 11in and 5.5in Hontoria guns, not '10-inch Armstrong' and '5.2-inch' weapons), ship types (the German large cruiser *Kaiser* is referred to as a 'battleship') and names (the Austro-Hungarian cruiser *Kaiserin und Königin Maria Theresia* is incorrectly named '*Kaiserin Maria Teresa*'). There is also an anecdote quoted that alleges that the decision to change the calibre of battleship guns from 13in to 12in after the *Kearsage* class was the result of British pressure, something that finds no confirmation in any of the standard accounts of US battleship development. These, however, are but minor niggles in what is an excellent book that could well serve as a standard work on the Spanish-American War at sea.

Aidan Dodson

Patrick Maurand and Jean Moulin
Les frégates anti-aériennes *Cassard* & *Jean Bart*

Marines Editions, Rennes 2016; softback, 128 pages, data tables and schematic drawings, many colour/B&W illustrations; price €29.00.
ISBN 978-2-357-43143-0

Les frégates furtives type 'La Fayette'

Lela Presse, Le Vigen 2018; hardback, 160 pages, data tables and schematic drawings, many colour/B&W illustrations; price €25.00.
ISBN 978-2-37468-009-5

These two books are the latest in a series of French-language monographs on ships currently in service with the *Marine Nationale*, and follow an earlier book by the same authors on the anti-submarine frigates built from the 1960s to the present day.

Cassard and *Jean Bart*, together with two sisters, were intended to replace the four Tartar conversions of the postwar T 47-type fleet escorts, whose American missile systems they were to have inherited. The two sister-ships were subsequently cancelled, but *Cassard* and *Jean Bart* were authorised under the 1978 and 1979 estimates and entered service in 1988 and 1991 respectively. The design was adapted from that of the C 70 antisubmarine corvette (later frigate), but with an all-diesel propulsion system which dispensed with the large air intakes associated with gas turbines in order to make space for the Tartar area defence missile system, which was later upgraded to fire the Standard SM-1MR missile. *Cassard* and *Jean Bart* remain in service, but are due to be replaced by an air defence variant of the new FREMM frigate in 2022–23.

The five frigates of the *La Fayette* class were the world's first 'stealth' frigates. Designed during the mid-1980s following the Falklands War, they introduced many of the features which have since become standard in modern warships: angled, 'blocky' superstructures to reduce radar signature, with minimal protrusions. They are also notable for making extensive use of the composite material CVR (GRP) for superstructures. The second ship, *Surcouf*, introduced modular construction to the French Navy, using prefabricated, fully-fitted blocks, thereby reducing building times. The ships have proved eminently well-suited to overseas deployment in defence of French interests in areas where there is a moderate threat, and the design has been widely exported, with variants built for Taiwan, Saudi Arabia and Singapore. However, their offensive capabilities remain limited to a single 100mm gun and MM40 anti-surface missiles, with over-the-horizon guidance provided by an AS 365 Panther helicopter. Due to budgetary constraints they received neither the Aster 15 area defence missile – they were fitted instead with the dated Crotale short-range SAM – nor the bow and towed sonars for which space was allocated in the original design. Designed for a service life of 30 years, they are currently receiving a low-cost modernisation, but may have to soldier on until the 2030s.

Because all seven of these ships remain in commission, official data continues to be restricted. Despite this, the French Navy gave every possible help to the authors, including personal visits to *Jean Bart* and *Guépratte* which have resulted in many on-board photographs of the various items of equipment. Given this constraint one could ask for nothing more than what is provided here. There are comprehensive accounts of both the technical aspects of the ships and their service careers, accompanied by informative data tables. A series of profile drawings by Jean Moulin show the ships' appearance at various stages of their careers and the modifications made to their weapon and sensor outfits.

Production values are high; both books are produced on good-quality paper with a clear, easy-to-read modern typeface and attractive page design. It would have been good to see some of the photographs of *Cassard* and *Jean Bart*, which are excellent throughout and mostly in colour, reproduced at a larger size, but this is a minor criticism. In the book on the *La Fayette* class the photos of the ships often occupy the full page, and many are stunning.

John Jordan

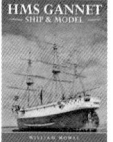

William Mowll
HMS *Gannet*, Ship and Model.

Seaforth Publishing, 2018; hardback, 128 pages, 285 illustrations, mainly in colour; price £25.00.
ISBN 978-1-5267-2628-5

The main theme of this book is the author's building of a ¹⁄₄₈ scale model of the sole surviving gunboat of the Victorian Royal Navy, an exercise which took three years. However he starts with the birth and operational career of the full-scale ship, one of many such vessels designed to police the British empire. Currently at Chatham's Historic Dockyard, HMS *Gannet* had a career which included not only policing duties but also early hydrographic surveys. She was subsequently employed as a drill ship for the reserves and a training ship; as TS *Mercury*, by then in private hands, she helped introduce a new generation to naval skills prior to the First World War.

This role continued until 1968 when she was handed back to the Royal Navy. Three years later she was purchased by the Maritime Trust and in due course arrived at Chatham for restoration, a process that didn't really get underway until 2002 when Tommi Nielsen of Gloucester was contracted to restore the hull and rigging to the level of completion that is now visible to the visitor to the Historic Dockyard. An attempt to dock *Gannet* in a hurry led to the discovery of a sluice valve stuck open by cement, causing the dock to unexpectedly empty. This did, however, demonstrate the great strength of the hull, which remained undamaged despite slipping off the blocks.

The major part of the book covers, step by step, the building of the author's model, illustrated by some 270 numbered pictures. Each is accompanied by a paragraph of description and explanation, with quotes from the ship's building specification dotted throughout. Some of the illustrations show the work underway on the restoration of the actual ship, with photographs from the ever-helpful Nielsen, although the majority cover specific model components and model-making techniques.

The author used the plank-on-frame method to reproduce the original composite structure of *Gannet*, ie wooden planks over iron frames (one of three surviving composite ships, strangely, given the short period of time that the technique was used), but his modelling work also includes metal casting, turning, fitting and rope work for the rigging, as well as the wood work.

This is a book full of great detail, clearly the result of much careful research and labour in making a fine model. The great pains that Mowll has gone to ensure authenticity are evident, and should encourage others to try their hand at modelling or, as in this reviewer's case, to try to get better at it.

W B Davies

David Goodey and Richard Osborne
Destroyer at War: The Fighting Life and Loss of HMS *Havock* from the Atlantic to the Mediterranean, 1939–1942

Frontline Books, Barnsley 2017; hardback, 293 pages, illustrated with 37 B&W photographs and 3 maps; price £25.00. ISBN 978-1-52670-900-7

Anyone with knowledge of the Second World War at sea will be instantly familiar with the Royal Navy's 'H' class destroyers, which were particularly prominent during the crucial first three years. *Destroyer at War* tells the story of one of these ships, HMS *Havock*, through a mixture of official accounts, newspaper reports and other documentation, together with eye-witness testimony, some written at the time in the form of diaries, other recalled later. The idea for the book came from a chance discussion between the two authors at a World Ship Society Naval Meeting in June 2013. Richard Osborne had just presented a talk on the loss of the destroyer; David Goodey's interest stemmed from the fact that his father had served aboard *Havock* and he had already researched the subject.

HMS *Havock* was built by William Denny of Dumbarton as part of the 1934 Programme and by the first week in February 1937 she had received her full crew at Chatham and was ready for departure for service with the Mediterranean Fleet. The ship was bombed *en route* to Malta while within the Spanish Civil War zone, providing a rude awakening for her inexperienced crew. In July of that year she was attacked in the same area, first by submarine-launched torpedoes and later by bombers.

In September 1939 *Havock* and her sister *Hotspur* were sent to the South Atlantic on convoy protection duties. It was during a manoeuvre in Pernambuco Harbour, Brazil, that she stripped blades on a turbine, exacerbating existing concerns about the reliability of her main engines, which would require 'heavy machine shop' docking to put right. It is a measure of the incessant demand for the services of *Havock* that this defect was never put right and she 'crabbed' by a few degrees while underway for the rest of her life.

By December 1939, the ship was in home waters and in April 1940 was despatched to Norway, taking part in the First Battle of Narvik. No sooner had she emerged, virtually unscathed, from that ultimately disastrous campaign than she was sent to Holland following the commencement of the German *Blitzkrieg* on the Low Countries on 10 May. Here intensive aerial assault exposed the inadequacy of her AA defences. On 14 May, while patrolling off the Dutch coast, *Havock* was ordered to Plymouth, for immediate onward passage to the Mediterranean. There was to be no let-up.

The ship was to spend practically two years in the Mediterranean until her loss in April 1942. By the time she foundered off the Tunisian Coast, she had won no fewer than eleven Battle Honours, most in the Mediterranean, including Matapan, Greece, Crete, Libya, the Malta Convoys and Sirte. Battered, damaged and finally patched up in a Malta Dockyard itself under furious bombardment, she was ordered to negotiate the dangerous passage to Gibraltar with a captain and crew many of whom were clearly suffering from PTSD.

David Goodey and Richard Osborne deserve the highest praise for bringing an often familiar train of events vividly to life through a skilful blend of well-researched and authoritative narrative and eye-witness accounts. Importantly, they regularly 'step back' from the immediate action in order to place *Havock*'s activities within the broader context of the war taking place elsewhere. Their account of the First Battle of Narvik is quite simply the best this reviewer has read. There continues to be a surfeit of often indifferent naval books about the Second World War; this is a splendid exception.

Jon Wise

Charles C Roberts, Jr
The Boat that Won the War: An Illustrated History of the Higgins LCVP
Seaforth Publishing, Barnsley, 2017; hardback, 128 pages, profusely illustrated in B&W and colour; price £25.00.
ISBN 978-1-5267-0691-1

Ron MacKay, Jr
The US Navy's 'Interim' LSM(R)s in World War II: Rocket Ships of the Pacific Amphibious Forces
McFarland, Jefferson, North Carolina, 2016; paperback, 344 pages, illustrated with B&W photographs and maps; price US$45.00.
ISBN 978-0-786-498598

Both these books describe amphibious vessels used in the Second World War, but the authors treat their subjects in different ways. Thousands of LCVPs (Landing Craft Vehicles and Personnel) were built and carried by larger amphibious ships for use in assaults over beaches. Capable of carrying 36 fully-equipped Marines or a light vehicle with its trailer or field gun, they were based on a design by Higgins Industries, but the vast scale of production required led to seven other manufacturers being brought into the programme, including a US Army Engineer Base Shop in Australia. LCVPs were mostly made of marine mahogany and, apart from a few early examples, had a single diesel engine which gave a top speed of 9 knots fully loaded. Typically they had a crew of four comprising a coxswain, motor mechanic and two deck hand/machine-gunners.

Charles Roberts' text is extremely detailed and is enhanced by numerous photographs, drawings and graphics. It must be considered the definitive description of how these iconic little landing craft were designed and built – the author's engineering knowledge was backed up by restoring a 1943 Higgins Boat for operation and display at his own World War II Museum. Given that LCVPs served in many theatres, it would not have been possible to describe every action in which they were used, but Roberts does describe how they were manned, deployed and operated during large-scale beach assaults. Diagrams from wartime manuals and photographs give a very good idea of how coxswains were taught to beach and retract the craft, tasks often made difficult by beach gradients and wave action. These little landing craft have rather been taken for granted in the past, but this book rectifies that shortcoming and is thoroughly recommended to readers with an interest in amphibious warfare.

In contrast, only twelve Landing Ships, Medium (Rocket) – LSM(R)s – were completed, all of them at Charleston Navy Yard during 1944 as modifications of the basic LSM design. Two General Motors diesel engines gave a maximum speed, fully loaded, of just over 13 knots with a range of 4,900 nautical miles at 12 knots. They had a complement of 6 or 7 officers and up to 83 enlisted men and were completed in two groups: eight with fin-stabilised rockets and four with spin-stabilised rockets. They also had a single 5in/38-calibre gun, two 40mm Bofors and two 20mm Oerlikons. All had SU-1 surface search radar and extensive communications facilities, which made them into formidable little fighting ships in the amphibious operations area.

Ron MacKay concentrates on the LSM(R)s' relatively brief operational service during the Okinawa campaign. Detailed descriptions of how the vessels' deployment from Charleston to Okinawa was administered give insight into an aspect of the Pacific War that has not previously been well covered, but the text's main strength lies in its detailed descriptions of their employment, which included rocket attacks on beach defences and support for destroyers on radar picket duty. The losses of some of their number to Japanese air attacks are well described, but unfortunately much of the text is written in the style of a USN Report of Proceedings and is both dry and repetitive. That said, this is recommended as a text book for anyone with an interest in the Pacific War in general and USN amphibious operations in the Okinawa Campaign in particular.

David Hobbs

John D Grainger
The British Navy in the Mediterranean
The Boydell Press, Woodbridge, Suffolk, 2017; hardback, 306 pages; price £65.00.
ISBN 978-1-78327-231-0

Author John Grainger has followed up a previous book covering British involvement in the Baltic Sea with an overview of the Royal Navy's involvement and influence in the Mediterranean from the 11th century to the present day. After a chapter detailing early maritime forays, principally during the Crusades, he focuses on the period from 1580 and the Levant Company's trade with the Ottoman Empire to the evacuation of Tangiers in 1683, moving on through the wars with France and Spain fought between 1688 and 1815, which established Britain's ascendancy in the Mediterranean and saw bases established at Gibraltar, Port Mahon (Minorca) and finally Malta.

Through the rest of the 19th century, the Royal Navy dominated the Mediterranean, supporting Greece and Italy in their struggles for independence and the Ottoman Turks in their efforts to resist Russian expansionism. British warships supported Britain's diplomatic interests around the region, and Gibraltar and Malta both grew in importance as naval bases to support the increasing numbers of warships. With the opening of the Suez Canal in 1869, the Mediterranean was no longer a closed basin but a thriving sea route between Europe and the Far East.

The final chapters cover the period that begins with the Great War. During this conflict, the Mediterranean was essentially of secondary strategic importance. In the interwar years, especially the 1920s, it became the key to Britain's imperial naval strategy, with a strong and efficient Mediterranean Fleet able to sail eastwards to reinforce the Far East in the event of war with Japan, or westwards in the event of war in the Atlantic. The changing political scene in 1930s Europe shifted the strategic focus back to the Atlantic.

The Mediterranean was far from a secondary theatre during the Second World War, seeing several fierce convoy battles until, with the invasions of Sicily and Italy in 1943, the need for large fleets declined. Postwar developments saw an economically exhausted Britain unsuccessfully trying to re-assert her previous influence. It also saw a gradual increase in the American naval presence, and the formation of NATO meant that there was little need for substantial British naval forces in the area or for major naval bases at Malta and Gibraltar. Today the Mediterranean remains an important sea route for merchant shipping, but it is the US Navy that dominates.

Grainger has done his job well and I would recommend this as a good overview of the topic. However, there are reservations. The principal one is the book's pricing: at £65.00 it is probably too expensive for the general readership that Grainger's 'broad-brush' approach would probably best engage, while academic readers may be disappointed by the lack of depth and detail.

The other issue is one of production. On occasions words and sentences seem out of place – possibly a symptom of careless cutting and pasting. The maps are small but are still take up a full page, and the key states that places used as British naval bases are underlined; however, no underlining is visible (although bases are shown as squares). The author's account includes an episode in which Gloster Gladiator fighters(!) are alleged to have dropped torpedoes against Italian ships in Bomba harbour. While generally of a minor nature, errors such as these are a particular disappointment in an otherwise well-presented and relatively expensive volume.

Andy Field

Richard Perkins (with introductions by Andrew Choong)
British Warship Recognition:
Volume IV: Cruisers 1865–1939, Part 2
Volume V: Destroyers, Torpedo Boats and Coastal Forces, 1876–1939

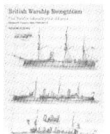

Seaforth Publishing, in association with the National Maritime Museum, 2017; large format hardbacks, 192 and 312 pages respectively, illustrated in full colour throughout; price £60.00 and £70.00.
ISBNs 978-1-4738-9149-4 and 978-1-5267-1112-0

Two more volumes in this series of facsimile reproductions of Richard Perkins' remarkable collection of recognition albums include what is by a substantial margin the largest so far, and hence the most expensive book in a series likely to stretch the budget of even the most dedicated warship buff.

Volume IV completes the coverage of cruisers, rather inconveniently split with volume III (reviewed in *Warship 2018*) at the mid-point of the First World War. The remainder of the wartime light cruisers are covered, heading backwards in time through scouts, protected cruisers, despatch vessels, corvettes and iron frigates. The sheer number of ships of the late Victorian era is impossible to ignore (as is how many ended up in bizarre guises performing all manner of harbour service duties). To further complicate matters for those using the books for reference, Perkins names each class by whichever ship came first alphabetically, although Seaforth have used the more conventionally accepted official class names in the contents listing at the front of the book. With many more ships per class than in the earlier volumes, there is more emphasis on minor differences between individual vessels: number and location of scuttles, positions of steam-pipes, and variations in rig. Much of this is covered by way of tables, and it has to be said that these, devised to Perkins' own requirements, do take a bit of getting used to. By way of example of the books' value, the light cruiser shown on pages 205–207 of *Warship 2017* can now be identified as *Calliope*, distinguished by the position of the yard on the mainmast (visible in the original image, when enlarged), while the foremast yards suggest that the cruiser on page 206 of *Warship 2013* is *Undaunted* rather than *Inconstant*.

In a slight departure from earlier volumes, in the cases of just a few ships their late-war camouflage schemes are depicted (though from one side only); no reason is given for this. A further one-off is that one page simply consists of a reproduction of a single clipping from the *Illustrated London News* that had clearly caught Perkins' eye; again this is unexplained.

Volume V, following the format of the original folios, is a step up in size, and hence in price. Weighing in at a shade under 3.5 kilograms, it really is an unwieldy beast, made more awkward still by the fact that a small number of the pages (all within the concluding sections on coastal forces and torpedo boats) have been reproduced at 90 degrees to the others, meaning that the book has to be rotated to be read.

The treatment also changes slightly in this volume. Coverage reverts to a more conventional chronological order, and this has the incidental but notable benefit of providing a comprehensive visual record of the development of the British destroyer. Meanwhile, the number of ships involved results in coverage that is rather more varied, with some vessels receiving several profiles to show appearance changes over their careers, while others are covered by just one for a number of very similar-looking ships, plus extensive accompanying tables and numerous detail drawings. This means that identification of, say, an undated photo of an unnamed 'V' and 'W' class destroyer will still be a challenge requiring time and application. On the other hand, it also gives an idea of just what an undertaking it must have been for Perkins to create all this in the first place, especially when dealing with the large number of war-built vessels of more-or-less standard design. Incidentally, this famous and numerous class throws up one interesting anomaly, in that while the 'official' cut-off date for the end of the books' coverage is 1939, in the case of *Wolsey* there is a profile dated 1941, after her conversion to an escort destroyer. This begs the intriguing question of how Perkins went about getting his information by this time. It remains a pity that so little is known about the man; what is known is outlined in the introductions to these volumes, which are an impressive testament to a unique diligence and industry.

Stephen Dent

Trent Hone
Learning War: The Evolution of Fighting Doctrine in the US Navy, 1898–1945
US Naval Institute Press, Annapolis 2018; hardback, 432 pages, maps, figures and B&W plate section; price US$34.95.
ISBN 978-1-68247-293-4

During the 19th century the US Navy was a traditionalist institution commanded by elderly career officers who obtained their posts through seniority, regardless of merit. In the first years of the 20th century a revolution took place, led by Rear Admiral Stephen B Luce and his 'insurgents'. Luce wanted to create a 'professional' navy, with 'educated' officers whose promotion was subject to selection. These proposals initially met with resistance from Congress, which feared the loss of civilian control over the military, but received the support of key individuals such as Secretary of the Navy Josephus Daniels. It is no coincidence that most of the admirals whose names are associated with the key actions and campaigns of the Pacific War of 1941–1945 graduated from the Naval Academy between 1903 and 1906.

Trent Hone's study of professional development in the US Navy from 1900 to the end of the Second World War is not always an easy read. The terminology introduced and defined in the early chapters ('enabling constraint', 'symmetry break' and 'heuristics') is more commonly associated with 'management speak' than naval history. However, it provides the necessary framework for the subsequent discussion and analysis of the evolution of doctrine in the US Navy. The key features of this evolution were tactical exercises devised by the Naval War College ('Fleet Problems'), the encouragement of collaborative decision-making through conferences in which every officer felt safe to give his views, regardless of seniority, and innovation fostered by an effective balance of low variability (experimentation) within enabling constraints (the set task or 'problem').

Ironically, the US naval doctrine that evolved proved just as unsuited to the realities of the Pacific War as that of the Imperial Japanese and Royal Navies, largely because the US Navy shared the IJN's obsession with the 'decisive fleet battle'. The cruiser and destroyer tactics, with their wedge- and 'V'-shaped attacking formations, did not survive the early battles of Guadalcanal, and the US surface navy suffered huge losses which might have crippled any lesser force. However, because the *processes* in place to develop doctrine were superior to those of other navies, the learning curve was steeper: within a year and a half the US Navy was getting the upper hand in the smaller battles, and was ready to embark on a massive counter-offensive against the island chains of the central Pacific using the new carrier task groups.

Trent Hone's analysis of the confused night actions in the Solomons is outstanding, as is his account of the carrier offensive in the Central Pacific; the Solomons campaign is illustrated with clear maps, that in the Central Pacific with tables showing the composition of each of the carrier task groups during the different phases of the operation. The author shows how the key to the carrier offensive was the adoption of a system of 'modular' task forces, with ships, task groups and commanders swapped and rotated so that the Japanese were given no time to regroup and consolidate. The adoption of a dual command structure in the summer of 1944, in which the Fifth Fleet staff under Spruance rotated with the staff of the Third Fleet under Halsey, meant that one commander was able to conduct an operation while the other planned the next, thereby keeping the Japanese off balance.

With the massive expansion of the US Navy during the war, both in terms of ships and personnel, it was inevitable that doctrine would congeal, that 'exploration' would give way to 'exploitation' and that variability at local command level would be superseded by standardised procedures. Hone sees this as regrettable but inevitable.

This is an excellent, thought-provoking book which is essential to understanding how the US Navy approached, and eventually prevailed in the War in the Pacific.

John Jordan

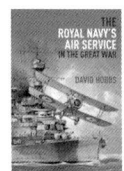

David Hobbs
The Royal Navy's Air Service in the Great War
Seaforth Publishing, 2017; hardback, 528 pages, illustrated with maps and numerous B&W photographs; price £35.00.
ISBN 978-1-84832-348-3

This book tells the story of British naval aviation from its very earliest days up to the end of the First World War, by which time the RNAS had been superseded by the newly-created Royal Air Force. It is a massive tale, and one for which 500-plus pages barely suffice as the RNAS steadily becomes ever bigger and more diverse in its activities.

The early part of the story features the classic 'characters' that invariably crop up in tales of the early days of aviation: those who learn to fly in 24 hours to win a bet, or who are tailed on the ground by their driver in a Daimler full of spares and supplies. Judging from the photographs, plenty of smoking also went on despite the proximity of large tanks of aviation fuel! The Admiralty, in contrast, and to their considerable credit, seems to have taken a careful and pragmatic approach to the whole business, recognising from the outset that there was considerable potential in this new medium and giving encouragement to those who wished to explore it, while at the same time not leaping in too enthusiastically.

The first seaplanes suffered from a range of unforeseen problems. Taking off had to be in exactly the right conditions – too rough and damage resulted, while too calm and the aircraft tended to fail to come 'unstuck' from the surface of the sea – and even then the motion was described as like that of a stick being dragged along park railings. The initial conversion of *Campania* was not an unqualified success: her high freeboard meant that hoisting seaplanes back on board in anything other than the calmest conditions resulted in them being bashed against the side. Aerial navigation posed challenges, and casualties from becoming lost were high; one pilot only made it back to his ship because he spotted a distant Zeppelin which turned out to be shadowing it! Engine failure was another persistent risk; many brave young men simply disappeared without trace because their machines weren't able to withstand the hostile conditions often encountered over the North Sea.

The chapters covering the development of shipboard aviation are particularly good, while others are simply lists of sorties and operations. Hobbs makes a valiant, and largely successful, effort to be even-handed when it comes to the controversial subject of the RNAS's replacement by the new RAF, pointing out that while in retrospect the service certainly could have fought its corner better in the political arena, it was also far too preoccupied with the business of fighting the real enemy.

A wide range of sources was consulted in the preparation of this volume, including both primary and secondary material. Perhaps the most important of these is an apparently unique surviving copy of an official Admiralty history of air operations in the Great War compiled in 1919 but never published. Unusually for a book from this publisher, there are numerous minor typos as well as instances of repetition and even of contradiction. Hopefully the worst of these will be corrected in any future reprint; in the meantime they should not be regarded as anything more than a minor blemish on an otherwise fascinating account.

Stephen Dent

Richard Larn
The Isles of Scilly in the Great War
Pen & Sword Books, Barnsley, 2017; paperback, 176 pages, numerous B&W illustrations; price £12.99.
ISBN 978-1-47386-766-5

Stephen Wynn
The Isle of Sheppey in the Great War
Pen & Sword Books, Barnsley, 2017; paperback, 160 pages, numerous B&W illustrations; price £12.99.
ISBN 978-1-47383-406-6

These two books are part of a series of titles produced by Pen and Sword Books entitled 'Your Towns and Cities in the Great War', intended to highlight the impact the conflict had on local communities in the UK. Scilly and Sheppey are both of obvious interest from a navalist's point of view. Both were strategically important. By 1914, Scilly, sited at the crossroads of no fewer than six major shipping routes, had historically monitored and defended the Western Approaches for over five centuries. Sheppey, with its naval dockyard town of Sheerness, similarly occupied a key position at the mouth of the River Medway leading to Chatham and on the Thames estuary. Both remain comparatively remote: even Sheppey, relatively near to London, leads nowhere.

Both works, according to the publishers, are written by experienced and knowledgeable authors. However here, sadly, the similarities end. Richard Larn had a long seagoing career in both the Royal and Merchant Navies, and is a native and resident Scillonian. He skilfully chronicles the myriad little naval actions fought against U-boats in the vicinity of the islands, showing not only how Scilly was at the forefront of attempts to counter this new form of *guerre de course* with aircraft and primitive hydrophonic detection equipment, but also the impact this war 'on their doorsteps' had on a tiny, isolated community.

By contrast, Stephen Wynn, a retired policeman living in Essex, has no apparent links with the Isle of Sheppey, thus there is an evident lack of empathy with the subject and, more significantly, a glaring absence of background research, as his sketchy bibliography (entirely composed of internet references) attests. At one point, he remarks that Sheppey became known locally as 'Barbed Wire Island' and, as one of only two designated military areas in Kent, its main claim to fame during the Great War was in relation to the establishment of the country's first military flying school at Eastchurch. The core reason why Sheppey was defended, which extended to restrictions relating to entry to and departure from the island, was due to the threat of invasion and the vulnerability of its key dockyard facility and naval base. If one read this book and knew nothing about the island's location, one would remain totally ignorant about the nature of its strategic significance within the wider context of the war itself.

Part of the blame for the failures of this book must lie with the publisher's directive that the series should record 'how each year of the war brought a change in the spirit of the populace as the huge battles taking place in Belgium, France, Gallipoli and elsewhere took their toll on the menfolk' – no mention of the war at sea and of the accompanying loss of life. Britain has drifted away from its identity as a maritime nation, and the Great War is now firmly perceived in the public eye as a land war fought exclusively on the Western Front. Thank goodness for Richard Larn's perceptive account of the impact of these terrible years on his own local area, of the rising importance of the submarine, and the use of aircraft to counter this threat, which has had a such a profound effect on naval warfare over the past 100 years.

Jon Wise

Ken Brown
U-Boat Assault on America
Seaforth Publishing, Barnsley 2017; hardback, 208 pages, B&W plate section and map; price £25.00.
ISBN 978-1-4738-8728-2

The subject of this new book is Operation 'Drumbeat' (*Paukenschlag*), the German U-boat operation against US shipping off New York which took place between 12 January and April 1942, in the aftermath of Pearl Harbor and the German declaration of war against the United States. Often described by the German U-boat crews as a 'Second Happy Time', the operation caught the United

States totally unprepared and sank 82 ships of half a million tons, including a large umber of oil tankers. The loss of oil threatened the Allied war effort and led to petrol rationing in the northeast USA.

Numerous accounts of Operation 'Drumbeat' have been published. However, new author Ken Brown aims to look at the broader context. He attempts to analyse the factors which propelled and inhibited the performance of the US Navy, the political background to the construction programmes and naval policies of the 1930s, and the role of New York City in the war. Only in the final chapter of the book does the author give a brief account of 'Drumbeat', its consequences for the US Navy, and the measures taken to defeat the U-boat assault.

There are some interesting and unusual observations here, particularly those concerning New York. Aspects include the officially-authorised links between the intelligence services and the Mafia aimed at controlling the waterfront and infiltrating possible sabotage attempts on behalf of the Axis, the activities of the pro-Nazi German-American Bund, and the irregular 'Coastal Picket Patrol' (comprising yachts from the Cruising Club of America) and 'Civilian Air Patrol' (using private aircraft armed with bombs and depth charges!) mobilised as a temporary measure to counter the German U-boats. One of the reasons for the heavy losses in coastal traffic was the failure of the American people to take the war seriously; although the U-boats found many of their targets by cruising on the surface off the city at night and picking off ships silhouetted against the shore lights, there was considerable local resistance to the imposition of a blackout because of fears that it would affect tourism.

The author has used a number of secondary sources for his narrative, and the early chapters in particular are effectively a compilation. The result is a picture which is relatively complete but not always coherent. The chapter on Roosevelt's Atlantic War Mobilisation of 1941 (Chapter 2) would have been better as a preparation for Chapter 9 (Operation Drumbeat). Although the author's aim of providing a broader context is laudable, there is a lot of material in chapters 3 to 8 which is only marginally relevant to the U-boat campaign off New York. Chapter 5, Diplomacy and Submarine Development Between the Wars, is particularly disappointing: the discussion of the interwar naval arms limitation treaties focuses largely on the impact on surface ships; the account of the limitations on the construction and employment of submarines is incomplete and confused, and there is no attempt to look at the types of submarine built by the major powers during the interwar period – the German-inspired 'cruiser'-type submarine which dominated US and IJN thinking during the 1920s does not even merit a mention. There are also errors: the names of the German cruisers *Magdeburg* and *Frankfurt* are misspelled (pages 64 and 84 respectively), and the 'super-Hoods' (G2) would have had nine (not twelve) 16in guns (page 84).

There is much to admire in this book. The chapters dealing with the city of New York are particularly good and look at 'cultural' issues which have rarely been highlighted in earlier histories of the campaign. However, the author's lack of expertise in submarine design, development and regulation is a weakness which lessens the book's impact.

John Jordan

Quintin Barry
The War in the North Sea: The Royal Navy and the Imperial German Navy 1914–1918

Helion & Co, Solihull 2016; hardback, 544 pages,
141 photographs & 24 reproductions of paintings, all B&W,
19 maps, 8 appendices, bibliography and index; price £29.95.
ISBN 978-1-911096-38-2

Quintin Barry, who has previously written several books on military history, has now tackled the entire naval war in the North Sea from 1914 to 1918. Beginning with four introductory chapters on the prewar years, he then devotes a single chapter to each of the major surface actions and eleven to the Battle of Jutland. The narrative is interspersed with chapters that describe the changes in the high commands on both sides – those on 'the Byzantine complexity of the management of the German navy' provide new insights – and also more extended campaigns like that of the Dover Patrol, the distant blockade, the battle against the U-boats and the eventual adoption of convoy. There are numerous monochrome illustrations, although the 24 small reproductions of paintings (mostly by Wyllie) are of limited historical value. The 19 specially-commissioned maps cover the most important actions (eleven for Jutland); regrettably there are no indications of the sources used.

Barry acknowledges his debt to Arthur Marder's *From the Dreadnought to Scapa Flow* and notes that, at the time of writing, several important new works were about to be published. He has been able to make reference to some that appeared before his own in the centenary year of Jutland (including, it should be stated, one by this reviewer) but he has in the main used older sources, including the British and German official histories, Naval Staff Monographs and published papers and memoirs of participants – there are, however, only a few references to the Jutland Official Despatches.

The main chapters are well-written and generally accurate but the introductory chapters are weaker. The description of the evolution of distant blockade is patchy and informed by only some post-Marder studies. On the 'dreadnought race', there are mistakes about the German Novelles, and the author fails to make it clear that, after the year 1911–12, Germany had to accept that she could no longer compete with British rates of construction. There are errors in ships' data, notably on the anti-torpedo boat guns of German dreadnoughts, which were not 4.1in (10.5cm) calibre. Moreover, Fisher's personnel reforms concerned not the status of stokers and seamen but of engineer and executive officers.

Only a few caveats must be made for the chapters dealing with the naval war. The description of the concluding events at the Dogger Bank relies on Admiral

Goldrick's earlier book rather than the more recent *Before Jutland*. Barry is rightly critical of Beatty's tactical leadership at Jutland, but does not mention how the late formation of the battlecruiser line compromised gunnery at the start of the Run to the South. Von Hase was not unfair in criticising Beatty for failing to follow the retreating German battlecruisers and predreadnoughts after 8.35pm. The British 4th Destroyer Flotilla is not credited with the torpedoing of *Rostock*, and Barry repeats the criticism of Commander Holloway Frost USN that the attack by the 12th Destroyer Flotilla showed 'an inexcusable lack of initiative' – a claim disputed by Marder, who considered Frost to be 'hypercritical … and with a tendency to be wise after the event'. The surface actions of the final quarter of 1917 are a reminder that, since Jutland, staff work had not improved at the Admiralty, and may even have deteriorated in the Grand Fleet under Beatty as C-in-C. Barry also reminds us that, but for faulty German intelligence, the last sortie of the High Seas Fleet in April 1918 might have resulted in serious British or even American losses.

Of the conclusions in the final 'Retrospective' chapter, that on the excellent design of German battlecruisers is unquestionable, but Barry goes too far in claiming that all German designs were superior. In the main, the light cruisers and destroyers were too lightly gunned, while some of the latter were too small for North Sea operations and coal firing was retained for too long. Of the principal commanders, Barry is clear about Beatty's shortcomings, but he appears to agree with Horsfield that Jellicoe lacked the power of decision. He nevertheless accepts that, faced with the supreme test of deciding how best to the deploy the Grand Fleet, Jellicoe made the correct choice. He rightly praises Hipper for his leadership in the Run to the South but, during the Run to the North, the latter did not display the 'prime quality of insubordination' by ignoring Scheer's order to chase. Scheer showed Nelsonic 'tenacity of purpose' in retreating towards Horns Reef, but also made tactical misjudgements, especially in his second action-turn-about.

Barry has written a useful single-volume account of the naval actions in the North Sea during the Great War which will appeal particularly to those who do not have the five volumes of Marder's great work.

John Brooks

Donald Stoker & Michael T McMaster (eds)
Naval Advising and Assistance: History, Challenges and Analysis
Helion, Solihull, 2017; hardback, 304 pages; price £49.95.
ISBN 978-1-91151-282-0

'Military advising', as opposed to 'naval advising', is a descriptive term with dubious connotations linked with the murky world of arms dealing, mercenaries and the exploitation of impoverished states by rich and powerful nations. '*Naval* Advising' on the other hand is a topic which has been overlooked by historians. This book seeks to address this omission. *Naval Advising and Assistance* consists of twelve essays by different authors, each of whom examines a particular naval 'mission' to countries ranging from Japan to Peru and Egypt to Poland undertaken during the 19th and 20th centuries.

Historically, as the editors point out in their introduction, such missions have been identified from the advising nation's perspective as aids to modernisation or nation building, as ways to increase political influence over the receiver nation, as ideological weapons, as elements of counterinsurgency, and as ways to make money. However, because the mission-despatching nations have tended to be principally Britain, France and the US, research has hitherto concentrated on information drawn from those nations' archives. Thus key documentary evidence concerning the motivations and reactions of the receiving nations have remained hidden. Fortunately, as this book demonstrates, this shortcoming can now be addressed as receiver nation records become more widely available.

Although it might be thought highly prestigious in terms of a nation's standing to be known to be acting in a naval advisory capacity, the reality is often very different. John Ferris reveals that the British naval aviation mission to Japan in the early 1920s ended in a sorry mess. It is a story which features espionage, fraud and naïve racial stereotyping. However, the receiver nation, Japan, profited enormously. As British naval aviation stuttered during the interwar years, its Japanese counterpart forged ahead on the back of British know-how; the disastrous loss of Singapore in 1942 was the ironic consequence.

Fabio De Ninno describes how naval missions to Persia, China and Spain brought mixed success for Italy in the 1930s. Mussolini believed that Italy, as a young and rising nation, needed to spread its sphere of influence, secure its prosperity and build a new fascist civilisation. Naval missions would assist in that aim. In the cases of Persia and China this meant building up the navies and the supporting infrastructure from scratch. Although there was some commercial gain for Italian shipyards in terms of warship construction for Persia, the aim of exerting regional political influence had faltered by the time the mission ended. Similarly the construction of a riverine navy in China barely materialised owing to the growth of the Germany–Italy–Japan axis in the latter part of the 1930s. Although the mission to Spain tangibly helped the Nationalists to win the Civil War, Franco turned out to be more influenced by Germany.

Ideological differences brought an end to other promising alliances. The Russians assisted the Chinese at the end of the Second World War to create a sea-going navy with the specific goal of re-capturing outlying islands, specifically Taiwan, from the Nationalists. Although successful for several years, Xiaobing Li demonstrates how growing differences between the two Communist nations caused a parting of ways at the end of the 1950s which retarded the subsequent development of the Chinese PLAN by several decades. Douglas Peifer, on the other hand, describes the successful evolution of what became the small but efficient East German Navy during the Cold War. The transformation from police force to fully-fledged *Volksmarine* and Warsaw Pact member was achieved through a progressive loosening of what had started as very strict Russian control and administration by the naval mission.

There are many other examples which illustrate the difficulties of setting up and, more vitally, maintaining a successful naval mission in a receiver country. More often than not success or otherwise depended on the personalities who headed the missions. This is a fascinating read for the naval historian, although the hefty price tag and the fact that individual warship names are few and far between might deter some potential readers.

Jon Wise

Robert C Stern
The Battleship Holiday: The Naval Treaties and Capital Ship Design
Seaforth Publishing, Barnsley 2017; hardback, 272 pages, many B&W illustrations; price £35.00.
ISBN 978-1-84832-344-5

On 12 November 1922, the American Secretary of State Charles Hughes, in his address to the delegates at the Washington Conference on the Limitation of Armament, proposed a ten-year 'naval holiday' in capital ship construction by the major powers: Britain, France, Italy, Japan and the USA. The events leading up to and the repercussions from this bold proposal are the central tenets of Robert Stern's comprehensive, if sprawling, study of the battleship in the 20th Century.

The author's intention in tackling what he admits is a topic that has been well covered is stated in the introduction. Stern contends that there are three major 'threads' to consider: diplomacy, technology and operational performance. He claims that most studies hitherto have concentrated on one or two of these threads and that his treatment covers all three.

He commences with an analysis of the clash between USS *Monitor* and CSS *Virginia* in Hampton Roads in 1862 and progresses through the pre-dreadnought and dreadnought eras before arriving at Jutland, the greatest of all battles involving these seaborne behemoths. Curiously, in view both of its significance with respect to a shift in world naval power and the technological lessons it afforded, the key Battle of Tsushima in 1905 warrants only passing references. The First World War, and principally Jutland, occupy three lengthy chapters of this large format book, one of which is entitled 'The Art and Practice of Main-Battery Fire Control in 1916'. Certainly the technological considerations of Stern's three-pronged coverage are addressed in lavish detail.

All this means that Secretary of State Hughes' suggestion about introducing a naval 'holiday' is not mentioned until page 100, nearly half-way through the book. Thereafter the deliberations and outcomes of the Washington Conference are examined minutely, likewise Jutland from the perspective of 'lessons learned'.

Stern's narrative then moves on to the setbacks encountered during the Geneva Conference of 1927 where neither the vexed question of capital ship tonnage nor the related problem of cruiser limitation was resolved. These issues were merely passed over to the two London Conferences of 1930 and 1935. Rightly, he also makes reference to the construction of the firstGerman *Panzerschiffe*, a milestone the significance of which was shortly to become apparent. While limited progress was made in 1930, Stern remarks that the 1935 London Naval Conference, the last in the series, 'was truly an exercise in futility'. The nadir was reached when the Japanese, who had grown progressively disenchanted since Geneva nearly a decade earlier, withdrew from the treaty system.

A lengthy chapter entitled 'The New Generation: 1934–1949' examines, in admirable detail and depth, the capital ship construction which finally emanated from the previous, mostly negative deliberations. This is followed by an overtly nationalistic study of seven Second World War battles or engagements involving capital ships. Stern focuses closely on the Second Battle of Guadalcanal, which merely serves to illustrate the obvious technical superiority of two modern US battleships over a 1915-vintage Japanese battlecruiser. By contrast, his other examples, which include Taranto, the River Plate, North Cape and Casablanca, all seem to imply that victory was achieved in each case due to a large degree of good fortune.

The last chapter recounts the long-drawn-out postwar demise of the battleship. Sadly, Stern fails to make more than a passing footnote reference to the fitting of guided weapons to the *Iowa* class or to the last active service undertaken by these survivors of the interwar naval treaties during the First Gulf War.

The Battleship Holiday is lavishly illustrated with a fine selection of photographs and line drawings. It has been extensively researched, drawing on a wide range of both primary and secondary sources. Robert Stern is a well-informed writer, with a particularly fine grasp of the technical aspects of his subject. It is a shame therefore that he is also unnecessarily self-indulgent in this book, allowing himself to digress too often from the complexities of a core subject matter which so dominated naval thinking in the interwar period.

Jon Wise

David C Evans (editor)
The Japanese Navy in World War II – In the Words of Former Japanese Naval Officers (Second Edition)
Naval Institute Press, Annapolis 2017; paperback, 568 pages, numerous plans and illustrations; price US$34.95.
ISBN 978-1-59114-568-4

This book is the first paperback version of a classic work initially published in 1969 and subsequently revised and expanded into the current, second edition in 1986. Moreover, many of its chapters had previously appeared in various issues of the US Naval Institute's *Proceedings* during the 1950s and 1960s.

As its title suggests, *The Japanese Navy in World War II*

is largely an account of IJN operations during the Pacific campaign from the perspective of its own personnel. However, the book also benefits from a general introduction by the editor of the first edition, US academic Raymond O'Connor, which puts the chapters that follow into context. Further clarity is provided by further short, explanatory introductions to each chapter, often adding a valuable US viewpoint to the Japanese accounts.

The 17 chapters – the second edition added a further five articles to an initial twelve – adopt a broadly chronological methodology, commencing with the initial attack on Pearl Harbor in December 1941 and extending to the loss of the battleship *Yamato* in April 1945. However, some take a more thematic approach. These include, for example, articles on why Japan's anti-submarine warfare efforts were unsuccessful and a concluding review of the reasons for the ultimate failure of the country's naval strategy.

The book covers a broad spectrum of subjects from viewpoints that range from the 'coal face' experiences of Japanese junior officers to the more strategic outlook of leading admirals such as legendary destroyer squadron flag officer Raizo Tanaka. The various authors had first-hand knowledge of the subjects they choose to write about, allowing them to provide considerable detail. Their accounts were also written relatively soon after the relevant events occurred, avoiding the risk of recollections being clouded by the passage of time. Effective translation means that the clarity of these recollections is retained in the English language.

The unparalleled Japanese perspective provided by *The Japanese Navy in World War II* made the book essential reading for students of the Pacific War when first published in the 1960s. This remains true today. However, it is worth noting that this reprint is unchanged from the second, 1986 edition. The editorial commentary does not therefore benefit from any of the research undertaken into Japanese wartime naval operations over the succeeding thirty years.

Conrad Waters

Jonathan North
Nelson at Naples: Revolution & Retribution in 1799
Amberley Publishing, Stroud 2018; hardback, 288 pages, 27 colour and B&W illustrations; 4 maps; price £20.00.
ISBN 978-1-44567-937-2

The British have a predilection for toppling national heroes from their pedestals. In 2017 the journalist Afua Hirsch suggested that the statue of Horatio Nelson should be removed from its column because of his support of slavery at a time in history when public opinion was beginning to move against that odious practice. Jonathan North likewise has the reputation of Britain's most famous naval hero firmly in his sights in his account of the events which took place in the Bay of Naples during the summer of 1799.

Revelations about Nelson's conduct on that occasion are not new. Robert Southey, Nelson's Victorian biographer, called it 'deplorable' and a 'stain' on the Admiral's memory. Bicentenary biographers including Roger Knight and John Sugden did not shy away from the subject, but did offer mitigating circumstances in their analyses of Nelson's conduct. Jonathan North, on the other hand, is wholly condemnatory of the naval hero.

So what did Nelson do? After the French Revolution, the Republican Army temporarily seized the Kingdom of Naples before a counter-revolution wrested back control, trapping hundreds of Neapolitans. A peace treaty was signed permitting the republicans to be transported to France and safety. At that point, Nelson's fleet entered the Bay of Naples, apprehended the would-be exiles who were already waiting in their transports and returned them to their royalist captors, in whose hands most experienced torture and death.

The larger part of *Nelson at Naples* comprises a most comprehensive summary of the events leading up to Nelson countermanding the terms of the surrender. Jonathan North draws on a wide range of sources of information to provide an international, rather than a British-dominated account of events. Having assembled the facts, he assesses the extent of Nelson's culpability in a final chapter titled 'The Controversy'.

Some historians, including Southey, have tried to deflect part of the responsibility from Nelson onto William and (in particular) Emma Hamilton, but North is quite clear that the Admiral alone used his power to sweep aside any opposition, despite severe misgivings from within the ranks of his own senior officers.

This is a well-written and committed account, and the author is unequivocal in his condemnation of Nelson's behaviour. Revisionist history is healthy as long as it remains balanced and unbiased itself; the reader will have to decide whether or not this is the case with *Nelson at Naples*.

Jon Wise

Philip Kaplan
Building for Battle: U-boat Pens of the Atlantic Battle
Pen & Sword Military, Barnsley 2017; hardback, 175 pages, 150 B&W & colour illustrations; price £25.00.
ISBN 978-1-52670-544-0

U-boat Pens is one of two books from this author under the 'Building for Battle' banner, the other being *Hitler's D-Day Defences*. The title implies a book about the construction of the U-boat pens, while the jacket synopsis more specifically suggests a study of the massive and strategically important pens of the Brittany coast in particular: Brest, Lorient, St-Nazaire, La Pallice and Bordeaux. However, the book does not deliver what is promised.

While it does include some information about these pens, it is much more of a general read about other aspects of the Second World War at sea. The first chapter briefly covers the origins of the U-boat service in the First World

War and touches on German preparations to build the pens, but there is very little detail about the planning or construction process and no plans or diagrams are included. Another chapter begins to describe the building of the bunkers but then goes on to recount the story of Operation 'Frankton' and the 'Cockleshell Heroes' raid on Bordeaux, which is not particularly relevant here. There is a chapter on the Type VII U-boat and another devoted to the U-boat captains, though this mainly covers the already well-known careers of Prien, Schepke and Kretschmer. Other chapters cover daily life aboard a U-boat, the overall U-boat campaign, and the Allied measures to combat the U-boats. Almost half of the final chapter consists of an extract from Herbert Werner's bestselling memoir *Iron Coffins*. Much of this has been written about in greater detail elsewhere and, more importantly, very little of it is specifically about the U-boat pens of the title.

The publisher has opted for an unusual format: a medium-sized hardback with glossy pages throughout and the text spread across two columns. However the text is very large, mostly black and bold, almost like that in a large-print title. Where it crosses a dark photograph it is reversed out in white, while quotes are printed in blue, meaning the colour of the text keeps changing as you read down the page, which looks distinctly odd.

There are many photographs and illustrations. However, while some of the photographs of the surviving U-boat pens are interesting not all the images are relevant to the text – a page of wartime propaganda posters or random photographs of merchant seamen, for example. The large text and the many illustrations make the book a fairly quick read which feels far too short on detail to be worth the hefty price tag. Someone new to the subject of the U-boat war might find this an interesting introduction to the topic in general but, ultimately, this book does not fulfil its title and synopsis.

John Peterson

Janusz Skulski and Stefan Draminski
Anatomy of the Ship: Battleships *Yamato* and *Musashi*
Osprey Publishing, Oxford 2018: hardback, 336 pages, 45 digitally colourised photos, 1020+ drawings, 359 CG 3D colour images, 17 tables; price £35.00.
ISBN 978-1-84486-317-4

Yamato and *Musashi* were the largest battleships of all time, with the heaviest naval guns (18.1in/46cm) and the thickest armour (25.6in/650mm armour on turret faces) ever put afloat. A third unit, *Shinano*, was completed as a carrier. Most Japanese naval documentation disappeared in August 1945; much was destroyed in American firestorms, some was taken home for safe custody by naval officers and constructors, to reappear slowly in the public domain. Accurate data and many details of ship performance and losses came from postwar investigations by the US Navy. The earlier published plans of the super-battleship pair were very crude: they were released by the USN after the war, appearing in *Jane's Fighting Ships* 1947–48. Fukui's *Nippon No Gunkan* (1956) had a not-entirely-accurate single fold-out sheet (a lift from quarterdeck to hangar was claimed). A 1975 publication, *Plans of Ships of the IJN – History of Shipbuilding in the Showa Era*, contained 28 drawings of *Yamato*.

With this dearth of information, Skulski's *The Battleship Yamato* (192 pages, Conway Maritime Press 1988, reprinted 1995), which contained more than 600 detailed and elaborate drawings, was particularly important. In his introduction, Skulski provided an interesting account of his drawing process: construction of a simple 1:133 model from existing documentation and reconstructions, then comparing photos of the model and of the actual ship for preliminary detailed drawings. These drawings were then refined in an iterative process to produce the superb drawings published in the book. Skulski's method worked well for the exterior of the ship.

The new book, which extends full coverage to both battleships, will inevitably be compared and contrasted with Skulski's original. It is twice the size, many of the extra pages being taken up by the co-author's computer-generated imagery. The first twenty pages describe the ships, the next sixteen summarise their careers. A section titled 'Primary Views' containing 30 CG colour illustrations then follows, preceding the main feature, the line drawings, totalling more than one thousand interspersed with the remaining CG images. The line drawings have been revised from the 1988 originals, where possible providing more upper deck and superstructure detail and showing both ships at different stages in the war. There is more space for illustrating smaller items of equipment such as ships' boats and embarked aircraft.

Numerous new photos have been located and added to the 22 from the earlier book; all have been colourised and are much sharper, although the purist may not approve of the colour. The very comprehensive data tables are rearranged but unchanged. While the table of sketch designs in the original book is omitted, the new book includes more detail on the loss of *Musashi* and a brief mention of *Shinano*. Two unexpected claims in the text have been made, seemingly from uncritical use of oral history and without authentication: a speed of over 30 knots (*Musashi*), and an estimate of hits on *Yamato* greatly in excess of the careful figures obtained after the war by painstaking interrogations; the estimate is from Gakken, a popular Japanese publishing conglomerate. The 350 new 3D colour illustrations, necessarily external views, look magnificent but do not add greatly to the value of the book.

Details of examination of wreckage and the supposed sequence of events during the sinkings are presented sketchily; more could have been said about the ships' resistance to action damage. Surprisingly, there are no references, no list of sources, no bibliography and no index, only sincere thanks to unnamed friends and particular gratitude to a few named individuals. This is a book for the specialist and, particularly, the model maker.

Ian Sturton

WARSHIP GALLERY
The Polish Navy against the *Luftwaffe* and *Kriegsmarine* 1–3 September 1939

Przemysław Budzbon presents a series of previously unpublished photographs of the Polish Navy before, during and after the German aggression of September 1939.

Poland emerged from the Great War with access to the Baltic. The 'Polish Corridor', carved out between the bulk of Germany and East Prussia, was a key strategic aspect to be considered in the event of war with Germany. The Polish military presence in the Corridor was seen in terms of prestige rather than purposeful effort, while the future role of naval forces was limited to the harassment of enemy traffic in the Baltic by submarines operating from neutral or friendly bases. In the event of war with the Soviet Union, the Polish Navy aimed to close the Gulf of Finland with minefields and submarines, with the additional mission of escorting friendly shipping carrying supplies.

The strategic dilemma led the Poles to limit their naval expenditures to submarines, light surface forces and mine warfare. By the summer of 1939 the Polish Navy comprised the destroyers *Wicher*, *Burza*, *Grom* and *Błyskawica*, the attack submarines *Orzeł* and *Sęp*, the minelaying submarines *Wilk*, *Ryś* and *Żbik*, the large minelayer *Gryf*, the minesweepers *Jaskółka*, *Rybitwa*, *Mewa*, *Czapla*, *Czajka* and *Żuraw*, the gunnery training ship *Mazur*, two elderly gunboats, a handful of auxiliaries, and nineteen aircraft.

The imminence of hostilities with Germany led the Polish Government to the conclusion that its surface forces would be of little value and, following the signature of the Anglo-Polish military pact in April 1939, on 30 August three destroyers were despatched to Britain.

During the 1920s the Polish naval forces comprised six ex-German torpedo boats – one was lost to a boiler explosion in 1925 – and a handful of lesser ships. This photograph was taken at Gdynia in 1932–34. From left to right: the torpedo boat *Mazur* (ex-*V-105*, ex-Dutch *Z-1*) following her recent conversion to a gunnery training ship, *Ślązak* (ex-German *A-59*), *Podhalanin* (ex-*Góral*, ex-German *A-80*) and *Krakowiak* (ex-German *A-64*). At the head of the line is *Wicher*, the first of the new French-built destroyers. (Muzeum Marynarki Wojennej collection)

Wicher was left to cover the eventuality that the enemy might restrict military action to the seizure of the Free City of Danzig or the Corridor. However, on 1 September 1939 Poland was subjected to a full-scale invasion. The fleet based in Gdynia and Hela under Rear Admiral Józef Unrug faced a force comprising the old battleship *Schleswig-Holstein* (moored in Danzig opposite the Polish military base on Westerplatte), three light cruisers, nine destroyers, ten submarines, 40 lesser ships and 449 aircraft.

It was the *Luftwaffe* that was given the primary rsponsibility for eliminating the Polish fleet. At 1400 on 1 September, in an air raid on Gdynia by *Ju 87* dive-bombers of IV.(St)/LG1, *Mazur* and the auxiliary *Nurek* were disabled. During the attack *Wicher*, four minesweepers and gunboats manoeuvred in the roadstead while firing on the German aircraft. They were waiting for *Gryf* which, hidden inside the Bay of Puck, was embarking mines from barges in preparation for a planned night minlaying operation.

At 1700 the force moved off slowly in line ahead to the northwest towards Hela, and was joined by *Gryf* with 300 mines on board. An hour later the column, headed by *Wicher*, was attacked by three squadrons of IV./LG1 totalling 32 *Ju 87* dive-bombers escorted by eight *Bf 109E* fighters of I.(J)/LG2. This would result in the first naval aerial battle of the Second World War that could be compared, albeit on a much smaller scale, to the battles fought two years later in the Mediterranean. The aircraft failed to obtain any direct hits during attacks which lasted 25 minutes, but near-misses and machine-gunning proved fatal. *Mewa* caught fire, was disabled and had to be taken on tow. *Gryf* had her captain killed, her steering gear damaged and a number of mines derailed or damaged. Her second-in-command decided to jettison the entire load of mines, and in consequence the Polish surface forces lost their *raison d'être*, which was the key German success of the day. Next morning *Wicher* and *Gryf* were berthed in Hela naval harbour, their use restricted to the role of floating batteries.

On the morning of 3 September, the German destroyers *Leberecht Maaß* (flying the flag of Rear Admiral Günter Lütjens) and *Wolfgang Zenker* made a sortie towards Hela. They were engaged around 0700 by

On the right of this photograph, taken in Gdynia naval harbour in the summer of 1931, can be seen the torpedo boat *Mazur* (ex-German *V-105*, ex-Dutch *Z-1*) following her conversion to a gunnery training ship. Alongside is the torpedo boat *Kujawiak* (ex-German *A-68*). Between the two ships, in the background, can be seen (from left to right): the minesweeper *Jaskółka* (ex-German *FM-27*), the gunboat *Komendant Piłsudski* (ex-Russian *Lun'*) and a naval tug. (Author's collection)

WARSHIP GALLERY

Bałtyk, the former French colonial cruiser *d'Entrecasteaux,* served as a naval accommodation hulk. Because of her size she was the most prominent feature of prewar Gdynia harbour. (Muzeum Marynarki Wojennej collection)

The destroyer *Wicher,* led by her sister *Burza,* leaving Gdynia during the mid 1930s. Built at the CNF shipyard in France as an export version of the *torpilleurs d'escadre* of the *1500 tonnes* type (see *Warship No 13*), they were classified as *Kontrtorpedowce* (from the French *contre-torpilleur*) in Poland. Although criticised for shortcomings in their design and build, they were the largest and most heavily armed ships of their type in the Baltic until the appearance of the German *Leberecht Maaß* in 1937. (Marek Twardowski collection)

Close-up of the submarine *Wilk* during the Festival of the Sea in Gdynia shortly after commissioning. She was completed in France in 1932 as one of a class of three minelaying submarines derived from the *Saphir* class. These submarines had not been refitted since their completion, which hampered their performance when the war begun. (Muzeum Marynarki Wojennej collection)

A photograph taken on 29 June 1939 during the 'Festival of the Sea', looking out onto the roadstead of Gdynia. It depicts the Polish Navy in its heyday, and almost all the ships which two months later were to face the German onslaught are present. The first row is headed by the gunboat *Komendant Piłsudski* followed by three *Jaskółka* class minesweepers with fishing boats in the foreground,

WARSHIP GALLERY

The four Polish-built minesweepers of the *Jaskółka* were commissioned in 1935–36; they are seen here at Hela during the late 1930s. From left to right: *Mewa*, *Czajka*, *Rybitwa* and *Jaskółka*. In September 1939 they were supplemented by two further units, albeit hastily pressed in service – see the article by Marek Twardowski in *Warship* No 15. (Muzeum Marynarki Wojennej collection)

Gryf, *Wicher* and the four-gun 6in Bofors battery. During a five-minute duel the Poles scored a hit on the leading destroyer, damaging her forward gun mounting. The German hit on *Gryf* knocked out one AA mounting, while the already-damaged *Mewa* was hit in the bow.

Two hours later an air raid by *Ju 87* dive-bombers from 4.(St)/186 *Trägergruppe* set fires on *Gryf*, which had to be abandoned. Following a near-miss *Mewa* broke her

some of which were mobilised at the end of August. Two submarines of the *Wilk* class head the second row, followed by *Sęp* and *Orzeł*. The sail training ships *Iskra* (naval) and *Dar Pomorza* (merchant navy) are at the head of the third row, followed by the gunnery training ship *Mazur*, the minelayer *Gryf* and (out of picture) the four destroyers of the *Wicher* and *Grom* classes. (Marek Twardowski collection)

Gryf docked in Gdynia in September 1938. Freshly arrived from the Le Havre shipyard of Augustin–Normand in France, she was the largest Polish warship of the day. The concept of combining minelaying capabilities with a heavy gun armament and spacious berths for cadets left little room for propulsion, and with a top speed of 20 knots she was incapable of offensive minelaying. (Photo by Antoni Ratajczak, from the author's collection)

Grom off Gdynia in the last weeks of peace. Built, together with her sister *Błyskawica*, by Samuel White in Britain (see *Warship No 4*), they were the most powerful ships in the Polish Navy. (Marek Twardowski collection)

Orzeł entered the war as the most combat-ready of all the Polish submarines. Built in the Netherlands (see *Warship No 42*), she is seen in June 1939 coming through the southern entrance of Gdynia harbour. (Marek Twardowski collection)

moorings and sank. The raid was repeated at 1510 when *Wicher* was sunk, while the burning *Gryf* took further hits and started to take on water. The wreck was demolished during a third raid of *He 59* floatplanes of 3.(M)/KüFlGr 706 at 1715.

By this time the Polish surface fleet had been practically annihilated, while torpedoes fired at 1939 by *U-30* in the direction of the liner *Athenia* immediately moved the focus of the war at sea to the North Atlantic.

The Aftermath

German forces cut off the Corridor the same day. Fighting around Gdynia ended on 19 September, while the Hela base, which was under close surveillance by the *Luftwaffe*, was unable to support the operations of submarines (only *Ryś* was based there, on 5 September). For the first week of hostilities the Polish submarines waited in vain around Hela for the expected armada of landing ships; they were later moved to the central

The torpedo boat *Mazur* has the dubious honour of being the first warship to be sunk during the Second World War. She received one hit in the bow, while the explosion of a second bomb blew her 75 metres away from the quay. Fully manned, with the boiler and engine room personnel closed up, *Mazur* took 46 men down with her. (Marek Twardowski collection)

The destroyer *Wicher* was sunk by one 250kg bomb which struck her bow and three 50kg bombs to starboard, amidships. Her wreck is seen lying alongside the southern mole of Hela Naval Harbour. (Narodowe Muzeum Morskie collection)

Baltic with the intention of harassing traffic. Led by elderly officers, they had no success, although the German minesweeper *M-85* and two fishing boats were lost to mines laid by *Wilk* and *Żbik*. All but one were interned in neutral harbours by 25 September. *Wilk* managed to flee to Britain where she was joined by *Orzeł*, which under her first lieutenant escaped from Tallin.

After 3 September the only Polish surface warships that remained operational in home waters were the minesweepers based in Jastarnia. They shelled German troops in the vicinity of Gdynia and laid a mine barrage off Hela, but an air raid at 0730 on 14 September by 4.(St)/186 *Trägergruppe* brought their activity to an end. *Jaskółka* and *Czapla* were sunk; the other three were slightly damaged; they were evacuated to Hela and later seized by the Germans. The Polish garrison on Hela lasted until 1 October. That effort had neither strategic nor operational significance, but was valuable to the morale of the defeated nation.

Acknowledgments:

Muzeum Marynarki Wojennej (Naval Museum), Narodowe Muzeum Morskie (National Maritime Museum), Adam Jarski and Marek Twardowski.

The wreck of the minesweeper *Czapla* at Jastarnia. She was struck by one 50kg bomb on the stern and suffered a near-miss of a larger 250kg bomb. Her 75mm gun was landed to augment the shore defences. (Narodowe Muzeum Morskie collection)